Lecture Notes in Computer Scien

T0238283

Commenced Publication in 1973
Founding and Former Series Editors:
Gerhard Goos, Juris Hartmanis, and Jan van Leeuwen

Yannis Ioannidis   Boris Novikov
Boris Rachev (Eds.)

Advances in Databases and Information Systems

11th East European Conference, ADBIS 2007
Varna, Bulgaria, September 29-October 3, 2007
Proceedings

 Springer

Volume Editors

Yannis Ioannidis
University of Athens
Department of Informatics and Telecommunications
Informatics Buildings, Panepistimioupolis, 15784 Ilissia, Athens, Greece
E-mail: yannis@di.uoa.gr

Boris Novikov
University of St.Petersburg
Department of Computer Science
28, Universitetsky Prospekt, Staryj Peterhof, 198504 St.Petersburg, Russia
E-mail: borisnov@acm.org

Boris Rachev
Technical University of Varna
Department of Computer Science and Technologies
1, Studentska Str., 9010 Varna, Bulgaria
E-mail: brachev@gmail.com

Library of Congress Control Number: 2007935199

CR Subject Classification (1998): H.1, H.2, H.3, H.4, H.5

LNCS Sublibrary: SL 3 – Information Systems and Application, incl. Internet/Web
and HCI

ISSN   0302-9743
ISBN-10   3-540-75184-X Springer Berlin Heidelberg New York
ISBN-13   978-3-540-75184-7 Springer Berlin Heidelberg New York

Springer is a part of Springer Science+Business Media

springer.com

© Springer-Verlag Berlin Heidelberg 2007
Printed in Germany

Typesetting: Camera-ready by author, data conversion by Scientific Publishing Services, Chennai, India
Printed on acid-free paper   SPIN: 12163120   06/3180   5 4 3 2 1 0

Preface

The series of East European Conferences on Advances in Databases and Information Systems (ADBIS) is an established and prestigious forum for the exchange of the latest research results in data management. It provides unique opportunities for database researchers, practitioners, developers, and users from East European countries to explore new ideas, techniques, and tools, and to exchange experiences with colleagues from the rest of the world. This volume contains the proceedings of the 11th ADBIS Conference, held in Varna, Bulgaria, September 29 – October 3, 2007. The conference included 3 keynote talks, 36 research papers in 13 sessions, and 2 tutorials. Twenty-tree of the research papers as well as papers or extended abstracts for the keynote talks are included here; the remaining papers appear in local proceedings.

Distinguished members of the database and information-retrieval communities delivered the three keynotes. Timos Sellis, an expert in the area of multidimensional indexing and data warehousing, analyzed the entire lifecycle of ETL workflows, from specification to optimized execution, offering solutions as well as future challenges. Gerhard Weikum, a leader of several efforts falling at the intersection of databases and information retrieval, discussed the emergence of several "Webs" and how these may be harvested and searched for knowledge. Finally, Paolo Atzeni, well-known for several contributions to database theory, addressed the perennial problem of schema and data translation in the context of emerging model management systems and outlined several research challenges that emerge.

The Research Program Committee consisted of 55 members and was chaired by Yannis Ioannidis (University of Athens, Hellas) and Boris Novikov (University of St. Petersburg, Russia). It accepted 36 papers (23 for the Springer proceedings and 13 for the local proceedings) out of 77 submissions coming from 29 countries. The reviewing process was administrated by the Conference Management System developed and supported by Yordan Kalmukov (University of Rousse). Boris Rachev (Technical University of Varna), Irena Valova (University of Rousse) and Yordan Kalmukov (University of Rousse) edited the proceedings.

The program and social activities of ADBIS 2007 were the result of a huge effort by many hundreds of authors, reviewers, presenters, and organizers. We thank them all for helping to make the conference a success. In particular, we want to thank Peter Antonov (Technical University of Varna) and Angel Smrikarov (University of Rousse) for the smooth local organization.

July 2007

Yannis Ioannidis
Boris Novikov
Boris Rachev

Organization

The 11th East-European Conference on Advances in Databases and Information Systems (ADBIS) was organized by members of the Department of Computer Sciences and Technologies at the Technical University of Varna, and the Department of Computing at the University of Rousse, Bulgaria, in cooperation with the Moscow ACM SIGMOD Chapter.

General Chair

Boris Rachev, Technical University of Varna, Bulgaria

Program Committee Co-chairs

Yannis Ioannidis, National and Kapodistrian University of Athens, Greece
Boris Novikov, University of St. Petersburg, Russia

Program Committee

Antonio Albano (Universita' di Pisa, Italy)
Periklis Andritsos (University of Trento, Italy)
Dmitry Barashev (University of Saint Petersburg, Russia)
Michael Böhlen (Free University of Bozen-Bolzano, Italy)
Stefan Brass (University of Halle, Germany)
Albertas Caplinskas (Institute of Mathematics and Informatics, Lithuania)
Barbara Catania (University of Genoa, Italy)
Chee-Yong Chan (National University of Singapore, Singapore)
Damianos Chatziantoniou (Athens University of Economics and Business, Greece)
Carlo Combi (University of Verona, Italy)
Jens-Peter Dittrich (ETH Zurich, Switzerland)
Pedro Furtado (University of Coimbra, Portugal)
Shahram Ghandeharizadeh (USC, USA)
Maxim Grinev (Institute for System Programming, RAS, Russia)
Oliver Guenther (Humboldt-Universitaet, Germany)
Hele-Mai Haav (Institute of Cybernetics, Estonia)
Piotr Habela (Polish-Japanese Institute of IT, Poland)
Mohand-Said Hacid (University Lyon 1, France)
Theo Härder (University of Kaiserslautern, Germany)
Leonid Kalinichenko (Institute of Informatics Problems, RAS, Russia)
Mikhail Kogalovsky (Market Economy Institute of RUS, Russia)
Georgia Koutrika (Stanford University, USA)

Organizing Committee Chairs

Peter Antonov, Technical University of Varna
Angel Smrikarov, University of Rousse

Organizing Committee Coordinators

Irena Valova, University of Rousse
Yordan Kalmukov, University of Rousse

Organizing Committee Members

Milko Marinov, University of Rousse
Yulka Petkova, Technical University of Varna
Silyan Arsov, University of Rousse
Antoaneta Ivanova, Technical University of Varna
Veneta Aleksieva, Technical University of Varna
Miroslava Strateva, Technical University of Varna
Polina Kalmukova, Graphic Designer

ADBIS Steering Committee Chair

Leonid Kalinichenko, Russian Academy of Science, Russia

ADBIS Steering Committee

Andras Benczur (Hungary)
Albertas Caplinskas (Lithuania)
Johann Eder (Austria)
Janis Eiduks (Latvia)
Hele-Mai Haav (Estonia)
Mirjana Ivanovic (Serbia)
Mikhail Kogalovsky (Russia)
Yannis Manolopoulos (Greece)
Rainer Manthey (Germany)
Tadeusz Morzy (Poland)
Pavol Navrat (Slovakia)
Boris Novikov (Russia)
Jaroslav Pokorny (Czech Republic)
Boris Rachev (Bulgaria)
Anatoly Stogny (Ukraine)
Bernhard Thalheim (Germany)
Tatjana Welzer (Slovenia)
Viacheslav Wolfengagen (Russia)

Table of Contents

Object-Oriented Systems

Indexing

Clustering and OLAP

Moving Objects

Query Processing

DB Architectures and Streams

XML and Databases

Distributed Systems

ETL Workflows: From Formal Specification to Optimization

Timos K. Sellis and Alkis Simitsis

[1] School of Electrical and Computer Engineering,
National Technical University of Athens, Athens, Hellas
timos@dblab.ece.ntua.gr
[2] IBM Almaden Research Center, San Jose CA 95120, USA
asimits@us.ibm.com

Abstract. In this paper, we present our work on a framework towards the modeling and optimization of Extraction-Transformation-Loading (ETL) workflows. The goal of this research was to facilitate, manage, and optimize the design and implementation of the ETL workflows both during the initial design and deployment stage, as well as, during the continuous evolution of a data warehouse. In particular, we present our results which include: (a) the provision of a novel conceptual model for the tracing of inter-attribute relationships and the respective ETL transformations in the early stages of a data warehouse project, along with an attempt to use ontology-based mechanisms to semi-automatically capture the semantics and the relationships among the various sources; (b) the provision of a novel logical model for the representation of ETL workflows with two main characteristics: genericity and customization; (c) the semi-automatic transition from the conceptual to the logical model for ETL workflows; and (d) the tuning of an ETL workflow for the optimization of the execution order of its operations. Finally, we discuss some issues on future work in the area that we consider important and a step towards the incorporation of the above research results to other areas as well.

1 Introduction

Successful planning and decision making in large enterprises requires the ability of efficiently processing and analyzing the organization's informational data. Such data is typically distributed in several heterogeneous sources and stored under different structures and formats. To deal with such issues, as well as performance issues, and to support the functionality of On Line Analytical Processing (OLAP) applications and Decision Support Systems (DSS), Data Warehouses (DW) are employed to integrate the data and provide an appropriate infrastructure for querying, reporting, mining, and other advanced analysis techniques. The procedure of designing and populating a DW has been characterized as a very troublesome and time consuming task with a significant cost in human, system, and financial resources [13].

In the past, research has treated DW as collections of materialized views. Although this abstraction may suffice for the purpose of examining alternative strategies for view maintenance, it can not adequately describe the structure and contents of a DW in real-world settings. A more elaborated approach is needed (a) to represent the population of

Y. Ioannidis, B. Novikov, and B. Rachev (Eds.): ADBIS 2007, LNCS 4690, pp. 1–11, 2007.

the DW with data stemming from a set of heterogeneous source datastores, and (b) to take into consideration that during their transportation, data may be transformed to meet the schema and business requirements of the DW. This procedure normally composes a labor intensive workflow and constitutes an integral part of the back-stage of DW architectures.

Hence, to overcome the above problems, specialized workflows are used under the general title *Extraction-Transformation-Loading* (ETL) workflows. ETL workflows represent an important part of data warehousing, as they represent the means in which data actually gets loaded into the warehouse. Their generic functionality and most prominent tasks include:

– the identification of relevant information at the source side,
– the extraction of this information,
– the transportation of this information to the Data Staging Area (DSA), where, usually, all transformations take place,
– the transformation, (i.e., customization and integration) of the information coming from multiple sources into a common format,
– the cleansing of the resulting data set, on the basis of database and business rules, and
– the propagation and loading of the data to the DW and the refreshment of data marts.

Figure 1 depicts a generic architecture of the DW environment.

Several research approaches have studied the modeling part of ETL processes. On the other hand, several commercial tools already exist in the market and all major DBMS vendors provide such functionality. However, each individual effort follows a different approach for the modeling and representation of ETL processes, making essential the adoption of an unified formal description of such processes. For a further discussion on the importance of ETL processes and on the problems existing due to the lack of a uniform modeling technique, along with a review of the state of the art in both research and commercial solutions, we refer the interested reader to [7].

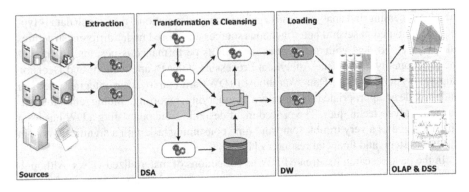

Fig. 1. Generic Architecture of Data Warehouse environment

In this paper, we present our work towards the modeling and optimization of ETL workflows. Section 2 presents our framework for the formal specification of ETL workflows. Section 3 describes our technique for the logical optimization of ETL workflows. Finally, Section 4 concludes our discussion with a prospect of the future.

2 Formal Specification of ETL Workflows

2.1 Identification of ETL Design Requirements Using Semantic Web Technology

During the initial phases of a DW design and deployment, one of the main challenges is the identification of the involved sources and the determination of appropriate inter-schema mappings and transformations from the data sources to the DW. Currently, research and commercial ETL tools mainly focus on the representation and design of ETL scenarios. The identification of the required mappings and transformations is done manually, due to the lack of precise metadata, regarding the semantics of the data sources and the constraints and requirements of the DW. Hence, such information is incomplete or even inconsistent, often being hard-coded within the schemata or metadata of the sources or even provided in natural language format after oral communication with the involved parties; e.g., business managers and administrators/designers of the DW. As a result, designing ETL processes becomes a very tedious and error-prone task. Given the fact that typical ETL processes are quite complex and that significant operational problems can occur with improperly designed ETL systems, developing a formal, metadata-driven approach to allow a high degree of automation of the ETL design, is critical in employing a Data Warehouse solution.

In our research, we have worked on the aforementioned problem. Earlier work argues that in the context of a DW application, ontologies constitute a suitable conceptual model for describing the semantics of the datastores and automatically identifying relationships among them using reasoning techniques [9,11]. The schema of a datastore describes the way that data is structured when stored, but does not provide any information for its intended semantics. Therefore, metadata are required to allow for the understanding, management, and processing of this data. Semantic Web technologies provide a means to formally specify the metadata, so that automated reasoning techniques can be used to facilitate further processing.

A graph-based representation, called *datastore graph*, is employed as a common model for the datastores. Graphs constitute a generic data model allowing the representation of several types of schemas, including relational and XML schemas, thereby allowing for both structured and semi-structured sources to be handled in a unified way. A graph representation, termed ontology graph, is introduced for the application ontology. Providing a visual, graph-based representation, with different symbols for the different types of classes and properties in the ontology, makes it easier for the designer to create, verify and maintain the ontology, as well as use it as a means of communication between different parties involved in the project.

Annotation of a datastore is accomplished by formally defining mappings between the nodes of the datastore graph and the ontology graph. These mappings can be represented as labels assigned to the nodes of the data store graph, i.e., the datastore is semantically described by the annotated datastore graph. The mappings may be specified either (semi-)automatically using results provided by related research efforts [2] or manually – e.g., by implementing drag-and-drop functionality between the visual representations of the corresponding graphs. In both cases, the time and effort required for establishing and maintaining the mappings significantly decreases with respect to common practice.

Based on the application ontology and the annotated datastore graphs, automated reasoning techniques are used to infer correspondences and conflicts among the datastores, thus, identifying relevant sources and proposing conceptual operations for integrating data into the DW.

Furthermore, the application ontology along with a common application terminology, can be used as a common language, to produce a textual description of the requirements of an ETL process. The verbalization of such requirements further facilitates the communication among the involved parties and the overall process of design, implementation, maintenance, and documentation. Recent results describe how a common application terminology can be established semi-automatically, using linguistic techniques [10]. In that work, a template-based technique is introduced to represent the semantics and the metadata of ETL processes as a narrative, based on information stored in the application ontology, which captures business requirements, documentation, and existing schemata. In addition, the customization and tailoring of reports to meet diverse information needs, as well as the grouping of related information to produce more concise and comprehensive output are considered.

The result of the above work is accompanied by a simple graphical model, which facilitates the smooth redefinition and revision efforts and serves as the means of communication with the rest of the involved parties [13]. A graph-based representation of the involved datastores and transformations is presented in a customizable and extensible manner. The transformations used in this model follow a high level description annotated with sufficient information for their ensuing formal specification in the logical level. (For a further analysis on this issue, we defer to subsection 2.3.)

2.2   Logical Modeling of ETL Workflows

A conceptual model for ETL processes serves as a suitable means for communications and requirements understanding in the early stages of a DW project during which, the time constraints of the project require a quick documentation of the involved data stores and their relationships, rather than an in-depth description of a composite workflow. For the ensuing stages of the project, a formal and more rigorous logical model is necessary.

In our research, we have extensively dealt with this challenge by presenting a formal *logical model* specifically tailored for the ETL environment [8,12,14]. The model concentrates on the flow of data from the sources towards the data warehouse through the composition of activities (transformations) and datastores. The core of the proposed model treats an ETL scenario as a graph of ETL activities having interconnected input and output schemata. This graph, which is referred to as Architecture Graph, can be used as the blueprints for the structure of an appropriate workflow in repository management, visualization, and what-if analysis tools. Activities, datastores, and their respective attributes are modeled as the nodes of the graph. Provider relationships that connect the input and output schemata of activities, are the edges of the graph. The operational semantics of ETL activities (or of a logical abstraction of them) are expressed in the declarative database language LDL++, which is a Datalog variant [15]. This language is chosen due to its expressiveness (e.g., it supports external functions, updates, and set-valued attributes) and suitability for ETL environment. (The rule-body pair is quite suitable to characterize input-output relationships in the Architecture graph.)

In addition, a principled way of transforming LDL++ programs to graphs, all the way down to the attribute level is proposed [8]. The resulting graph provides sufficient answers to the measurement of the quality of ETL workflows, as well as to what-if and dependency analysis in the process of understanding or managing the risk of the environment [14]. Moreover, due to the obvious, inherent complexity of this modeling at the finest level of detail, abstraction mechanisms to zoom out the graph at higher levels of abstraction are considered (e.g., visualize the structure of the workflow at the activity level). The visualization of the Architecture Graph at multiple levels of granularity allows the easier understanding of the overall structure of the involved scenario, especially as the scale of the scenarios grows.

The facilitation of the ETL design is realized through an extensible set of template activities [12]. In the framework proposed, there is a discrimination between logical and physical activities. Logical activities can be grouped in four main categories: filters (e.g., selection, not null, unique value), unary operations (e.g., aggregation, projection, function), binary operations (e.g., union, join, diff), and composite operations (e.g., slowly changing dimensions, format mismatch, data type conversion). Physical activities usually apply in datastores (e.g., transfer, table management, and transactional operations). The template activities proposed do not constitute a complete set of operations, rather they are generic templates of abstract operations that are core operations in practically every frequently used ETL transformation.

Furthermore, a template language allows the construction of any desirable template activity and an instantiation mechanism is responsible for the instantiation of these templates. This feature provides the logical model with two characteristics: (a) genericity, the model becomes powerful to capture ideally all the cases of ETL activities, and (b) extensibility, the model supports the extension of its built-in functionality with new, user-specific, templates.

An example template activity corresponding to the 'difference' operator is depicted in Figure 2. This operator is used for the comparison of two datasets. For example, during the extraction of a dataset from a source, a usual procedure is to compare the two snapshots of the extracted data – each one corresponding to the previous and current extraction, respectively – in order to discriminate the newly inserted/updated/deleted data. Here, assume that during the extraction process we want to detect only the newly inserted rows. Then, if PK is the set of attributes that uniquely identify rows (in the role of a primary key), the newly inserted rows can be found from the expression $\Delta_{<PK>}(R_{new}, R)$. The formal semantics of the difference operator are given by the following calculus-like definition: $\Delta_{<A_1...A_k>}(R, S) = \{x \in R | \neg \exists y \in S : x[A_1] = y[A_1] \wedge \cdots \wedge x[A_k] = y[A_k]\}$. The template activity in Figure 2, expressed in LDL++, contains an intermediate predicate in order to enhance intuition. The semijoin predicate is used so that all tuples that satisfy it should be excluded from the result. Note also that

```
a_out([i<arityOf(a_out)](A_OUT_$i$,)[i=arityOf(a_out)](A_OUT_$i$))<-
    a_in1([i<arityOf(a_in1)](A_IN1_$i$,)[i=arityOf(a_in1)](A_IN1_$i$)),
    a_in2([i<arityOf(a_in2)](A_IN2_$i$,)[i=arityOf(a_in2)](A_IN2_$i$)),
    ~semijoin([i<arityOf(a_in1)](A_IN1_$i$,)
    [i=arityOf(a_in1)](A_IN1_$i$)),
    [i<arityOf( a_out )](A_OUT_$i$= A_IN1_$i$,)
    [i=arityOf( a_out )](A_OUT_$i$= A_IN1_$i$)

semijoin([i<arityOf(a_in1)](A_IN1_$i$,)[i=arityOf(a_in1)](A_IN1_$i$))<-
    a_in1([i<arityOf(a_in1)](A_IN1_$i$,)[i=arityOf(a_in1)](A_IN1_$i$)),
    a_in2([i<arityOf(a_in2)](A_IN2_$i$,)[i=arityOf(a_in2)](A_IN2_$i$)),
    [i<arityOf( @COMMON_IN1 )](@COMMON_IN1[$i$]=@COMMON_IN2[$i$],)
    [i=arityOf( @COMMON_IN1 )](@COMMON_IN1[$i$]=@COMMON_IN2[$i$])
```

Fig. 2. An example template activity: difference activity

there are two different inputs, which are denoted as distinct by adding a number at the end of the keyword *a_in*. A simpler version of this template could be produced using macros.

2.3   Bridging the Two Worlds: From Conceptual to Logical Modeling

In the previous subsections, we have presented a conceptual (subsection 2.1) and a logical (subsection 2.2) models suitable for the ETL environment. In this subsection, we bridge the different levels of our framework by briefly presenting a semi-automatic transition from conceptual to logical models for ETL processes. (For further details, we refer the interested reader to [3,4].) By relating a logical to a conceptual model, we exploit the advantages of both worlds. On one hand, there exists a simple model, sufficient for the early stages of the data warehouse design. On the other hand, there exists a logical model that offers formal and semantically-founded concepts to capture the characteristics of an ETL process.

During the transition from one model to the other, several issues should be confronted. First, the correspondence between the two models should be identified. Since the conceptual model is constructed in a more generic and high-level manner, each conceptual entity is mapped to a logical entity; however, in general the opposite does not hold.

There are constructs of the conceptual model that can be directly mapped to the logical model; e.g., datastores and attributes. But, there exist at least two issues where the mapping between the two models is not straightforward: the mapping of the transformations and the finding of the execution order of a logical workflow.

Regarding the former issue, a logical activity is determined by its schemata and its operational semantics. An activity has input and output schemata, a set (schema) of parameters, and the schemata that describe which attributes are generated or projected out by its application. The schemata of an activity are specified by a topological sort of the conceptual design. (Recall from subsection 2.1 that the latter is represented as a graph.) This procedure is done either directly (input, output, and parameter schemata) or indirectly (generated and projected out schemata); e.g., the generated schema consists of attributes that belong to the output but not to the input schemata.

For the determination of the operational semantics special information should have been provided in the conceptual model. This information is placed in a conceptual construct called 'note' that captures designer's comments. For example, a note may contain information about the transformation type, e.g, $addAttribute$, and the transformation expression, e.g., $Date = SysDate()$. With such information, in the logical level a template activity $addAttribute$ is chosen, which apart from the schema mappings provided by the topological sort of the conceptual design, it also contains an expression: $@OutField = @Value$ that represents the name and the value, respectively, of the new attribute that will be created. After the instantiation of variables in the template expression, the resulting activity creates a new attribute called $Date$ that contains values returned by the function $SysDate()$. When such information is not available from the conceptual design, then the designer of the logical model should specify the internal semantics of the activity.

The conceptual model is not a workflow; instead, it simply identifies the mappings and the transformations needed in an ETL process. The placement of the transformations into the conceptual design does not directly specify their execution order. However, the logical model represents a workflow and thus, it is very important to determine the execution order of the activities. A method has been proposed for the determination of a correct execution order of the activities in the logical model [3,4]. Briefly, the main idea is that the transformations of the conceptual design are grouped into strata of order-equivalent transformations (intuitively, of transformations that are swappable). This stratification provides a semantically correct execution order of logical activities. Still, the resulting workflow is not unique. Different placement of order-equivalent activities is feasible; the final choice depends on the quality and the execution time of the workflow. A discussion on the latter follows in the next section.

3 Optimization of ETL Workflows

An ETL workflow usually follows a quite complex design that applies numerous operations in large volumes of data. Frequently, an ETL workflow must be completed in a certain time window and thus, it is necessary to optimize its execution time. Despite the significance of such optimization, so far the problem has not been considered in research literature [6]. Even more surprisingly, although the leading commercial tools provide advanced GUI's for the design of ETL scenarios, they do not provide/suggest/enforce any optimization technique to the created scenarios. On the contrary, a designed workflow, hopefully optimized manually by the designer, is propagated as is to the DBMS for execution and hence, in practice the DBMS undertakes the task of optimization.

This optimization policy is not enough because DBMS optimizers can interfere only in portions of a scenario and not in its entirety. This is due to the fact that ETL workflows cannot be considered as (multi-)queries. It is not possible to express all ETL operations in terms of relational algebra and then optimize the resulting expression as usual. The traditional logic-based algebraic query optimization may be blocked in an ETL scenario, basically due to existence of data manipulation functions. In addition, the cases of functions with unknown semantics – 'black-box' operations – or with 'locked' functionality – e.g., external call to a dll library – are quite often. Another complication in ETL workflows is that they may involve processes running in separate environments, usually not simultaneously and under time constraints. Hence, it is more realistic to consider and treat ETL workflows as complex transactions and not as complex queries.

In our research, we have taken a novel approach to the problem and we have set up a theoretical framework for the optimization of ETL scenarios by taking into consideration the aforementioned particularities [5,6]. The ETL workflows optimization problem is modeled as a state search problem. An ETL workflow is considered as a state. Given an initial scenario, if the placement of the operations contained in that scenario changes, then a new state is produced. If such change does not affect the global semantics of the scenario, then the state produced is equivalent to the initial one. Equivalent states comprise the state space. According to a cost model, each state may have a different cost. The goal is to develop algorithms that may find the optimal scenario in terms of the execution cost.

The optimization technique proposed lies in the logical level. Its main advantage is that it is not bounded by the operational semantics of ETL operations. On the contrary, the major concern of this method is the schemata of ETL operations. As a simple example, assume two operations a_i and a_j. Intuitively, if the parameters needed for the execution of the latter are not affected by the execution of the former operation then the positions of these two operations in the scenario can be interchanged.

A finite set of transitions that can be employed on a state is introduced. The application of a transition to a state is proved that results in another equivalent state. In our context, equivalent states are assumed to be states that based on the same input produce the same output. Practically, this is achieved in the following ways: (a) by transforming the execution sequence of the operations, i.e., by interchanging the positions of two operations in the workflow; (b) by replacing common tasks in parallel flows with an equivalent task over a flow to which these parallel flows converge; or (c) by dividing tasks of a joint flow to clone operations applied to parallel flows that converge towards the joint flow. The transitions used are the following.

- *Swap*. This transition can be applied to a pair of unary operations and interchange their sequence. Swap concerns only unary operations, e.g., selection, checking for nulls, primary key violation, projection, and function application.
- *Factorize* and *Distribute*. These operations involve the interchange of a binary operation - e.g., union, join, and difference - and at least two unary *homologous* operations that have the same functionality, but are applied over different data flows that converge towards the involved binary activity. The Factorize transition replaces the two unary operations with a new one placed right after the binary operation. Distribute is the reverse transition. Intuitively, Factorize and Distribute essentially model the swap between unary and binary operations.
- *Merge* and *Split*. These two transitions are used to 'package' and 'unpackage' a pair of operations without changing their semantics. Merge indicates that a set of operations is grouped according to the constraints of the ETL workflow; thus, for example, a third operation may not be placed between two merged operations. Split un-groups a set of operations previously grouped by a merge transition. The use of these two transitions ensures that specific design constraints remain intact and also proactively reduces the search space without sacrificing any of the design requirements.

In our work, we have experimented with three search algorithms. First, an exhaustive approach has been used to construct the search space in its entirety and to find the optimal ETL workflow. (This was used as a base metric for comparison with solutions constructed by the other algorithms.) The size of the search space increases exponentially with respect to the size of the initial ETL scenario (i.e., user scenario). Therefore, this approach is feasible only in a limited number of ETL scenarios, which contain a relative small number of ETL operations. For larger scenarios, as those mostly used in practice, this straightforward approach can not be applied mostly due to efficiency reasons and more elaborated methods should be used.

Another approach been tested uses a set of heuristics that stem from the experimentation with a large number of ETL scenarios. This heuristic algorithm reduces the search space efficiently without any significant loss in the quality of the solution proposed. A greedy version of the heuristic algorithm turns out that works satisfactory for small and medium sized ETL workflows. Experimental results on both algorithms suggest that the benefits of the proposed method can be significant. In general, the improvement of the performance of the initial ETL workflow exceeds 70% in reasonable times for the DW environment. (Execution times of the algorithms vary from few seconds up to 10 minutes depending on the size of the initial ETL scenario.)

It is noteworthy to mention that the above optimization approach may be applied to more general workflows (not just ETL). To the best of our knowledge, typical research efforts in the context of workflow management are concerned with the management of the control flow in a workflow environment. This is due to the complexity of the problem and its practical application to semi-automated, decision-based, interactive workflows where user choices play a crucial role. Therefore, our proposal for a structured management of the data flow, concerning both the interfaces and the internals of operations appears to be complementary to existing approaches for the case of workflows that need to access structured data in some kind of data store or to exchange structured data between operations.

4 Conclusions

In this paper, we have presented our work on a framework towards the modeling and optimization of ETL workflows. A conceptual model suitable for the early stages of a DW project has been described. The main challenges in that phase include the identification of the appropriate source datastores and the determination of the inter-attribute mappings along with the appropriate (abstract) transformations that constitute the ETL process. Also, a logical model that provides formal foundations for ETL workflows, has been presented. The main characteristics of that model are its genericity and extensibility. Towards the provision of an end-to-end solution for ETL processes, we have discussed the mapping between the two aforementioned models.

The facilitation of the design of ETL scenarios is one side of the coin. Although, the graph-based modeling approach that we have followed, is elegant and powerful in terms of measuring the quality and performing what-if and risk analysis, still, the optimization of the execution time of such workflows is a real challenge. For that reason, we have presented a logical optimization strategy that is not bounded by the operational semantics of the individual operations that constitute an ETL workflow. As a thorough experimental analysis has shown, the proposed algorithms significantly improve the performance of ETL workflows in reasonable times for DW environments.

Concerning our future directions, a first goal is the generalization of the techniques introduced to more generic workflows (not only ETL). Towards that, we already have some preliminary results in flows of web services. Another issue is to consider the case of ETL in more generic types of data, instead of the traditional relational data. We have already extended our techniques for semi-structured data, but, still, there are more challenges: spatial, biomedical, multimedia data and so on. The application of ETL technology in different, no static environments – e.g., p2p networks – is also of interest.

Recently, the term 'active' or 'real-time' DW was coined to capture the need for a DW containing data as fresh as possible. The periodic population of a DW belongs to the past. Modern necessities demand that OLAP and DSS should handle updated data in an on-line fashion. Thus, approximation and streaming techniques should be taken into consideration too. In that sense, traditional ETL should adapt to new on-line environment. As a preliminary effort towards this direction, we have proposed a new join operator, namely MeshJoin, which is able to handle the join of a large static relation with a stream of data [1].

Acknowledgements. We would like to thank our colleagues that have contributed in various ways to the result of this work: Panos Vassiliadis (University of Ioannina), Dimitrios Skoutas (National Technical University of Athens), Spiros Skiadopoulos (University of Peloponnese), and Manolis Terrovitis (National Technical University of Athens).

References

1. Polyzotis, N., Skiadopoulos, S., Vassiliadis, P., Simitsis, A., Frantzell, N.-E.: Supporting streaming updates in an active data warehouse. In: Proceedings of the 23rd IEEE International Conference on Data Engineering (ICDE '07), IEEE Computer Society Press, Los Alamitos (2007)
2. Rahm, E., Bernstein, P.A.: A survey of approaches to automatic schema matching. VLDB Journal 10(4), 334–350 (2001)
3. Simitsis, A.: Mapping conceptual to logical models for ETL processes. In: Proceedings of the ACM 8th International Workshop on Data Warehousing and OLAP (DOLAP '05), pp. 67–76. ACM Press, New York (2005)
4. Simitsis, A., Vassiliadis, P.: A method for the mapping of conceptual designs to logical blueprints for ETL processes. Decision Support Systems (DSS) (to appear)
5. Simitsis, A., Vassiliadis, P., Sellis, T.K.: Optimizing ETL processes in data warehouses. In: Proceedings of the 21st International Conference on Data Engineering (ICDE '05), pp. 564–575 (2005)
6. Simitsis, A., Vassiliadis, P., Sellis, T.K.: State-space optimization of ETL workflows. IEEE Transactions on Knowledge and Data Engineering 17(10), 1404–1419 (2005)
7. Simitsis, A., Vassiliadis, P., Skiadopoulos, S., Sellis, T.K.: Data Warehouses and OLAP: Concepts, Architectures and Solutions. In: Wrembel, R., Koncilia, C. (eds.) Data Warehouse Refreshment, IRM Press (2006)
8. Simitsis, A., Vassiliadis, P., Terrovitis, M., Skiadopoulos, S.: Graph-based modeling of ETL activities with multi-level transformations and updates. In: Tjoa, A.M., Trujillo, J. (eds.) DaWaK 2005. LNCS, vol. 3589, pp. 43–52. Springer, Heidelberg (2005)

9. Skoutas, D., Simitsis, A.: Designing ETL processes using semantic web technologies. In: Proceedings of the ACM 9th International Workshop on Data Warehousing and OLAP (DOLAP '06), pp. 67–74. ACM Press, New York (2006)
10. Skoutas, D., Simitsis, A.: Flexible and customizable NL representation of requirements for ETL processes. In: Proceedings of the 12th Int'l Conf. on Applications of Natural Language to Information Systems (NLDB '07), pp. 433–439 (2007)
11. Skoutas, D., Simitsis, A.: Ontology-based conceptual design of ETL processes for both structured and semi-structured data. Int'l Journal of Semantic Web and Information Systems (to appear)
12. Vassiliadis, P., Simitsis, A., Georgantas, P., Terrovitis, M., Skiadopoulos, S.: A generic and customizable framework for the design of ETL scenarios. Information Systems 30(7), 492–525 (2005)
13. Vassiliadis, P., Simitsis, A., Skiadopoulos, S.: Conceptual modeling for ETL processes. In: Proceedings of the ACM 5th International Workshop on Data Warehousing and OLAP (DOLAP '02), pp. 14–21. ACM Press, New York (2002)
14. Vassiliadis, P., Simitsis, A., Terrovitis, M., Skiadopoulos, S.: Blueprints and measures for ETL workflows. In: Delcambre, L.M.L., Kop, C., Mayr, H.C., Mylopoulos, J., Pastor, Ó. (eds.) ER 2005. LNCS, vol. 3716, pp. 385–400. Springer, Heidelberg (2005)
15. Zaniolo, C.: LDL++ Tutorial. UCLA (1998) Available at:
    http://pike.cs.ucla.edu/ldl/

Harvesting and Organizing Knowledge from the Web

Gerhard Weikum

Max-Planck Institute for Informatics
Saarbruecken, Germany
weikum@mpi-inf.mpg.de

Extended Abstract

Information organization and search on the Web is gaining structure and context awareness and more semantic flavor, for example, in the forms of faceted search, vertical search, entity search, and Deep-Web search. I envision another big leap forward by automatically harvesting and organizing knowledge from the Web, represented in terms of explicit entities and relations as well as ontological concepts. This will be made possible by the confluence of three strong trends: 1) rich Semantic-Web-style knowledge repositories like ontologies and taxonomies, 2) large-scale information extraction from high-quality text sources such as Wikipedia, and 3) social tagging in the spirit of Web 2.0. I refer to the three directions as Semantic Web, Statistical Web, and Social Web (at the risk of some oversimplification), and I briefly characterize each of them.

Semantic Web: Although the Semantic Web in its originally envisioned glorious form is still a very elusive goal, the vision itself has created a significant momentum towards creating ontologies and representing knowledge in more rigorous formats than text (see, e.g., [5,7]). These include general-purpose ontologies and thesauri such as SUMO, OpenCyc, ConceptNet, or WordNet, as well as domain-specific ontologies and terminological taxonomies such as GeneOntology, SNOMED, or UMLS. While each of these collections alone may be viewed as fairly partial, connecting them and combining them with "softer" knowledge sources such as Wikipedia could be a powerful way of organizing more and more knowledge in rigorous representations that allow effective querying and reasoning. Richly annotated natural-language corpora such as multilingual thesauri, word-sense-tagged texts, or even representations in logic-based frames start becoming an interesting asset as well.

Statistical Web: Information-extraction (IE) technology - entity recognition and learning relation patterns - has made enormous progress and become much more scalable in recent years [1,10] and also much less dependent on human supervision [3,4,8]. Much of this progress comes from major advances in the underlying fields of natural language processing (NLP) and statistical learning, but there is also a much better understanding of algorithmic efficiency and how to engineer large-scale IE. To be clear, all these technologies will remain computationally

Y. Ioannidis, B. Novikov, and B. Rachev (Eds.): ADBIS 2007, LNCS 4690, pp. 12–13, 2007.

expensive, but the gloomy picture of such issues being "AI-complete" and practically hopeless is gone.

Social Web: There is a growing amount of "low-hanging fruit" that allows us to harvest knowledge without any rocket science. A large extent of this comes from the Web 2.0 trends, or more specifically, the human contributions to the emerging Social Web (aka. Human Semantic Web) in the form of tagging (and thus semantically annotating) Web pages, passages or phrases in pages, images, videos, etc. and creating so-called folksonomies (e.g., [6,10]). Another big contributor is the strong proliferation of high-quality knowledge repositories with some explicit structure that is suitable for entity, relation, and topic recognition. Probably, Wikipedia is the best example. Although it is still primarily hyperlinked text, the link structure, the thematic categories to which articles are manually assigned, and the templates that are used for authoring certain types of articles (e.g., about music bands) provide enormous benefits for semantic tagging. Several recent projects have made excellent use of Wikipedia and similar sources for building explicit knowledge bases and connecting these with other sources (e.g., [2,9]).

Each of the three directions - Semantic Web, Statistical Web, Social Web - poses interesting research themes. I believe that connecting these different kinds of implicit and explicit knowledge sources opens up synergies and great opportunties towards the vision of large-scale knowledge management and search. The talk will present various approaches in each of three areas, discuss their strengths and weaknesses, and point out ideas on a combined methodology.

References

1. Agichtein, E., Sarawagi, S.: Scalable Information Extraction and Integration. Tutorial Slides, KDD 2006,
   http://www.cs.columbia.edu/~eugene/kdd2006_tutorial/KDD06Tutorial.pdf
2. Auer, S., Lehmann, J.: What have Innsbruck and Leipzig in common? Extracting Semantics from Wiki Content. In: ESWC 2007 (2007)
3. Banko, M., Cafarella, M.J., Soderland, S., Broadhead, M., Etzioni, O.: Open Information Extraction from the Web. In: IJCAI 2007 (2007)
4. Etzioni, O., Cafarella, M.J., Downey, D., Popescu, A.-M., Shaked, T., Soderland, S., Weld, D.S., Yates, A.: Unsupervised Named-Entity Extraction from the Web: An Experimental Study. Artif. Intell. 165(1) (2005)
5. Fellbaum, C. (ed.): WordNet: An Electronic Lexical Database. MIT Press, Cambridge (1998)
6. N. Koudas (ed.): IEEE Data Engineering Bulletin, Special Issue on Data Management Issues in Social Networks, 30(2) (June 2007)
7. Staab, S., Studer, R. (eds.): Handbook on Ontologies. Springer, Heidelberg (2004)
8. Suchanek, F.M., Ifrim, G., Weikum, G.: Combining Linguistic and Statistical analysis to Extract Relations from Web Documents. In: KDD 2006 (2006)
9. Suchanek, F., Kasneci, G., Weikum, G.: YAGO: A Core of Semantic Knowledge Unifying WordNet and Wikipedia. In: WWW 2007 (2007)
10. Suciu, D. (ed.): IEEE Data Engineering Bulletin, Special Issue on Web-Scale Data, Systems, and Semantics 29(4) (December 2006)

Schema and Data Translation: A Personal Perspective

Paolo Atzeni

Dipartimento di Informatica e Automazione
Università Roma Tre, Italy
`atzeni@dia.uniroma3.it`

Abstract. The problem of translating schemas and data form a model to another has been under the attention of database researchers for decades, but definitive solutions have not been reached.

Motivation for the problem comes from the variety of sources available in modern systems, which often use different approaches (and data models) for the organization of information.

The topic is discussed here by first setting the context with reference to the recent proposal for model management system, which considers an even wider set of requirements. Then, definitions are given for the problem of schema and data translation and for the related one concerning data exchange. Some side technical issues are then discussed: how schemas, models and mappings are described, and what is the relationship between source and target schemas in terms of information capacity. Finally, a specific proposal for data translation is discussed in some detail.

1 Introduction and Motivation

It is widely acknowledged that the need to transform, integrate and exchange data is common to many application contexts. In fact, we often use different systems to handle data, with different models, and we therefore need to translate data and their description from one to another. The problem has been considered for decades, but definitive solutions are not yet available (Abiteboul et al.[1], Haas [20]). The problem is relevant at the schema level, during the specification or design phase, and at the data level, when we have databases, and we want to translate them into some other system, which may be similar (for example, relational to relational) or completely different (for example, XML to relational or viceversa).

The recent developments in the Internet world have added more motivation to these needs, as it has become possible, at least in principle, to implement communication between systems at any level, without significant limitations in the amount of data exchanged and on the length of the interaction. Therefore bigger amounts of data, more heterogeneous than ever, have become available for multiple uses, which include on-line "multidatabase" integration (Kim et al.[24]), as well as datawarehousing (Chaudhuri and Dayal [16]) or peer-to-peer interaction (Halevy et al.[21]).

Moreover, the available information need not be in database form, but there could be interest in integrating it in some way: it might be contained in sources of

Y. Ioannidis, B. Novikov, and B. Rachev (Eds.): ADBIS 2007, LNCS 4690, pp. 14–27, 2007.

various kind, such as Web pages, XML documents, mail messages, spreadsheets, annotation over multimedia content, and many others. Also, there is often the need to reorganize and present information and data according to the user preferences and profile, or to the context and environment: in general, there are often adaptation and personalization requirements both in data representation and in query answering (De Virgilio and Torlone [18], Koutrika and Ioannidis [25]).

From all the above arguments, we derive the importance of a general need, according to which we have data at various places, in various forms and we need to exchange, replicate, integrate, transform, and adapt it. The more data is available, the more probably it is heterogeneous.

In current practice, translation problems are often tackled by means of ad-hoc solutions, for example by writing code for each specific application, but this is clearly very heavy and hard to maintain. Therefore, a major feature of any significant attempt to the problem would be *generality*: we need approaches that are maintainable and scale. Generality requires high-level descriptions of families of problems (not just individual problems); in the database field (as well as in other areas of computer science) high-level descriptions are made of *metadata*, descriptions of structures (and possibly of their meaning).

In this paper, we discuss a number of issues related to schema and data translations, with emphasis on an approach based on generality.

We will first set the framework, by devoting Section 2 to the major points of "model management," a high-level approach to problems that require the management of descriptions of application artifacts (Bernstein [10]). Then, in Section 3, we will present a general discussion of the major issues of interest, schema and data translation, and data exchange, and of a number of related points, such as information capacity dominance and equivalence and the techniques used to represent schemas and mappings in heterogeneous environments. In Section 4 we will be more specific, by presenting our project in the area, which refers to the actual development of one of the model management operators, ModelGen, as a general approach to schema and data translation. Finally in Section 5 we briefly mention concurrent projects and discuss future directions.

2 Model Management

Model management (Bernstein et al. [10,11,12]) is an approach to problems that require access, integration, and exchange of bulk heterogeneous data. It is based on the representation and management of schemas and mappings between them. The basic idea is to provide a high-level programming interface, based on a set of operators, which are defined in a generic way (that is, independently of any specific data model). It involves operators that can be composed to form scripts that give the lines along which the problems of interest can be solved. A basic set of model management operators includes the following¿

- *Match*: given two schemas, it returns a mapping between the two;
- *Merge*: given two schemas and a mapping between them, it returns an "integrated" schema;

- *Diff*: given a schema and a mapping, it returns the "subschema" that is not involved in the mapping;
- *Compose*: given two mappings, it returns their composition.

In order to show a possible application of the operators, let us refer to an example problem (Bernstein and Rahm [14]). Assume we have a database DB_1 and a datawarehouse DW_1 that contains data from DB_1; therefore, we can say that there is a mapping m_1 between DB_1 and DW_1. Assume now that we want to extend the datawarehouse with data from another database DB_2, whose schema is similar to that of DB_1, but with some differences. By using model management operators, we could proceed as follows:

1. match DB_1 and DB_2, to find the mapping, say m_2, between them;
2. compose m_2 and m_1, to obtain the mapping m_3 between DB_2 and DW_1 that describes how DB_2 (or probably a portion of it) can be mapped to the existing data warehouse;
3. use Diff to find the portion DB_2' of DB_2 that is neglected by m_3;
4. by means of a human intervention, design a new portion DW_2' of the dataware-house that can handle the data in DB_2' as well as a mapping between DB_2' and DW_2';
5. match DW_1 and DW_2' and then merge them, to obtain the new datawarehouse DW_2.

Obviously, the problem specifications that are obtained in this way are not easy to implement. Indeed, a lot of research has been devoted recently to both the precise definition and implementation of the various operators and to the actual clarification of the features of the mapping definition languages as well as of the generality of the data models of interest. Two important observations are useful. First of all, the operators can be defined with respect to schemas, but it is even more important (and more challenging) to have extended versions that are applied to instances as well. The second issue is that the involved schemas can refer to different models. For example (Bernstein [10, Sec.4.3]) let us assume that we have a conceptual (say, ER) schema C_1 and its relational implementation R_1 and assume that the implementation is changed (for some reason) to R_2; we would like to change the specification in a suitable way. Again, we could use Match, Compose, and Diff to find the "new" portion R_2' of the implementation, but then we would need to translate it to the conceptual model; indeed, there exist known techniques to "reverse engineer" a relational schema to the ER model, and we could apply them, but techniques change somehow if we have different data models involved. Indeed, a specific operator is included in model management for this purpose:

- *ModelGen*: given a (source) data model M_1, a (source) schema S_1 (in data model M_1) and a (target) data model M_2, it produces a schema S_2 in M_2 that suitably corresponds to S_1; the data level extension, given also a database D_1 over schema S_1, generates a corresponding database D_2.

3   Schema and Data Translation, a Long Standing Issue

3.1   Schema Translation and Data Exchange

The specification of the modelGen operator has been considered in the database literature since the beginning. As soon as different data models were conceived, for example the hierarchical, the network and the relational one, the translation of schemas and data from one to another was studied; for example, the seminal paper on the ER model (Chen [17]) discusses how ER schemas can be translated to the various logical models. Many other contributions followed, including Lien [27] and Markowitz and Shoshani [30]. However, most of, if not all, the approaches are data model specific, in the sense that they work for a specific pair of models, a source and a target one.

A problem that can be related to schema translation, and that has been deeply studied in the literature recently goes under the name of *data exchange* (Miller et al.[19,20,35,38]): given a source schema S_1 and a target schema S_2, find a suitable transformation that maps each instance I_1 of S_1 to a corresponding instance I_2 of S_2.

The two approaches, schema translation and data exchange, can be seen as complementary (Bernstein and Melnik [12, Sec.3]), as being the two ways to obtain a transformation (an "executable mapping") between a source and a target: in schema translation only the source schema is given, whereas in data exchange they are both given. Data exchange can be seen as a process in two steps. First correspondences between schema elements are found (for example by using some form of the Match operator mentioned in the previous section); then correspondences are used as the basis to produce a complete translation (for example, a set of SQL queries). The second step is in turn usually divided in two substeps, with the first one that produces an intermediate, declarative specification of the mapping in terms of constraints, and the second one that actually "compiles" the constraints into an executable form.

Another point of view for complementarity between translation and data exchange is the following: data translation (that is, the translation of schemas *and* data) can be seen as composed of schema translation (the schema level of ModelGen) followed by a data exchange step that can take advantage of the two schemas, the given source one and the target one generated by schema translation, as well as of the schema level mapping, also produced by ModelGen.

3.2   Information Capacity in Schema Translation

Schema and data translation and data exchange share an important requirement, which we have kept in an implicit form so far: what is the relationship that should hold between the source and target schemas (and instances)? We have said, in a superficial way, that we want a target schema and database that "suitably correspond" to the source ones. Now, what does "suitably correspond" means? Indeed, the comparison of database schemas is also a long studied problem, under various keywords that include "schema equivalence" and "information capacity dominance" (Atzeni et al.[3], Hull [22], McGee [31], Miller et al.[33,36]). To make a long story short, we can say that the requirement we would like to enforce is, at the instance level, the fact that the source instance can be precisely reconstructed from the target one and, at the schema level, the fact that, for each instance of the source schema, there is an instance of the target schema from which it can be reconstructed. Then, a number of issues arise, beginning with the expressive power of the language used to establish the correspondence (Atzeni et al.[3], Hull [22]), which can lead to different notions of dominance and equivalence. In general, this theory is indeed elegant, but has mainly negative results and involve difficulties in the proper interpretation of results, as some tricky cases can emerge.

It is interesting to see the relationship that exists between information capacity dominance and the basic properties of mappings between the sets of instances of two schemas [36,33]. Dominance corresponds to a mapping between source and target instances that is a total injective function; the function need not be surjective, and therefore dominance requires a bijection between the set of source instances and a subset of the target ones. This is indeed a common requirement when we want to be sure that the source can be properly implemented in terms of the target. General observations that can be made when trying to study information capacity in translation settings, especially heterogeneous, is that general characterizations are difficult, and desirable properties need not hold: for example, in schema translation, we could have a source and target data model with different expressive power, so that there is no equivalent nor dominated schema that can be obtained as a result of an intuitive translation, and some ad-hoc step may be needed; similarly, in data exchange, it could be the case that the source and target schema (which are given in this case) are not really comparable, and therefore some piece of information gets lost in the transformation. However, information capacity can be very useful in guiding the process and in avoiding some clearly wrong choices (Miller et al.[33]).

3.3   Models, Schemas, Mappings

Let us briefly comment on the artifacts of interest for our problems. We have seen in Section 2 that model management operators manipulate schemas and mappings, and we have kept referring to them in Section 3. Now we want to briefly discuss the possible ways to represent schemas and mappings, especially in heterogeneous settings.

The two major issues are related to the choice of representing models in an explicit way and to how details are represented, or just omitted, in order to look for simplicity.

The choices are mainly related to the specific problem dealt with. Some pieces of work that deal with transformations consider rather simple formalisms (for example, a tree based model, Abiteboul et al.[2]), that include nesting but without much detail, so that both flat models (such as the relational one) and nested ones (for example in the XML world) can be represented as special cases. They make use of graphs, which can become more complex, with typing and constraints (Miller et al.[33]), and include some pieces of metalevel information (Melnik et al. [32]), which describes the types of the various elements.

The precise description of models is important if we want to handle translations that explicitly refer to them. By following a database approach, we can think that models abstract schemas in the same way as schemas abstract instances. Therefore, we could introduce the notion of a "metaschema" that corresponds to the description of a model (Mark and Roussopoulos [29]). Then, an issue arises: which are the elements that can appear in a model? In order to answer the question we can start from the observation (Hull and King [23]) that the constructs used in most known models can be expressed by a limited set of generic (i.e. model-independent) *metaconstructs*: lexical, abstract, aggregation, generalization, function. As a consequence, it becomes possible to fix a set of metaconstructs (a *metamodel*) and to define a model by specifying the constructs (with the corresponding metaconstructs) it involves. We will give some more details on this in the next section, as this is the basis for our approach.

Another proposal, with similar goals but different features is that of Barsalou and Gangopadhyay [9], who proposed an extensible model, with three levels and a set of meta-types; specialization and refinement can be used to extend the model, as meta-constructs are organized in an inheritance hierarchy (lattice) of predefined concepts.

It is worth noting that there are contexts where the distinction between schemas and instances is not so sharp as it is with databases; then, more elaborate techniques would be needed; this could be the case with a "semistructured" use of XML, or for some features in the semantic Web world, for example with RDF [28] or Topic Maps [15]; in particular, the possibility of multiple levels of "instance of" would definitely change the picture. This can turn out to be interesting if the translations involve "schematic heterogeneity", that is, the possibility of transforming schema elements into data and viceversa (Lakshmanan et al. [26], Miller [34]).

The notion of mapping has been considered in even more different ways. At one extreme, we have specifications that are just loose correspondences between elements. For example in some graph based representations of schemas (for example, see Melnik et al [32]), mappings can be specified as binary relations between nodes (which represent schema elements). It has also been argued for the need to "reify" mappings (Bernstein [10]), that is, to represent a mapping as an additional schema, with binary relations between it and the actual schemas involved in the mapping. More sophisticated approaches are needed in the context of data exchange. The most popular project (the Clio project [19,35,38]) assumes one starts with simple correspondences and wants to obtain executable code. More recently, Bernstein and Melnik [12] have commented on the distinction between "approximate" mappings (essentially correspondences) and "engineered" mappings (executable specifications with robust implementations).

4   Model Independent Schema and Data Translation

In this section, we describe our long term project on schema and data translation, an implementation of the ModelGen operator. There have been two main phases in our activities, the first mainly concentrated on schema translation, which led to the MDM tool (Atzeni and Torlone [4]), and the second focusing on a flexible, customizable environment for schema *and* *data* translation, the MIDST tool (Atzeni et al. [7,8]). Let us discuss the main issues of each in turn.

MDM introduces the notion of a *metamodel*[1] as composed of a set of generic metaconstructs [4,7]; this follows the observation we mentioned in Sec. 3.3 that constructs in the various models are similar to one another (Hull and King [23]). Then, a model is defined by its constructs and the metaconstructs they refer to. For example

- the relational model involves (i) aggregations of lexicals (the tables), with some properties, for example the indication, for each component (a column), of whether it is part of the key or whether nulls are allowed; (ii) foreign keys defined over components of aggregations;
- a simplified OR model has (i) abstracts (tables with system-managed identifiers); (ii) lexical attributes of abstracts (for example Name and Address), each of which can be specified as part of the key; (iii) reference attributes for abstracts, which are essentially functions from abstracts to abstracts.

This approach involves also the notion of the *supermodel*, a model that has constructs corresponding to all the metaconstructs known to the system. Thus, each model is a specialization of the supermodel and a schema in any model is also a schema in the supermodel, apart from the specific names used for constructs. The translation of a schema from one model to another is defined in terms of translations over the metaconstructs. The supermodel acts as a "pivot" model, so that it is sufficient to have translations from each model to and from the supermodel, rather than translations for every pair of models. Thus, a linear and not a quadratic number of translations is needed. Moreover, since every schema in any model is an instance of the supermodel, the only needed translations are those within the supermodel with the target model in mind; a translation is performed by eliminating constructs not allowed in the target model, and possibly introducing new constructs that are allowed.

Each translation in MDM is built from elementary transformations, which are essentially elimination steps. So, a possible translation from the ER model to the relational would have two elementary transformations (i) one that eliminates many-to-many relationship (introducing new entities instead), and (ii) a second that replaces entities and one-to-many relationships with tables and foreign keys (the traditional steps in translating from the ER to the relational model). Essentially, MDM handles a library of elementary transformations and uses them to implement complex transformations.

The major limitation of MDM with respect to our problem is that it considers schema translations only and it does not address data translation at all.

---

[1] Some authors (including Bernstein [10]) would use the term "metametamodel" instead, as they use the term model for what we call schema here, and metamodel for what we call (data) model.

Our current project, *MIDST (Model Independent Data and Schema Translation)*, improves MDM with respect to the following aspects:

- It has a *dictionary* that includes three parts (i) the meta-level that contains the description of models, (ii) the schema-level that contains the description of schemas; (iii) the data-level that contains data for the various schemas. The first two levels are described in detail in Atzeni et al. [6] and the third in the subsequent paper [7]. The dictionary is "exposed" and so it can be the basis for specifying translations.
- The elementary translations are also visible and independent of the engine that executes them. They are implemented by rules in a Datalog variant with Skolem functions for the invention of identifiers; this enables one to easily modify and personalize rules and reason about their correctness.
- The translations at the data level are also written in Datalog and, more importantly, are generated almost automatically from the rules for schema translation. This is made possible by the close correspondence between the schema-level and the data-level in the dictionary.
- Mappings between source and target schemas and data are generated as a by-product, by the materialization of Skolem functions in the dictionary.

Implementations of the tool have been recently demonstrated, first for the schema level and then for the schema and data level (Atzeni et al. [5,8]).

Let us comment on the major features of MIDST and on the main ideas upon which it is based.

The translation of schemas leverages on the organization of our dictionary, which is based on a relational approach. A schema is described in the dictionary as a set of schema elements, with references to both its specific model and the supermodel [6]. For example, an entity of an ER schema is described both in a table, say ER_ENTITY, referring to the ER model and in a supermodel table SM_ABSTRACT, corresponding to the abstract metaconstruct to which the entity construct refers. Similarly, a class of a UML diagram gives rise to a tuple in a specific table UML_CLASS and to one in SM_ABSTRACT again, because classes also correspond to abstracts. Indeed, our translation process includes steps ("copy rules") that guarantee the alignment of the two representations.

The supermodel's structure is pretty compact. In our relational implementation, it has a table for each construct. We currently have a dozen constructs, which are sufficient to describe a large variety of models. Translation rules are expressed using supermodel constructs. Therefore, they can translate any construct that corresponds to the same metaconstruct, without having to rewrite rules for each construct of a specific model. Therefore, we concentrate here on the portion of the dictionary that corresponds to the supermodel, as it is the only one really relevant for translations.

Fig. 1 shows an excerpt from the dictionary of MIDST, with two tables that represent "abstracts" (the supermodel term for "entity" or "class") and their attributes. The tuple with OID 201 in table SM_ATTRIBUTEOFABSTRACT belongs to schema 1 (in the same way as all other elements shown, as sOID always

SM_ABSTRACTS		
OID	sOID	Name
101	1	Employees
102	1	Departments
...	...	...

SM_ATTRIBUTEOFABSTRACT						
OID	sOID	Name	IsKey	IsNullable	AbsOID	Type
201	1	EmpNo	T	F	101	Integer
202	1	Name	F	F	101	String
203	1	Name	T	F	102	String
204	1	Address	F	F	102	String
...	...	...	...	...	...	...

Fig. 1. A small portion of the model-generic dictionary

equals 1). It has some properties, including Name, with value "EmpNo," and IsKey with value true, and a reference AbsOID with value 101: these say that in schema 1 there is an attribute EmpNo, which is part of the key of entity EMPLOYEES (the abstract with OID 101).

As in the MDM approach, translations in MIDST are built by combining elementary translations. Each elementary translation is specified by means of a set of rules written in a Datalog variant with Skolem functors for the generation of new identifiers. Elementary translations can be easily reused because they refer to the constructs in supermodel terms, and so each of them can be applied to all constructs that correspond to the same metaconstruct. The actual translation process includes an initial step for "copying" schemas from the specific source model to the supermodel and a final one for going back from the supermodel to the target model of interest.

We illustrate the major features of our rules by means of an example, which refers to the translation from the ER model to the relational one. Specifically, we show the simple rules that generate a table (an "aggregation of lexicals," in supermodel terms) for each entity ("abstract"), and its columns for the attributes of the entity. The following rule translates each abstract into an aggregation of lexicals, with the same name, and it is pretty intuitive:

```
SM_AGGREGATIONOFLEXICALS(
          OID: #aggregationOID_1(oid),
          sOID: target,
          Name: name)
    ← SM_ABSTRACT(
          OID: oid,
          sOID: source,
          Name: name)
```

A major feature of our rules is the use of Skolem functors to enforce correlation between the constructs created by different rules: functors are used to create new OIDs for the elements the rule produces in the target schema.[2] The head of the rule above has one functor: #aggregationOID_1 which generates the OID

---

[2] A brief comment on notation: functors are denoted by the # sign, include the name of the construct whose OIDs they generate (here often abbreviated for convenience), and have a suffix that distinguishes the various functors associated with a construct.

SM_INSTOFABSTRACT		
OID	dOID	AbsOID
1001	1	101
1002	1	101
...	...	...
1005	1	102
1006	1	102

SM_INSTOFATTRIBUTEOFABSTRACT				
OID	dOID	AttOID	i-AbsOID	Value
2001	1	201	1001	134
2002	1	202	1001	Smith
2003	1	201	1002	201
...	...	...	...	...
2011	1	203	1005	A
...	...	...	...	...

Fig. 2. Representation of an instance

for the new construct. If we consider the rule that generates columns for tables from attributes of entities, then the same functor would be used to correlate the columns with the appropriate table:

SM_COMPONENTOFAGGREGATION(
        OID: #componentOID_1(attOid),
        sOID: target,
        Name: name,
        IsKey: isId,
        IsNullable: isN,
        AggrOID: #aggregationOID_1(absOid),
        Type: type)
← SM_ATTRIBUTEOFABSTRACT(
        OID: attOid,
        sOID: source,
        Name: name,
        IsIdent: isId,
        IsNullable: isN,
        AbsOID: absOid,
        Type: type)

The main novel aspect of MIDST is the management of translations of actual data, derived from the translations of schemas. This is made possible by the use of a dictionary for the data level, built in close correspondence with the schema level one. We give the main ideas here, referring the interested reader to the original paper (Atzeni et al. [7]) for more information.

Data are described in a portion of the dictionary whose structure is automatically generated and is similar to the schema portion. A portion of the representation of an instance of the schema described by the dictionary in Fig. 1 is shown in Fig. 2. We do not have space to comment on the details, but we can say that each table in Fig. 2 has a row for each value or instance of construct in the schema: a row for each instance of each entity and a row for each value of each attribute.

The above representation for instances is clearly an "internal" one, into which or from which actual database instances or documents have to be transformed.

We have developed import/export features that can upload/download instances and schemas of a given model. This representation is somewhat onerous in terms of space, so we are working on a compact version of it that still maintains the close correspondence with the schema level, which is its main advantage.

The close correspondence between the schema and data levels in the dictionary allows us to automatically generate rules for translating data, with minor refinements in some cases. Indeed, the data level rule for the first rule shown above is the following:

I_SM_AGGREGATIONOFLEXICALS(
        OID: #i_aggregationOID_1(i_oid),
        dOID: i_target,
        AggOID: #aggregationOID_1(oid))
   ← I_SM_ABSTRACT(
        OID:i_oid,
        dOID: d_source,
        AbsOID: oid),
      SM_ABSTRACT(
        OID: oid,
        sOID: source,
        Name: name)

Without entering into the details, we can say that the rule generates data-level objects: it produces a data-level object (in this case, an instance of an aggregation) for each data-level object of the source database that is an instance of the source schema-level object (in the example, for each instance of the abstract that generates the aggregation). Let us note that the body of the data-level rule contains a generated data-level part as well as the body of the schema-level rule, which is needed to maintain the selection condition specified at the schema level. In this way the rule translates only instances of the schema element selected within the schema-level rule.

As we discussed in Sec.3.2, correctness is usually modelled in terms of dominance or equivalence of information capacity. However, we saw that formal characterizations and positive results are very rare. Therefore, given the genericity of our approach, it seems hopeless to aim at showing correctness in general. However, this is only a partial limitation, as we are developing a platform to support translations, and some responsibilities can be left to its users (specifically, rule designers, who are expert users), with system support. We briefly elaborate on this issue.

We follow the argument that can be called "axiomatic" (Atzeni and Torlone [4]): we assume the basic translations to be correct, a reasonable assumption as they refer to well-known elementary steps developed over the years. It is the responsibility of the rule's designer to specify basic translations that are indeed correct. So given a suitable description of models and rules in terms of the involved constructs, complex translations can be proven correct by induction.

In MIDST, we have the additional benefit of having schema level transformations expressed at a high-level, as Datalog rules: we can automatically detect which constructs are used in the body and generated in the head of a Datalog rule and then derive the signature. Since models and rules are expressed in terms of the supermodel's metaconstructs, by induction, the same can be done for the model obtained by applying a complex transformation.

For correctness at the data level, we can reason in a similar way. The main issue is the correctness of the basic transformations, as that of complex ones would follow by induction.

The validity of the approach, given the unavailability of formal results, has been evaluated by means of an extensive set of test cases, which have produced positive results. We defined a set of significant models, extended-ER, XSD, UML class diagrams, object-relational, object-oriented and relational, each in various versions (with and without nested attributes and generalization hierarchies), and tested translations over them.

5 Conclusion

There are other recent approaches to the same probelms, including those by Papotti and Torlone [37] and by Bernstein et al.[13]. Atzeni [7] and Bernstein and Melnik [12] provide brief comparisons and discussions.

As Bernstein and Melnik [12] recently noted, an ambitious goal not yet reached by current proposals is the generation of executable mappings that can be embedded in the run-time components of the actual systems: this is clearly not possible with our current tool, which provides off-line translation. This direction is definitely worth attention.

A second important direction we are working at is the development of a more general tool that allow for the integration of translation steps with other generic model management operators, in such a way that the scenarios discussed in Sec.2 can be supported.

References

1. Abiteboul, S., et al.: The Lowell database research self-assessment. Commun. ACM 48(5), 111–118 (2005)
2. Abiteboul, S., Cluet, S., Milo, T.: Correspondence and translation for heterogeneous data. Theor. Comput. Sci. 275(1-2), 179–213 (2002)
3. Atzeni, P., Ausiello, G., Batini, C., Moscarini, M.: Inclusion and equivalence between relational database schemata. Theoretical Computer Science 19(2), 267–285 (1982)
4. Atzeni, P., Torlone, R.: Management of multiple models in an extensible database design tool. In: Apers, P.M.G., Bouzeghoub, M., Gardarin, G. (eds.) EDBT 1996. LNCS, vol. 1057, pp. 79–95. Springer, Heidelberg (1996)
5. Atzeni, P., Cappellari, P., Bernstein, P.A.: Modelgen: Model independent schema translation. In: ICDE, Tokyo, pp. 1111–1112. IEEE Computer Society Press, Los Alamitos (2005)
6. Atzeni, P., Cappellari, P., Bernstein, P.A.: A multilevel dictionary for model management. In: Delcambre, L.M.L., Kop, C., Mayr, H.C., Mylopoulos, J., Pastor, Ó. (eds.) ER 2005. LNCS, vol. 3716, pp. 160–175. Springer, Heidelberg (2005)

7. Atzeni, P., Cappellari, P., Bernstein, P.A.: Model-independent schema and data translation. In: EDBT, pp. 368–385 (2006)
8. Atzeni, P., Cappellari, P., Gianforme, G.: Midst: model independent schema and data translation. In: Sigmod, pp. 1134–1136 (2007)
9. Barsalou, T., Gangopadhyay, D.: M(dm): An open framework for interoperation of multimodel multidatabase systems. In: ICDE, pp. 218–227. IEEE Computer Society Press, Los Alamitos (1992)
10. Bernstein, P.A.: Applying model management to classical meta data problems. In: CIDR, pp. 209–220 (2003)
11. Bernstein, P.A, Halevy, A.Y., Pottinger, R.: A vision of management of complex models. SIGMOD Record 29(4), 55–63 (2000)
12. Bernstein, P.A., Melnik, S.: Model management 2.0: manipulating richer mappings. In: Sigmod, pp. 1–12 (2007)
13. Bernstein, P.A., Melnik, S., Mork, P.: Interactive schema translation with instance-level mappings. In: VLDB, pp. 1283–1286 (2005)
14. Bernstein, P.A., Rahm, E.: Data warehouse scenarios for model management. In: ER, pp. 1–15 (2000)
15. Biezunski, M., Bryan, M., Newcomb, S.R. (eds.): ISO/IEC 13250:2000 Topic Maps: Information technology – document description and markup languages (1999), http://www.y12.doe.gov/sgml/sc34/document/0129.pdf
16. Chaudhuri, S., Dayal, U.: An overview of data warehousing and olap technology. SIGMOD Record 26(1), 65–74 (1997)
17. Chen, P.P.: The entity-relationship model: Toward a unified view of data. ACM Transactions on Database Systems 1(1), 9–36 (1976)
18. De Virgilio, R., Torlone, R.: Modeling heterogeneous context information in adaptive web based applications. In: Wolber, D., Calder, N., Brooks, C., Ginige, A. (eds.) ICWE, pp. 56–63. ACM Press, New York (2006)
19. Fuxman, A., Hernández, M.A., Howard Ho, C.T., Miller, R.J., Papotti, P., Popa, L.: Nested mappings: Schema mapping reloaded. In: VLDB, pp. 67–78 (2006)
20. Haas, L.M.: Beauty and the beast: The theory and practice of information integration. In: Schwentick, T., Suciu, D. (eds.) ICDT 2007. LNCS, vol. 4353, pp. 28–43. Springer, Heidelberg (2006)
21. Halevy, A.Y., Ives, Z.G., Madhavan, J., Mork, P., Suciu, D., Tatarinov, I.: The Piazza peer data management system. IEEE Trans. Knowl. Data Eng. 16(7), 787–798 (2004)
22. Hull, R.B.: Relative information capacity of simple relational schemata. SIAM Journal on Computing 15(3), 856–886 (1986)
23. Hull, R.B., King, R.: Semantic database modelling: Survey, applications and research issues. ACM Computing Surveys 19(3), 201–260 (1987)
24. Kim, W., Choi, I., Gala, S.K., Scheevel, M.: On resolving schematic heterogeneity in multidatabase systems. In: Modern Database Systems, pp. 521–550. ACM Press, Addison-Wesley (1995)
25. Koutrika, G., Ioannidis, Y.E.: Personalized queries under a generalized preference model. In: ICDE, pp. 841–852. IEEE Computer Society Press, Los Alamitos (2005)
26. Lakshmanan, L.V.S., Sadri, F., Subramanian, I.N.: SchemaSQL - a language for interoperability in relational multi-database systems. In: VLDB, pp. 239–250 (1996)
27. Lien, Y.E.: On the equivalence of database models. Journal of the ACM 29(2), 333–362 (1982)
28. Manola, F., Miller, E. (eds.): RDF primer, http://www.w3.org/TR/2004/REC-rdf-primer-20040210/

29. Mark, L., Roussopoulos, N.: Integration of data, schema and meta-schema in the context of self-documenting data models. In: Proceedings of the 3rd Int. Conf. on Entity-Relationship Approach (ER'83), pp. 585–602. North-Holland, Amsterdam (1983)
30. Markowitz, V.M., Shoshani, A.: On the correctness of representing extended entity-relationship structures in the relational model. In: SIGMOD, pp. 430–439 (1989)
31. McGee, W.C.: A contribution to the study of data equivalence. In: IFIP Working Conference Data Base Management, pp. 123–148 (1974)
32. Melnik, S., Rahm, E., Bernstein, P.A.: Rondo: A programming platform for generic model management. In: Halevy, A.Y, Ives, Z.G., Doan, A. (eds.) SIGMOD Conference, pp. 193–204. ACM Press, New York (2003)
33. Miller, R.J, Ioannidis, Y.E., Ramakrishnan, R.: Schema equivalence in heterogeneous systems: bridging theory and practice. Inf. Syst. 19(1), 3–31 (1994)
34. Miller, R.J: Using schematically heterogeneous structures. In: SIGMOD, pp. 189–200 (1998)
35. Miller, R.J., Haas, L.M., Hernández, M.A.: Schema mapping as query discovery. In: VLDB, pp. 77–88 (2000)
36. Miller, R.J., Ioannidis, Y.E., Ramakrishnan, R.: The use of information capacity in schema integration and translation. In: VLDB, pp. 120–133 (1993)
37. Papotti, P., Torlone, R.: Heterogeneous data translation through XML conversion. J. Web Eng. 4(3), 189–204 (2005)
38. Popa, L., Velegrakis, Y., Miller, R.J., Hernández, M.A., Fagin, R.: Translating Web data. In: VLDB, pp. 598–609 (2002)

A Protocol Ontology for
Inter-Organizational Workflow Coordination

Eric Andonoff, Wassim Bouaziz, and Chihab Hanachi

IRIT/UT1
Université Toulouse 1, 2 Rue de Doyen Gabriel Marty
31042, Toulouse Cedex 9, France
{Eric.Andonoff,Wassim.Bouaziz,Chihab.Hanachi}@univ-tlse1.fr

Abstract. As coordination is a central issue in Inter-Organizational Workflow (IOW), it is quite natural to model it as a specific entity. Moreover, the structure of the different IOW coordination problems is amenable to protocols. Hence, this paper show how these protocols could be modelled and made accessible to partners involved in an IOW. More precisely, the paper proposes a coordination protocol ontology for IOW and explains how workflow partners can dynamically select them. This solution eases the design and development of IOW systems by providing autonomous, reusable and extendable coordination components. This solution also supports semantic coordination through the use of the protocol ontology, and by making protocols shared resources exploitable in both design and execution steps.

1 Introduction

The aim of *Inter-Organizational Workflow* (IOW) is to support the cooperation between distributed and heterogeneous business processes running in different autonomous organizations. The different organizations involved in an IOW need to put resources and skills in common, and coordinate their respective business processes in order to reach a common goal, corresponding to a value-added service. Thus, IOW is a key technology for helping participating organizations to face the emergence of the open and dynamic worldwide economy [1].

Coordination of these different distributed, heterogeneous and autonomous business processes is a fundamental problem in IOW. By coordination, we mean all the work needed to put all these processes together in order to fulfil the common goal.

This coordination problem can be investigated in the context of two different scenarios: loose IOW and tight IOW [2]. In this work, we focused on *loose IOW* which refers to occasional cooperation between organizations, free of structural constraints, where the organizations involved and their number are not pre-defined but are selected at run time and in an opportunistic way.

Coordination in loose IOW raises several specific problems corresponding to four sequential stages at run time: (i) finding partners able to realize a workflow service (business process), (ii) negotiating of a workflow service between partners according to criteria such as due time, precision, visibility of the process evolution or way of

Y. Ioannidis, B. Novikov, and B. Rachev (Eds.): ADBIS 2007, LNCS 4690, pp. 28–40, 2007.

doing it, (iii) contracts signature between partners and (iv) synchronization of the distributed and concurrent execution of the IOW' workflow services [3].

More precisely, finding partners consist in selecting one or several provider organizations able to execute a workflow service, which is needed for a requested organization. Thus, after finding partners, a workflow service requested is connected to different workflow service providers. But connecting is not enough to definitively select the partner that is going to execute the requested workflow service: a negotiation step, in terms of due time, price, visibility of the service evolution and way of executing the service, is necessary to evaluate and select the best workflow provider. Contract specification follows up negotiation and permits the formal description of both the requested workflow service and the conditions of its execution. Finally, the execution of the requested workflow service implies its synchronization with the other workflow services which belong to the organizations involved in the IOW.

In fact, whatever the coordination problem considered in loose IOW[1], it follows a recurrent schema. After an informal interaction, the participating partners commit to follow a strict coordination/interaction protocol. This protocol rules the conversation by a set of laws which constraint the behaviour of the participating partners, assigns roles to each of them, and therefore organizes their cooperation.

Since coordination protocols constitute well identifiable and recurrent coordination patterns in IOW, it is useful to isolate them and to study, design and implement them as first-class citizen entities so as to allow the different Workflow Management Systems (WfMS) involved in an IOW to share protocols and reuse them at run-time. Doing so, we apply the principle of separation of concerns, recognized as a good design practice from a software engineering point of view [4], in order to separate individual and intrinsic capabilities of each workflow system from what relates to IOW coordination. The application of this principle has historically led to the advent of new technologies in Information Systems: data, user interfaces and more recently business processes have been successively pushed out of applications and led to specific software to handle them. Following this principle in IOW, we argue that coordination protocols have to be pushed out of IOW applications.

Following this principle, we have proposed to define a *Protocol Management System* (PMS) [5]. The PMS is a server of coordination protocols whose aim is to help the different WfMS involved in an IOW to deal with their coordination problems: finding partners, negotiation between partners, contract specification and execution. Given a coordination problem, the PMS helps a WfMS to choose an adequate coordination protocol and to deploy it. More precisely, the PMS offers three main services: the description of coordination protocols useful in the IOW context, their selection and their execution.

While our previous work reported in [5] specifies the PMS architecture (its components and their interactions), the present paper addresses two new and related issues: how to describe IOW coordination protocols and how to dynamically select them?

Moreover, this paper addresses these two problems in the light of the Semantic Web. On the one hand, the Web constitutes a perfect computing context for IOW, notably since it provides many facilities to ease syntactic communication and

---

[1] In the following of the paper, the term IOW will correspond to loose IOW.

interoperability between different organizations. On the other hand, the semantic Web approach, and more precisely ontologies, constitutes a complete solution for sharing knowledge [6] and solving semantic interoperability issues. Indeed, ontologies permit to describe and access common business vocabulary and shared coordination protocols in an explicit, machine readable and sharable way, thus facilitating semantic communication and interoperability between organizations. Thus, considering this computing context, we contribute to make IOW applications less restrictive and more dynamic, and automated IOW coordination, possible.

The main contribution of the paper is a coordination protocol ontology for IOW. This ontology defines, at a meta-level, the general concepts of IOW coordination protocols and classifies them according to the coordination problem they have to solve (finding partners, negotiation between partners...). Moreover, the paper shows how to exploit this ontology in order to dynamically select a coordination protocol.

To the best of our knowledge, this approach is novel in the context of IOW. Indeed, as discussed in section 5, existing propositions in the IOW literature do not provide a framework to deal with the whole set of IOW coordination problems, but rather provide ad hoc solutions to address a specific IOW coordination problem and implement a single coordination protocol [7], [8]. In our proposition, we do not want to force an organization to commit to a single coordination protocol, but to allow it to dynamically choose the most suitable coordination protocol, according to the coordination problem to be solved and according to the type of interaction the organization is able to participate in.

We have used Protégé-2000 [9] and OWL [10] to define this ontology. We also have used the nRQL (new Racer Query Language) language [11] to exploit the OWL ontology and, consequently, to dynamically select IOW coordination protocols.

This paper is organized as follows. Section 2 and 3 present the coordination protocol ontology. Section 2 defines the general concepts shared by all coordination protocols. Section 3 gives a classification of coordination protocols useful in the IOW context and also introduces the main axioms of the coordination protocol ontology. Section 4 shows how an organization involved in an IOW can dynamically select an adequate coordination protocol. Finally, section 5 compares our proposition with related works and concludes the paper.

2   The Coordination Protocol Ontology

The coordination protocol ontology gives a declarative and explicit representation of protocols that are useful in the IOW context. This ontology defines the invariant structure shared by all the protocols. It is defined around inter-related concepts, each of them highlighting a different aspect of a coordination protocol. Figure 1 below gives a graphical representation of these general concepts using the notions of classes (visualised as rectangles), class properties (visualised as ellipses and linked to their classes with dotted lines) and relationships between classes (visualised as oriented full lines). These concepts have been specified in Owl using Protégé-2000 (see Appendix 1).

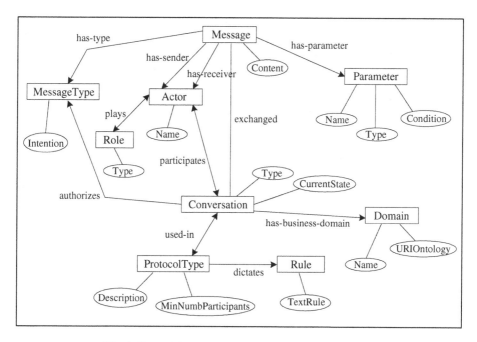

Fig. 1. General Concepts for IOW Coordination Protocols

The *ProtocolType* class abstracts the different types of coordination protocols that are useful in the IOW context. A *Conversation* is an occurrence of a Protocol Type. A protocol type is used in a conversation and coordinates its execution by authorizing types of messages and by enforcing the set of rules which constraint the behaviour of the participating actors. The *Conversation* class describes the conversations through their type (finding partners, negotiation between partners...) and their current state (open, closed), while the *MessageType* class defines the set of messages authorized in the context of a coordination protocol type. The *Intention* property of the MessageType class is a text that indicates the intention of messages of a given type. The *Rule* class describes the rules that govern a coordination protocol. Finally, a protocol type also includes a description in natural language for documentation purpose, and the minimal number of partners which have to be involved in the protocol.

The *Conversation* class is linked to the *Domain* class, which defines the business vocabulary of the conversation. The Conversation class is also linked to the *Actor* class, which defines the actors participating, i.e. playing a role, in a conversation. The different roles are described in the *Role* class. In the IOW context, whatever the protocol being considered, we have mainly three roles (Provider, Requester, Moderator) described through the value of the *Type* property. A partner involved in an IOW may be a *Provider* or a *Requester* of a workflow service. A *Moderator* manages a single conversation and has the same lifetime as the conversation it manages. It plays the type of coordination protocol underlying the conversation, grants roles to providers and requesters involved in the conversation, and ensures that any message that takes place in the conversation is compliant with the rules of the type of protocol it plays [12].

Moreover, several messages are exchanged during a conversation between the actors of the conversation. These messages are described in the *Message* class. Each message corresponds to a message type and has a sender, a receiver, a content, and eventually input or output parameters along with their associated conditions (preconditions or post-conditions). These parameters are described in the *Parameter* class.

To illustrate the meta-model concepts, let us take the example of the Contract Net Protocol (CNP) type [13]:

- CNP involves at least two actors: a single manager and one or several contractors.
- The requester role is played by the manager while the provider one is played by the contractors,
- The governing rules are the following: at the beginning the manager has a task (a workflow in our context) to subcontract, submits (MessageType) the specification of the task to contractors, wait for their bids and then awards (MessageType) the contractor having the best bid. The protocol is considered as finished when the contractor has been chosen. The protocol must also ensure that the manager cannot be a contractor.

We have specified this protocol ontology using Protégé-2000 and derived the graphic specification onto OWL using a specific transformation Plug-In [9]. We have chosen OWL as a target language for three reasons. The first one is that OWL offers a formal semantics, which eases the deduction between concepts of OWL specifications. The second one is that it exists languages to query OWL ontologies in order to exploit the hierarchy of concepts. This feature is very important to dynamically select IOW coordination protocols. The third reason is that OWL is a standard de facto for ontologies.

3   Classification of IOW Coordination Protocols

In addition to the previous general concepts, we also need additive information to handle coordination protocols better. We propose a classification of IOW coordination protocols so as to distinguish them and to easily select them according to the coordination problem to be dealt: finding partners, negotiation between partners... This classification only takes the coordination protocols that are useful in the IOW context into account. However, because of space limitation, we only have presented in the paper the useful protocols for the two first coordination problems: finding partners and negotiation between partners. This section identifies coordination protocols which are adequate to the IOW context. It then presents a classification of these protocols. Finally, basing on this classification, it introduces the main axioms of the coordination protocol ontology.

3.1   Identification of IOW Coordination Protocols

Regarding *finding partners*, the two main protocols proposed in the Multi-Agent Systems (MAS) and Web services literature are the Matchmaker and the Broker protocols [14], [15]. The Matchmaker Protocol is a protocol which connects a requester looking for a specific service to providers able to fulfil its need. More precisely, upon a service demand from a requester, the Matchmaker returns one or several providers able to execute a workflow service. The providers are supposed to have already advertised their workflow services to the matchmaker. The Broker protocol is nearly the same as the Matchmaker one. But, unlike the Matchmaker which directly connects a requester to providers, the Broker does not permit to directly solicit the providers which offer the requested service: the Broker is always an intermediary between the requester and the providers.

Regarding *negotiation between partners*, we have examined the Heuristic, Argumentation, Auction, Multi-Attribute Auction, Contract-Net, Iterative Contract-Net protocols proposed in MAS [16]. The criteria used to determine the negotiation protocols that suit loose IOW the best are the behaviour of the partners involved in the negotiation (competitive or cooperative), the number of partners (one or several), the number of negotiation rounds, and the question whether a multi-attribute negotiation is possible or not.

Table 1. Negotiation Protocol and Selection criteria

Protocol	Negotiation Behaviour	Number of partners	Number of rounds	Multi-attribute Negotiation
Auction	competitive	2 .. *	1 .. *	not possible
Multi-Attribute Auction	competitive	2 .. *	1 .. *	possible
Argumentation	competitive	1 .. *	1 .. *	possible
Heuristic	competitive	1 .. *	1 .. *	possible
Contract-Net	competitive or cooperative	1 .. *	1	possible
Iterative Contract-Net	competitive	1 .. *	1 .. *	possible

According to table 1 above, the Multi-Attribute Auction, Argumentation, Heuristic and Iterative Contract-Net protocols fit the IOW requirements better since (i) the partners involved in an IOW are competitive, (ii) the number of partners is one or several, (iii) the number of rounds is one or several, and (iv) the negotiation concerns several attributes (due time, price, visibility of the evolution of the service, way of doing it…).

Regarding the negotiation behaviour, we argue that the competitive behaviour better fit the IOW requirements since the different providers selected at the end of finding partners compete with each others with the aim to be chosen by the provider.

3.2 A Hierarchy of IOW Coordination Protocols

According to the identified IOW coordination protocols and using the sub-class concept for their classification, we present in Figure 2 our hierarchy of IOW coordination protocol which refines the ProtocolType class introduced in section 2. Thus, types of protocols are specialized according to their objective into two abstract classes: *FindingPartner* and *Negotiation*. Those classes are in turns recursively specialized until obtaining leaves corresponding to types of IOW coordination protocols like for instance Matchmaker, Broker, Argumentation, Heuristic... Each of these types of protocols may be used to deal with one of the coordination problems identified in the context of IOW. These different defined classes also feature new properties which are specific to each one. Thus, the *Objective* property, added in the ProtocolType class, indicates the coordination problem for which the protocol may be used (finding partners, negotiation between partners). Moreover, if we consider the problem of finding partners, we can make the following observation. As explained before, the Matchmaker protocol differs from the Broker protocol by the fact that it implements a peer-to-peer execution mode with the provider: the identity of the provider is known and a direct link is established between the requester and the provider at run time. Consequently, the *P2Pexecution* property defined in the FindingPartner class, will be used to choose one of these two types of protocol. In the same way; the Negotiation class is provider with additional properties (*OpenDialog*, *Explanations* and *Bid*) which will be used to choose one type of negotiation protocol among Argumentation, Heuristic, Multi-Attribute Auction and Iterative Contract Net.

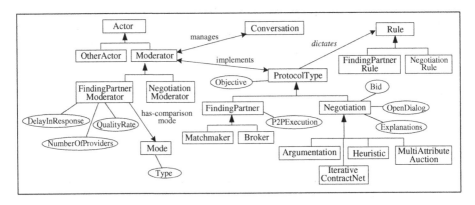

Fig. 2. Hierarchy of Coordination Protocols

Moreover, the Rule class is refined into two sub-classes: *FindingPartnerRule* and *NegotiationRule*, respectively describing rules for finding partners (i.e. rules which are available for both Matchmaker and Broker protocol types) and negotiation between partners (i.e. rules which are available for both Argumentation, Heuristic, Iterative Contract Net and Multi Attribute Auction protocol types) [17].

Finally, the Actor class is refined in order to distinguish actors playing the role of *Moderator* from actors playing the role of Requester or Provider. The *Moderator* and *OtherActor* sub-classes are introduced. More precisely, actors playing the role of Moderator (for short, moderators) manage a conversation, and consequently implement the type of coordination protocol underlying the considered conversation and ensure that any message that takes place in the conversation is compliant with the rules of the type of protocol it plays. The Moderator class is in turn refined into two new sub-classes: *FindingPartnerModerator* and *NegotiationModerator*. The instances of the first class are moderators for finding partners (i.e. matchmakers or brokers) while the instances of the second class are moderators for negotiation between partners (i.e. morderators implementing an heuristic, argumentation, iterative contract-net or multi-attribute auction protocol type. Each of these two classes specifies new properties which help in choosing a moderator. For instance, if we consider a moderator for finding partners, one can be interested in the comparison modes it supports (*has-comparison-mode* relationship). These comparison modes can be Plug-In, Exact, Subsume or Nearest-Neighbour as suggested in [14]. One can also be interested in its quality rate to compare it to other moderators (*QualityRate* property), in the minimum number of providers it is able to manage (*NumberOfProviders* property), and in the delay in its response (*DelayInResponse* property). Finally, one can also be interested in finding the conversation the moderator manages (*manages* relationship).

3.3 Axioms of the Coordination Protocol Ontology

To fully define the Coordination Protocol Ontology, it is also necessary to specify axioms that constrain the concepts of the ontology (classes, properties and relationships). These axioms are described using the Protégé Axiom Language (PAL). We give below a brief overview of the axioms we have defined for this ontology:

- A1: Each protocol type has a minimal number of participants greater than 2. The statement of this axiom is the following:

(forall ?ProtocolType (> (allowed-slot-value MinNumbParticipant ?ProtocolType) 2))

- A2: For each conversation, the type of its moderator is compliant with the type of the protocol type underlying the conversation. In the specific case of *finding partners*, the previous sentence can be described by the following axiom:

(defrange ?pt :FRAME Conversation used-in) -- pt: Protocol Type of a conversation
(forall ?Conversation (=> (instance-of (used-in ?pt)FindingPartner)
     (exists ?FindingPartnerModerator
          (and (implements ?FindingPartnerModerator ?ProtocolType)
          (manages ?FindingPartnerModerator ?Conversation))))

- A3: Each conversation has an actor who plays the role of moderator and implements the ProtocolType used in the conversation.

(defrange ?pt :FRAME Conversation used-in) -- pt: Protocol Type of a conversation
(forall ?Conversation
     (=> (exist ?Actor (and ((participates ?Actor ?Conversation)
          (plays ?Actor ?Role) (allowed-slot-value TypeRole Role 'Moderator'))))
     (and ((instance-of (?Actor) Moderator) (implements ?Actor ?pt)))))

- A4: Each negotiation is ruled by negotiation rules.

(for all ?Negotiation
(=> (dictates ?Negotiation ?Rule) (Exist (instance-of (?Rule) NegotiationRule))))

4   Dynamic Selection of IOW Coordination Protocols

As the IOW coordination ontology is specified, we have to show how to exploit it so as to help an organization involved in an IOW to dynamically select a coordination protocol. This section explains how this dynamic selection is done according to the considered IOW coordination problem that is to be dealt with, and to the type of interaction the organizations involved in the IOW able to participate in.

On the one hand, this selection consists in identifying the type of protocol to execute in concordance with the aim to be reached, and, on the other hand, it consists in choosing an actor/moderator implementing the identified type of protocol in the context of a conversation. The protocol ontology is at the basis of this selection and it is specified in OWL. So, we have used the nRQL [11] query language for OWL ontologies in order to query the OWL specification and select relevant instances of the ontology. nRQL is implement by the Racer system which is known as an highly effective and efficient optimized OWL-DL query processor.

The nRQL language is probably the most successful language to directly query an OWL ontology. Its main rival for querying OWL ontologies is OWL-QL[18]. But, as it is, this language is insufficient to be chosen to query our protocol ontology.

A nRQL query is composed of a query head and a query body. The query body consists of the query expression whereas the query head corresponds to variables mentioned in the body that will be bound to the result. Examples of nRQL queries are given below in this section.

On the one hand, and as indicated before, nRQL is first used to select a type of protocol to execute according to the IOW coordination problem to be dealt with. To do this, the query must navigate through the type of protocols hierarchy and select possible models of protocols. For instance, if the aim of a requester organization is to find a partner, the requester will obtain a set of *FindingPartner* protocols: Matchmaker and Broker. The nRQL query below implements this example and we also give the result of its execution.

```
Q1: (retrieve (?x) (?x (= |Objective| "FindingPartners")))
R1: (?x Matchmaker) (?x Broker)
```

The value of the other properties can be used to select a specific protocol type. For instance, if a query specifies a peer-to-peer execution, it will obtain the Matchmaker protocol type otherwise the Broker will be suggested to it. The nRQL query below implements this example and we also give the result of its execution.

```
Q2: (retrieve (?x) (and (= |Objective| "FindingPartners")
                        (= |P2Pexecution| True) ) )
R2: (?x Matchmaker)
```

On the other hand, nRQL is used to select an actor/moderator implementing the selected type of protocol. Then, properties of the FindingPartnerActor class, which correspond to properties of moderator implementing protocols, must be considered for this selection. For instance, to choose a Matchmaker able to compare workflow services in the Travel business domain and able to consider several providers, the following query must be submitted to the OWL protocol ontology:

```
Q3: (retrieve (?x) (and (?y (= |Name| "Travel")) (?z   ?y |has-
business-domain|) (?x ?z |manages|) (?x |FindingPartners| imple-
mentsfp)))
```

The result of this query is one or several existing actors/moderators of open conversations implementing the Matchmaker protocol (type).

5 Discussion and Conclusion

This paper has presented a novel approach to IOW coordination. The idea is to apply to IOW coordination the principle of separation of concerns, widely recognized as a good design practice from a software engineering point of view. Thus, coordination protocols involved in an IOW are isolated from the participating Workflow Management System (WfMS) and left to a Protocol Management System (PMS) [5]. The aim of the PMS is to support coordination between different WfMS involved in an IOW helping them to deal with their coordination problems: finding partners, negotiation between partners, contract specification and execution.

This paper has addressed two issues that must be tackled to really be able to implement a PMS: how to describe IOW coordination protocols and how to dynamically select them. More precisely, this paper has proposed an IOW coordination protocol ontology including a description of the general concepts needed to define coordination protocols and a classification of the main coordination protocols involved in the IOW context. This paper has also explained how to dynamically select these protocols using n-RQL. These propositions have two main advantages:

- *Easy design and development of IOW systems.* The principle of separation of concerns adopted in this paper eases and speeds up the design, development and maintenance of IOW systems. Following this practice, coordination protocols have been thought and designed as *autonomous*, *reusable* and *extendable components* independently from any specific workflow system behaviour. Thus, being relieved of coordination tasks, each partner of an IOW can focus on its own activity and the workflow can be adapted, maintained or reused without impact on the coordination protocols it uses when involved in the IOW.
- *Semantic coordination of IOW systems.* The proposed ontology permits to describe and access coordination protocols in an explicit, machine readable and sharable way. Thus, this ontology facilitates communication and semantic inter-operability between organizations involved in an IOW. Moreover, this ontology makes a dynamic selection of a coordination protocol possible: indeed, as explained in the paper, an organization is not forced to commit to a single coordination protocol, but dynamically chooses the coordination protocol which is the most suitable according to the coordination problem to solve and according to the type of interaction it is able to participate in. Consequently, this ontology is a contribution to make automated coordination possible.

Related works may be considered according to two complementary points of view: IOW and protocol. Regarding the IOW point of view, the main works addressing the coordination problem are [3], [7], [8], [19], [20]. All these works provide middleware-based solutions to deal with one of the two following problems: finding partners or negotiation between partners. They also exploit agent approach as an enabling technology to both model coordination and implement flexible and adaptative workflow processes. Regarding the finding partners issue, [7], [8] define a matchmaker to find and coordinate agents implementing workflow processes. [19], [20] both mix Web services and agents to implement flexible workflow processes running on the web. While [19] implements a matchmaker for finding partners, [20] implements a broker.

Regarding negotiation, [3] is the only work dealing with the negotiation issue. It defines an agent-based architecture including two specific mediators, a moderator implementing a coordination protocol, and conversation server recording information about open negotiations. Unfortunately, none of these works adopt a comprehensive approach to deal with all the coordination problems in a coherent and uniform framework. They also lack an engineering perspective aiming at proposing solutions covering the coordination protocol engineering life cycle including the design, modelling, selection and execution of such protocols. In our proposition, we adopt this engineering approach. Certainly, we do not cover the entire coordination protocol life cycle but the proposed ontology explains how to design and model IOW coordination protocols and we also provide a solution to dynamically select a coordination protocol. Finally, none of the previous work addresses automated IOW coordination.

Regarding the protocol point of view, three main works are close to our [12], [21], [22]. [21] defines an ontology to support negotiation in E-Commerce. More precisely, it defines a conceptual model to describe the general concepts of negotiation protocols. [12] defines a protocol conceptual model and shows how to transform it onto a corresponding Petri Net, thus obtaining an executable specification. Unfortunately, these two works neither identify and address IOW protocols, nor provide solutions to the dynamic selection of protocols. Moreover, [12] does not come within the web context. At the opposite, [22] defines an ontology of protocols devoted to business processes, and address their coordination through their composition. It also explains how to compile them into executable rules. However, [22] does not address the selection issue and does not provide a protocol classification useful to help the workflow partners in selecting the appropriate protocol according to the execution context.

Our research efforts are now focused on protocol execution. Indeed, we only have provided solutions to the design, the modelling and the dynamic selection of IOW coordination protocols, but we need to investigate their instantiation and execution. Thus, we will cover the all of coordination protocol engineering life cycle.

References

1. van der Aalst, W.: Inter-Organizational Workflows: An Approach Based on Message Sequence Charts and Petri Nets. Systems Analysis, Modelling and Simulation 34(3), 335–367 (1999)
2. Divitini, M., Hanachi, C., Sibertin-Blanc, C.: Inter Organizational Workflows for Enterprise Coordination. In: Omicini, A., Zambonelli, F., Klusch, M., Tolksdorf, R. (eds.) Coordination of Internet Agents, pp. 46–77. Springer, Heidelberg (2001)
3. Andonoff, E., Bouzguenda, L.: Agent-Based Negotiation between Partners in Loose Inter-Organizational Workflow. In: 5th Int. Conf. on Intelligent Agent Technology, Compiègne, France, pp. 619–625 (2005)
4. Ghezzi, C., Jazayeri, M.D., Mandrioli, D.: Fundamentals of Software Engineering. Prentice-Hall International, Englewood Cliffs (1991)
5. Andonoff, E., Bouaziz, B., Hanachi: Protocol Management Systems as a Middleware for Inter-Organizational Workflow Coordination. In: 1st IEEE Int. Conf. on Research Challenge in Information Science, Ouarzazate, Marocco, pp. 85–96 (2007)

6. Struder, R., Benjamin, V., Fensel, D.: Knowledge Engineering, Principle and Methods. Data and Knowledge Enginnering 25(1-2), 161–197

7. Zeng, L., Ngu, A., Benatallah, B., O'Dell, M.: An Agent-Based Approach for Supporting Cross-Enterprise Workflows. In: 12th Int. Database Conf., Bond, Australia, pp. 123–130 (2001)

8. Blake, B.: Agent-Based Communication for Distributed Workflow Management using JINI Technologies. Artificial Intelligence Tools 12(1), 81–99 (2003)

9. Standford University, Protégé-2000. Available at http://protege.stanford.edu/

10. World Wide Web Consortium, OWL Web Ontology Language. Available at http://www.w3.org/TR/owl-features/

11. Haarslev, V., Moeller, R., Wessel, M.: Querying the Semantic Web with Racer and nRQL. In: 3rd Int. Work. on Applications of Description Logic, Ulm, Germany (2004)

12. Hanachi, C., Sibertin-Blanc, C.: Protocol Moderators as active Middle-Agents in Multi-Agent Systems. Autonomous Agents and Multi-Agent Systems 8(3), 131–164 (2004)

13. Foundation for Intelligent Physical Agents, FIPA Contract Net Interaction Protocol Specification. Available at http://www.fipa.org/specs/fipa00029/

14. Klusch, M., Fries, B., Sycara, K.: Automated Semantic Web Service Discovery with OWLS-MX. In: 5th Int. Conf. on Autonomous Agents and Multi-Agents Systems, Hakodate, Japan, pp. 915–922 (2006)

15. Maximilien, M., Singh, M.: A Framework and Ontology for Dynamic Web Service Selection. Internet Computing 8(5), 84–93 (2004)

16. Jenning, N., Faratin, P., Lomuscio, A., Parsons, S., Sierra, C., Wooldridge, M.: Automated Negotiation: Prospects, Methods and Challenges. Group Decision and Negotiation 10(2), 199–215 (2001)

17. Bartolini, C., Preist, C., Jennings, N.: Architecting for Reuse: A Software Framework for Automated Negotiation. In: 3rd Int. Work. on Agent-Oriented Software Engineering, Bologna, Italia, pp. 88–100 (2002)

18. Fikes, R., Hayes, P., Horrocks, I.: OWL-QL: a Language for Deductive Query Answering on the Semantic Web. Web Semantics 2(1), 19–29 (2004)

19. Andonoff, E., Bouzguenda, L., Hanachi, C., Sibertin-Blanc, C.: Finding Partners in the Coordination of Loose Inter-Organizational Workflow. In: 6th Int. Conf. on the Design of Cooperative Systems, Hyeres, France, pp. 147–162 (2004)

20. Aberg, C., Lambrix, C., Shahmehri, N.: An Agent-Based Framework for Integrating Workflows and Web Services. In: 14th Int. Work. on Enabling Technologies: Infrastructure for Collaborative Enterprises, Linköping, Sweden, pp. 27–32 (2005)

21. Tamma, V., Phelps, S., Dickinson, I., Wooldridge, M.: Ontologies for Supporting Negotiation in E-Commerce. Engineering Applications of Artificial Intelligence 18(2), 223–236 (2005)

22. Desai, N., Mallya, A., Chopra, A., Singh, M.: Interaction Protocol as Design Abstractions for Business Processes. Transactions on Software Engineering 31(12), 1015–1027 (2005)

Appendix 1: Protégé-2000 Screenshot of the Coordination Protocol Ontology

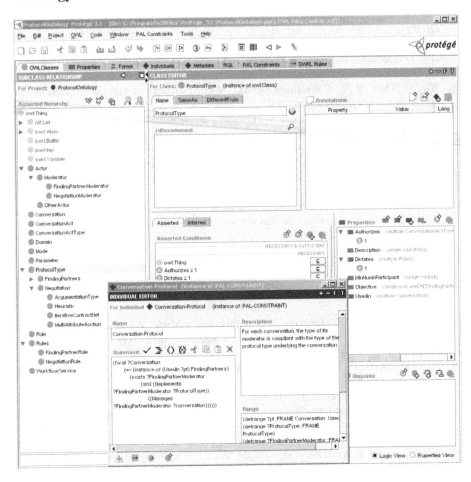

Preventing Orphan Requests by Integrating Replication and Transactions

Heine Kolltveit and Svein-Olaf Hvasshovd

Norwegian University of Science and Technology

Abstract. Replication is crucial to achieve high availability distributed systems. However, non-determinism introduces consistency problems between replicas. Transactions are very well suited to maintain consistency, and by integrating them with replication, support for non-deterministic execution in replicated environments can be achieved. This paper presents an approach where a passively replicated transaction manager is allowed to break replication transparency to abort orphan requests, thus handling non-determinism. A prototype implemented using existing open-source software, Jgroup/ARM and Jini, has been developed, and performance and failover tests have been executed. The results show that while this approach is possible, components specifically tuned for performance must be used to meet real-time requirements.

Keywords: Replication, transactions, non-determinism, orphan requests.

1 Introduction

Fault-tolerance is an important property of real-time and high availability applications. By moving from a centralized to a distributed system, the probability of a total system failure decreases, while the probability of a partial failure increases. A partial failure that is not dealt with correctly could easily jeopardize both consistency and availability of a system.

The *availability* of a system is defined as the fraction of the time that the system performs requests correctly and within specified time constraints [1], while *consistency* is the property that guarantees that the system will behave according to the functional requirements, and applies both to internal (state-changes) and external (output) behavior. Traditionally, transactions are used to ensure consistency [1], while replication provides availability [2]. The two most common types of replication are passive [3] and active [4].

A replicated system must be able to handle the problems occurring due to replicated invocations. A *replicated invocation* is a request from a replicated client to a (possibly replicated) server [5]. The actual problem and solution depends on whether the client is deterministic or not. For deterministic clients, it is only a matter of detecting duplicates, making sure that each request is only executed once and returning the same answer to all duplicate requests. Many clients, however, are not deterministic. Two common sources for non-determinism are multi-threading and timeouts, but others also exist [6]. These

Y. Ioannidis, B. Novikov, and B. Rachev (Eds.): ADBIS 2007, LNCS 4690, pp. 41–54, 2007.

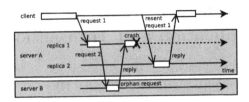

Fig. 1. An orphan request caused by non-determinism in server A

clients are said to be *non-deterministic* and the problem of orphan requests may arise if one of them crashes.

An *orphan request* is a request that is received and processed by a server, but it is no longer valid, normally because of a client failure. Figure 1 illustrates an orphan request. Replica A_1 receives a request from the client and invokes service B. A is then said to be a *client* of server B. A_1 then fails before it can reply to the client. Replica A_2 is chosen as the new primary and receives the retransmitted request from the client. However, since service A is non-deterministic, the re-sent request 1 might not lead to an identical request 2 to be sent to B_1. Consequently, the request from A_1 to B_1 is an orphan request and its results must be removed. If such a request is not handled, it may cause inconsitencies. Even worse, they might spread to other parts of the system, making the whole system inconsistent.

The main contribution of this paper is an integration of transactions and replication where possible orphan requests caused by non-determinism are aborted. Standard atomic commitment protocols, like 2-Phase Commit (2PC) [7], can not guarantee to remove all orphan requests. The approach suggested by the authors solves this by allowing the transaction manager to break replication transparency and therefore see the individual replicas of the transaction participants instead of the whole replica group. As long as at least one replica of the transaction manager is available it also renders 2PC non-blocking [1,8]. The problem of orphan requests in replicated systems have been handled before by integrating transactions and replication (e.g. [5,9,10]), but these approaches are either ineffective (extra messages in the critical path of the transaction), do not support state in all tiers or make unrealistic assumptions regarding the detection of orphans. The approach adopted in this paper does not have these weaknesses and at the same time it supports checkpointing at any time and restart of crashed replicas.

A prototype based on existing open-source group communication, Jgroup/ARM [11], and transaction implementations, Jini Transaction Service [12], is also presented. The prototype is performance evaluated to see if it can meet the stringent requirements of real-time systems.

The rest of this paper is organized as follows: The system model is presented in Section 2. Section 3 describes other approaches related to the integration of replication and transactions, while supporting non-determinism. A detailed description of how the integration is performed is given in Section 4. The method developed in Section 4 has been implemented in a test system and the tests performed on this system are presented in Section 5 and discussed in Section 6. Finally, Section 7 concludes the paper.

2   System Model

The system consists of a set S of fail-crash processes connected through unreliable channels without network partitions. The processes or *nodes* communicate by sending messages. A group G, which is a subset of S, implements a service that can be invoked by clients. These may be replicated. A node in G is called a member or *replica* of that service. Group membership is controlled by a *group membership service* that provides an interface for changes in G, implements a failure detector, notifies members of changes in G and controls that a request is sent to the correct replica(s) [13]. At any given time members of a group have a *view* of the group which is the set of the agreed upon members. Any replica that crashes is eventually excluded from the view, and any restarted and recovered replica is eventually included in the view.

Passive replication is used, i.e. for each group, there is a primary replica that receives, processes and replies to requests. The state of the backups are updated by periodically performing checkpoints [1], while forced writes of log records are propagated as a part of the atomic commitment protocol. The most common atomic commitment protocol, 2PC, is used.

Replicas are stateful, but have no persistent state. If a replica crashes and is restarted, the state is retrieved from one of the other replicas of the same group. The approach here assumes that there is always at least one replica that has not crashed, therefore the state will never be completely lost. Because there is no persistent state, it is not possible to distinguish a restarted node from a new node. All replicas are assumed to be non-deterministic, thus they can all produce orphan requests.

A request is assumed to be eventually received. They are periodically retransmitted from clients until a reply is received, and duplicates are filtered at each server. Thus, if a primary replica fails and a view with a new primary is installed, the new primary will receive the request when it is re-sent.

3   Related Work

Replication and transactions have historically been two separate techniques for achieving fault-tolerance. For instance, CORBA's transaction service (OTS) [14] and replication service (FT-CORBA) [15] are not integrated. A study by Little and Shrivastava [16] looks at two systems, one with transactions and no group communication, and one with group communication and no transactions. Their conclusion is that group communication can be useful for transactions, especially for supporting fast fail-over and active replication.

Many projects deal with replication and transactions, but only a few of these present a proper integration of the two concepts in a non-deterministic environment. Systems that support non-deterministic execution must be able to control its effects. ITRA [17] is an approach that handles the effects by replicating the result of each non-deterministic operation to the backups. ITRA supports replicated transactions by replicating the start, join-operations, prepare (including all operations), commit and abort operations. However, this is not an optimal integration since it incurs an unnecessary high overhead. In our approach only prepare, commit and abort operations are replicated.

Frølund and Guerraoui [18] presents a complete integration of replication and transactions for three-tier applications. However, it supports only stateless middle-tier servers, forcing all state to be stored in the end-tier databases.

Pleisch et al. [5,19] describes two schemes to handle non-determinism; one optimistic and one pessimistic. The first allows a subtransaction to be committed before its parent, while the latter forces the subtransaction to wait for the commit of the parent. By sending information about how to undo the changes to the backups before invoking a server, orphan subtransactions can be terminated in the pessimistic case, and compensated in the optimistic case. This inserts, however, extra messages in the critical path during failure-free execution.

A CORBA related approach [9] restarts execution of a failed subtransaction on a backup and aborts subtransactions where a parent transaction has failed. This integration, however, assumes that standard distributed commit protocols can be used and does not handle the intricate details of transaction completion in failure scenarios.

4   Integration of Transactions and Replication

This section presents an integration of replication and transactions. The goal is to support non-deterministic execution with minimal overhead caused by the integration in a failure-free scenario and minimal change in the application servers (transaction participants).

4.1   Replicating the Transaction Manager

To ensure availability, all single points of failure must be avoided. This is especially important for the transaction manager (TM) because it is a central component involved in distributed transactions. If the TM becomes unavailable, the most widely used atomic commitment protocol, 2PC, may block. By using replication 2PC becomes non-blocking [1,8].

The most important job of the TM is to make the decision to unilaterally abort or commit each transaction. Such a highly critical decision does not favor active replication, since every replica will have to behave deterministically. In practice, TMs are non-deterministic since they rely on timeouts in failure scenarios. This adds non-determinism since it cannot be guaranteed that all replicas timeout at exactly the same time [6]. Also, active replication does not scale well since executing the same processes on every replica waste resources which could have been used to serve other requests.

As can be seen in Figure 2, the protocol for a passively replicated TM is the same as for the non-replicated before the atomic commitment protocol is

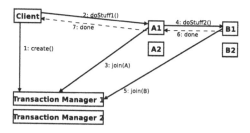

Fig. 2. The execution and join phase of a transaction

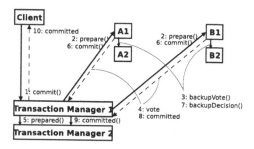

Fig. 3. The successful termination of a transaction

initiated by the client in Figure 3. However, a TM that supports 2PC must be able to persistently store the decision to commit or abort the transaction as the final part of the prepare phase [1,20,21]. In a non-replicated environment the decision is made persistent by force-writing a record to disk. The round-trip transmission time may be a lot shorter than the time needed to write to the disk. A solution where the prepare decision is persistently saved by sending it to the backups (message 5 in Figure 3) is faster and therefore preferred. In addition, it gives better availability since the prepared transactions can be committed by the backup in case of a primary failure. If a local disk was used, currently prepared transactions may be blocked until the TM has recovered.

A "transaction completed" message is sent to the backups, as indicated by message 9 in Figure 3. This is done instead of the lazy write to the log in the normal non-replicated 2PC [21]. Hence, the replicated nature of the TM is used to provide both availability and persistence of the decision.

4.2   Replicating the Transaction Participants

The transaction participants should be replicated for the same reason as any other component of a distributed system; to avoid single points of failure. To be able to handle non-determinism, passive replication is used.

Passive replication is subject to fail-overs. A *fail-over* happens when a primary replica fails and another replica of the same service is elected as the new primary by the group membership service. Consequently, any re-sent or new requests will

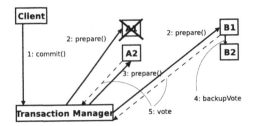

Fig. 4. A failure of a primary, and the consequent fail-over

be handled by the new primary. If any operation has been executed after the previous checkpoint, the new primary might not be fully updated, and care must be taken to avoid inconsistencies caused by orphan requests. Figure 4 shows a fail-over of a prepare request from A_1 to A_2.

There are only two ways to cause orphan requests. A failure of the client of the transaction, or a failure of a primary server which acts as a client to another server. The effects of an orphan request can be guaranteed to be removed by aborting all transactions that interact with a replica or client that fails. If the client of the transaction fails, the TM will not receive a commit message from the client. Without the commit message, it can use a timeout to safely abort the transaction. If a primary fails, the TM may still receive a commit message. However, if the TM can determine whether a transaction has been caught in a fail-over or not, it can abort potential orphan requests. The problem is then reduce to the detecting failed primary participants of the transaction.

Normally, replication of a server is hidden from the clients of that service, i.e. *replication transparency.* The unpredictable effects of non-determinism, however, can be controlled at the loss of replication transparency for the TM, by sending the prepare message to the primary only. If a participating primary replica of an active transaction fails, the TM can abort the transaction. Note that it is only the prepare message that does not fail-over. Since the primary persistently stores the vote to the backups during the prepare phase, the backups are then updated and an abort or commit message is allowed to fail-over.

Intuitively, by sending the prepare message only to the primary replicas that joined the transaction, primary failures should be detected: Failed primaries will not be able to reply and the transaction is aborted and possible orphan requests are rolled-back by the transactional abort mechanism. This is true for single fail-overs. Figure 5 illustrates this: When the TM does not get a reply from the failed primary replica, A_1, it eventually times out and aborts the transaction (as indicated by arrow 5). Contrast this to Figure 4 where the prepare request is allowed to fail-over and the orphan request to service B is not handled.

This protocol has one flaw: If the crashed replica is restarted and a second fail-over back to the original primary occurs, the TM may not be able to notice the failures. This is illustrated in Figure 6. The grey ovals are the current view of the group and the numbers inside are the view identifier. Two fail-overs of

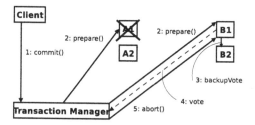

Fig. 5. A failure of a primary, without the fail-over

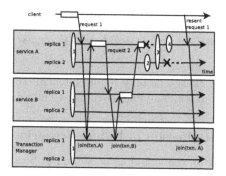

Fig. 6. A double fail-over

group A cause the TM not to notice the fail-over, since the last join message will be identical to the first and looks like a re-send due to a communication error. Therefore, request 2 is an orphan and the transaction should have been aborted. The TM can, however, see the difference of the two join messages if the view identifier was piggybacked on the join message (join(txn,group,viewID), instead of join(txn,group)). In the example, the two join messages would have viewID 1 and 4, respectively. Hence, the new protocol will be able to resolve double fail-overs correctly as well.

If a checkpoint was taken after the first request of the transaction, the new primary is aware of the transaction after a fail-over. If the prepare message was sent to the new primary, orphan requests may not be correctly aborted. Figure 7 illustrates this. A checkpoint is take after A_1 has joined the transaction. The checkpoint does not include information about request 2, which becomes an orphan request after replica A_1 fails. If the TM sends a prepare message to replica A_2 it would be able to vote yes, and the effects of the orphan request to B_1 could cause inconsistencies.

By simply allowing the TM to break replication transparency and therefore be able to avoid the automatic fail-over of the prepare message to a replicated service, it provides a way for orphan requests to be handled easily and correctly. This approach allows checkpointing at any time, thus reducing the time required to bring the state of the new primary up to date and only log records need to

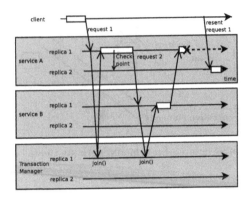

Fig. 7. A checkpoint and a possible orphan request in service B

be shipped as a part of 2PC. The only requirement for server applications is to implement the 2PC interface. The underlying system adds the view identifier.

4.3   Transaction Termination in Failure Scenarios

When the TM is passively replicated as presented in Section 4.1, a transaction may be unable to terminate, therefore blocking other transactions from completing. Consider the following case: A transaction has been created and some or all of the participants have joined it. Then the primary TM fails before the prepare phase has completed. This will leave the new primary with no knowledge of the transaction. When the client asks the TM to commit the transaction, the TM will reply that the transaction is unknown, and the client will assume that it has aborted. However, the transaction participants will still hold their locks on the items accessed by the transaction. Without proper termination of these transactions, the locks could be held forever, blocking other transactions from completing.

The locks held by a failed transaction can be removed by the client if it keeps control of the participants accessed by each transaction. Thus, when the client gets a reply from the TM that the transaction it tried to commit is unknown, the client can abort it. Because of possible nested invocations each participant must be able to tell which other participants it has caused to join the transaction, and so on. However, if one of the participants also fails, the participants invoked by that participant do not get the abort message.

A better way to remove the locks is to use a timeout. Each participant can periodically poll the TM to get the status of each active transaction. If the TM replies that the transaction is unknown, the transaction can be safely aborted. Also, this takes care of the scenario where the client fails.

This approach causes transaction commitment to be non-blocking as long as at least one of the TMs is available. When combined with the avoidance of failover for the prepare message and piggybacking the view identifier, all failures are correctly handled to avoid inconsistencies. Potential orphan requests are rolled-back and all types of non-determinism are supported.

5 Implementation and Testing

This section gives an overview of the prototype implementation, as well as the environment used for testing and the results of tests executed on the prototype. A presentation of the prototype is given in Section 5.1 and the environment for testing and the tests executed are presented in Section 5.2.

5.1 Prototype Implementation

A prototype of the transaction manager that do not allow fail-over of the prepare request was implemented, along with the transaction participants. The servers were implemented as Jini [22] services, and replicated using the Jgroup/ARM system [11].

The system where the tests are executed consists of four conceptual entities: A client, a transaction manager (TM) and two different banks. The banks and the transaction manager are implemented as Jini services that can be discovered and registered by the Jini registry, *Reggie* [12], or the group-enabled registry, *Greg* [23]. The transaction manager is based on the non-replicated *Mahalo* [12] as well as the actively replicated *Gahalo* [24].

5.2 The Test Environment

Figure 8 shows the system model. The grey ovals represent entities, while the white boxes are nodes where replicas of servers or the client execute. A single physical node may execute more than one service. The arrows in the figure represent the direction of the invocations.

The life cycle of the transaction used for testing is as follows:

1. A transaction is initiated by the client, and created by the TM.
2. The client invokes the withdraw operation of Bank$_A$, which joins the transaction.
3. The client invokes the deposit operation of Bank$_B$, which joins the transaction.
4. The client initiates 2PC, which is controlled by the TM.

As modeled in the figure, the TM can have up to four replicas, and the two banks can have up to two replicas each. These limitations are due to the fact that there were only five nodes available for executing the tests.

A Dual AMD MP 1600+ running at 1.4GHz powered each node. A 100Mbit Ethernet connected them and each had 1024 MB of RAM. The tests were executed using Linux kernel 2.6 and Java version 1.5.0.

All tests were carried out by executing 500 transactions and measuring the elapsed time at the client between transaction initiation and transaction completion. This is referred to as the *response time* of a transaction. Similarly, the response time of an invocation is the time passed between calling the remote method of the client and the return of the method call.

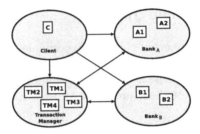

Fig. 8. A model of the system used for testing

Table 1. A summary of the response times for the test runs in Section 5.2

Test run	Description	Average (ms)	Standard Deviation (ms)	Delay (%)
	Nonreplicated system			
1	1 TM and 2 banks	47	10	0
	Passive replication of the TM			
2	2 passive TMs and 2 banks	77	17	64
3	3 passive TMs and 2 banks	92	19	96
4	4 passive TMs and 2 banks	106	21	126
	Fully replicated system			
5	1 TM and 2x2 banks	75	16	60
6	2 passive TMs and 2x2 banks	148	25	215
7	3 passive TMs and 2x2 banks	164	27	249

Failure-free performance tests were performed using the following configurations:

1. A non-replicated transaction manager and non-replicated banks.
2. Two passively replicated transaction managers and non-replicated banks.
3. Three passively replicated transaction managers and non-replicated banks.
4. Four passively replicated transaction managers and non-replicated banks.
5. A non-replicated transaction manager, with two replicas of each bank.
6. Two passively replicated transaction managers, with two replicas of each bank.
7. Three passively replicated transaction managers, with two replicas of each bank.

In addition, the response times during fail-overs were measured.

6 Comparing the Test Results

Table 1 summarizes the results of the test runs made in Section 5.2. The response time averages and standard deviations are presented[1]. It should be noted that

---

[1] The first 50 transactions of each test run are disregarded in this discussion because of extra startup cost.

these numbers only apply for these test runs and they should not be interpreted as any general response time guarantee, but rather as properties of the specific test run. However, they can be used as a reference for comparisons between the individual test runs.

The response time of a transaction was measured at the client and is the time elapsed from transaction creation to transaction termination.

The following sections presents a summary of the results from the test runs and compares passive replication and the non-replicated case. Finally, the fail-over delay is examined.

6.1   Cost of Replication

Replication increases the overhead of a service. The results of test runs 1–4, as presented in Table 1, clearly support this assumption. The average response time degrades when adding more replicas of a transaction manager, and the variance of the results increase. The numbers seem to indicate that replicating the TM causes about 50 percent longer response times, while each added replica on top of that increases the response time of about 30 percent of the non-replicated case.

The standard deviation seems to change similarly to the average response time. Table 1 shows a significant leap for the deviation of the response times when the TM is replicated and then scales linearly when adding the third and fourth replica.

Replication of the transaction participants (test run 5) has similar effects as when only replicating the transaction manager (test run 2). There is a 50 percent increase in the response time and about the same for the standard deviation. For test runs 6 and 7 the overhead increases more. Replicating the TM as well (test run 6) doubles the average response time. The cost of executing a fully replicated system with 2 replicas of each server is three times higher than executing a non-replicated one. If 3 replicas of the TM are executed (test run 7), the response time is three and a half times higher than in the non-replicated case.

A closer inspection reveals that for the fully replicated case (test run 6) the group management threads and layers causes an overhead that delays around 60 percent of the invocations, usually for around 10 – 20 ms. The average delay for a transaction just by running the group management threads was found to be 40 – 50 ms. The time to update the backups was found to be around 22 ms for the commit decision at the TM and 12 ms for each of the other updates. When added together these contribute to an average response time of 150 ms which differs with only 1.3% from the measured total time.

To make the non-replicated case fault-tolerant the log could be force written to disk instead of updating the backups. To force write a record to the disk takes approximately 20 ms. A successfully committed transaction requires three log forces and one lazy log write as part of 2PC [21]. Thus, the completion time for a fully fault-tolerant non-replicated system has a response time of around 107 ms. However, this solution does not provide high availability since it has single points of failure.

6.2 Fail-Over Delay

The observed client-side fail-over delay for the transaction test was found to be as much as 360–490 ms. However, the fail-over delay for a simpler application running on top of the same system was found to be between 200 and 250 ms. These measurements are closer to the real time between a failure and the continuation of the service by a new primary.

Gray and Reuter [1] distinguish five classes of transaction-oriented computing, with various properties and requirements. According to this classification the fail-over delay found here will be sufficient for batch processing, time-sharing (not widely used anymore), client-server and transaction-oriented processing. The last class, real-time processing, however, will probably require client-observed fail-overs of less than 200 ms, depending on the application.

7 Conclusion and Further Work

Many applications require high availability and strong consistency. Since system components fail from time to time, a system must be able to tolerate faults. Well known fault-tolerance techniques include transactions and replication. They are widely used and extensively studied as separate concepts and their efficiency has been well proven. However, to support both availability and consistency in a non-deterministic environment, the techniques should be integrated.

This paper addresses the issue of integrating replication and transactions without enforcing replica determinism. This is a highly desirable property since it allows any kind of application to be built on top of the system. The main contribution is an approach where the transaction manager is allowed to break replication transparency to ensure that no orphan requests survive. Thus, full support for non-determinism in general is achieved.

Tests were performed on a prototype built using existing open-source software. The tests show that transactions can be executed in a passively replicated environment with a 200 percent response time increase. Also, a failure of the primary will cause a fail-over delay of about 400 ms on average for the transaction manager. Measurements on a smaller application, however, indicate that the real fail-over time is probably closer to 200 ms. While these results show that the approach is possible, real-time systems have more stringent performance demands. The response time for a transaction in this prototype is too large for most real-time systems, e.g. telecommunications [25]. However, this is a property of this specific implementaion and not the approach, since both Jini and Jgroup are not tuned for real-time performance. For real-time systems the entire implementation should be focused on performance and the use of existing open-source software may not be suitable. Also, other transaction models (e.g. hierarchic [21]) than a centralized transaction manager receiving join-messages from all participants should be invesitgated.

For a real world application the advantage of increased availability must be weighed against the cost of replication. If the system cannot tolerate the downtime caused by a restart of a machine, replication should be used. On the other hand, if the increased response time cannot be tolerated, but a few minutes of unavailability once in a while can be, replication should not be used.

The system developed in this paper is a prototype where several shortcuts have been made to get a working system for basic testing. To be of any practical use, it must be able to restart crashed replicas, initiate new ones and update the new replicas with the current state. The Jgroup/ARM system has support for automatically performing these actions, but it has not yet been implemented in this prototype. Also, the system must be able to handle all failure scenarios during 2PC to be able to terminate all transaction in the presence of failures.

References

1. Gray, J., Reuter, A.: Transaction Processing: Concepts and Techniques. Morgan Kaufmann, San Francisco (1993)
2. Helal, A.A., Bhargava, B.K., Heddaya, A.A.: Replication Techniques in Distributed Systems. Kluwer Academic Publishers, Dordrecht (1996)
3. Budhiraja, N., Marzullo, K., Schneider, F.B., Toueg, S.: Distributed systems. In: Mullender, S. (ed.) Distributed systems, 2nd edn., pp. 199–216. ACM Press, Addison-Wesley, Reading (1993)
4. Schneider, F.B.: Replication management using the state machine approach, pp. 169–197. ACM Press/Addison-Wesley Publishing Co. (1993)
5. Pleisch, S., Kupšys, A., Schiper, A.: Preventing orphan requests in the context of replicated invocation. In: Proceedings of the 22nd International Symposium on Reliable Distributed Systems, Florence, Italy, pp. 119–128. IEEE Computer Society Press, Los Alamitos (2003)
6. Poledna, S.: Replica determinism in distributed real-time systems: A brief survey. Research Report 6/1993, Technische Universität Wien, Institut für Technische Informatik, Treitlstr. 1-3/182-1, 1040 Vienna, Austria (1993)
7. Gray, J.: Notes on data base operating systems. In: Operating Systems, An Advanced Course, pp. 393–481. Springer, London, UK (1978)
8. Reddy, P.K., Kitsuregawa, M.: Reducing the blocking in two-phase commit protocol employing backup sites. In: Proc. of CoopIS (1998)
9. Felber, P., Narasimhan, P.: Reconciling replication and transactions for the end-to-end reliability of CORBA applications. In: On the Move to Meaningful Internet Systems, 2002 - DOA/CoopIS/ODBASE 2002 Confederated International Conferences DOA, CoopIS and ODBASE 2002, pp. 737–754. Springer, Heidelberg (2002)
10. Frølund, S., Guerraoui, R.: Implementing e-transactions with asynchronous replication. Dependable Systems and Networks, 449–458 (2000)
11. Montresor, A.: System Support for Programming Object-Oriented Dependable Application in Partitionable Systems. PhD thesis, University of Bologna, Italy, Technical Report UBLCS-2000-10 (2000)
12. Sun Microsystems Inc.: Jini Technology Core Platform Specifications. 2.1 edn. (2005)
13. Coulouris, G., Dollimore, J., Kindberg, T.: Distributed Systems: Concepts and Design. 3rd edn. Addison-Wesley Longman Publishing Co., Inc. (2001)
14. Object Management Group: Transaction Service Specification, OMG Technical Committee Document formal/03-09-02 (2003)
15. Object Managment Group: Fault Tolerant CORBA, OMG Technical Committee Document formal/04-03-21 (2004)

16. Little, M.C., Shrivastava, S.K.: Integrating group communication with transactions for implementing persistent replicated objects. In: Krakowiak, S., Shrivastava, S.K. (eds.) Advances in Distributed Systems. LNCS, vol. 1752, pp. 238–253. Springer, Heidelberg (2000)
17. Dekel, E., Goft, G.: ITRA: Inter-tier relationship architecture for end-to-end QoS (2001)
18. Frølund, S., Guerraoui, R.: Transactional exactly-once. Technical report, Hewlett-Packard Laboratories (1999)
19. Pleisch, S., Kupšys, A., Schiper, A.: Replicated invocations. Technical report, Swiss Federal Institute of Technology (EPFL) (2003)
20. Bernstein, P.A, Hadzilacos, V., Goodman, N.: Concurrency control and recovery in database systems. Addison-Wesley Longman Publishing Co., Inc. (1986)
21. Mohan, C., Lindsay, B., Obermarck, R.: Transaction management in the R* distributed database management system. ACM Trans. Database Syst. 11, 378–396 (1986)
22. Arnold, K., Scheifler, R., Waldo, J., O'Sullivan, B., Wollrath, A.: The Jini Specification, 2nd edn. Addison-Wesley Longman Publishing Co., Inc. (2001)
23. Montresor, A., Davioli, R., Babaoğlu, Ö.: Jgroup: Enhancing Jini with group communication. In: Proceedings of the ICDCS Workshop on Applied Reliable Group Communication (2001)
24. Moland, R.: Replicated transactions in Jini. Master's thesis, University of Stavanger (2004)
25. Hvasshovd, S.O., Torbjørnsen, Ø., Bratsberg, S.E., Holager, P.: The ClustRa telecom database: High availability, high throughput, and real-time response. In: VLDB, pp. 469–477 (1995)

Discretization Numbers for Multiple-Instances Problem in Relational Database

Rayner Alfred[1,2] and Dimitar Kazakov[1]

[1] University of York, Computer Science Department, Heslington,
YO105DD York, United Kingdom
{ralfred,kazakov}@cs.york.ac.uk
http://www-users.cs.york.ac.uk/~ralfred
[2] On Study Leave from Universiti Malaysia Sabah,
School of Engineering and Information Technology,
88999, Kota Kinabalu, Sabah, Malaysia
ralfred@ums.edu.my

Abstract. Handling numerical data stored in a relational database is different from handling those numerical data stored in a single table due to the multiple occurrences of an individual record in the non-target table and non-determinate relations between tables. Most traditional data mining methods only deal with a single table and discretize columns that contain continuous numbers into nominal values. In a relational database, multiple records with numerical attributes are stored separately from the target table, and these records are usually associated with a single structured individual stored in the target table. Numbers in multi-relational data mining (MRDM) are often discretized, after considering the schema of the relational database, in order to reduce the continuous domains to more manageable symbolic domains of low cardinality, and the loss of precision is assumed to be acceptable. In this paper, we consider different alternatives for dealing with continuous attributes in MRDM. The discretization procedures considered in this paper include algorithms that do not depend on the multi-relational structure of the data and also that are sensitive to this structure. In this experiment, we study the effects of taking the *one-to-many* association issue into consideration in the process of discretizing continuous numbers. We implement a new method of discretization, called the *entropy-instance-based* discretization method, and we evaluate this discretization method with respect to C4.5 on three varieties of a well-known multi-relational database (Mutagenesis), where numeric attributes play an important role. We demonstrate on the empirical results obtained that entropy-based discretization can be improved by taking into consideration the multiple-instance problem.

Keywords: Discretization, Entropy-based, Semi-supervised clustering, Genetic Algorithm, Multiple Instance.

1 Introduction

Most multi-relational data mining deals with nominal or symbolic values, often in the context of structural or graph-based mining (e.g. ILP) [1]. Much less attention has

Y. Ioannidis, B. Novikov, and B. Rachev (Eds.): ADBIS 2007, LNCS 4690, pp. 55–65, 2007.

been given to the area of discretization of continuous attributes in a relational database, where the issue of *one-to-many* association between records has to be taken into account. Continuous attributes in multi-relational data mining are seldom used due to the difficulties in handling them particularly when we have a *one-to-many* association in a relational database. In fact, most data mining tools ignore the *multiple-instance* problem and treat all of the positive instances as positive examples and all of the negative instances as negative examples. For example, in a relational database, usually each target table record refers to one or more instances in another table through a foreign key. This relationship between tables in a relational database is called *non-determinate* or known as a *multiple-instance* problem.

Handling continuous attributes in multiple tables is different from handling attributes from a single table due to several factors. Firstly, discretization and aggregation of attributes stored in relational database need to use the structure (*schema*) of the relational database and to find out how attributes stored in non-target and target tables are related to each other. Next, by taking into consideration the occurrence of multiple instances in the non-target table, it makes discretization and aggregation more complex, since most traditional methods of discretization ignore the *multiple-instance* problem. And finally, entropy-based discretization in a relational database is not a straight-forward task as it has been done in a single table. In this paper, we have implemented a few discretization methods, including our new method of discretization called the entropy-instance-based discretization, embedded in DARA algorithm [1,2,3]. In DARA algorithm, we employ several methods of discretization in conjunction with C4.5 classifier, as an induction algorithm. We then evaluate the effectiveness of each discretization method with respect to C4.5, on three varieties of a well-known multi-relational database (Mutagenesis) [4].

The paper is structured as follows. In section 2, we explain the pre-processing sequence of steps, called DARA [1,2,3] that transforms the data representation of a relational database, with a possibly high degree of *one-to-many* associations into a vector space model. This model is applicable to clustering operations as a means of aggregating multiple instances. In Section 3, we define three different methods, implemented in DARA, of discretizing continuous attributes in relational database. Section 4 describes the experimental setup and evaluation. We draw conclusions in Section 5.

2   Data Transformation Using Dynamic Aggregation of Relational Attributes (DARA)

In a relational database, records are stored separately in different tables and are then associated through the matching of primary and foreign keys. A single record, R, stored in a main table can be associated with a large volume of records stored in another table, as shown in Fig. 1, in an example of a *one-to-many* association.

Let R denote a set of m records stored in the target table, and S denote a set of n records $(T_1, T_2, T_3, \ldots, T_n)$, stored in the non-target table. Let S_i be in the subset of S, $S_i \in S$, and associated with a single record R_a stored in the target table, $R_a \in R$. Thus, the association of these records can be described as $R_a \rightarrow S_i$. Since a record can be characterized on the basis of a series of terms or records associated with it, we use the

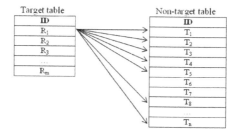

Fig. 1. A *one-to-many* association between target and non-target relations

vector space model to cluster these records, as described by Salton *et al.* [7]. In the vector space model, a record is represented as a vector or 'bag of terms', i.e., by the terms it contains and their frequency, regardless of their order. These terms are encoded and represent instances stored in the non-target table referred to by a record stored in the target table. The non-target table may have a single attribute or multiple attributes and the process of encoding the terms to transform them into a vector space model is as follows:

Case I: *Non-target table with a single attribute*

- Step 1) Compute the cardinality of the attribute domain in the non-target table. In case of datasets with continuous and discrete values, discretizes the continuous values first (using the methods described in Section 3) and take the number of bins as the cardinality of the attribute domain.
- Step 2) To encode values, find the appropriate number of bits, n, that can represent all different values for the attribute domain, where $2^{n-1} <$ |Attribute Domain| $\leq 2^n$. For example, if the attribute has 5 different values (London, New York, Chicago, Paris, Kuala Lumpur), then we just need 3 ($5 < 2^3$) bits to represent each of these values (001, 010, 011, 100, 101).
- Step 3) For each encoded term, increment the corresponding counter in the bag of terms or just add the term to the bag of terms if it is not in the bag; the resulting bag of terms can be used to describe the characteristics of a record associated with them.

Case II: *Non-target table with multiple attributes*

- Step 1) Repeat step 1) and step 2) from Case I, for all attributes
- Step 2) For each instance of a record stored in the non-target table, concatenate p number of columns' values, where p is less than or equal to the total number of attributes. For example, let $F = (F_1, F_2, F_3,..., F_k)$ denote k field columns or attributes in the non-target table. Let $F_1 = (F_{1,1}, F_{1,2}, F_{1,3}, ..., F_{1,n})$ denote n values that are allowed to be used by field/column F_1. So, we can have instances of a record in the non-target table with these values $(F_{1,a}, F_{2,b}, F_{3,c}, F_{4,d}..., F_{k-1,b}, F_{k,n})$, where $a \leq |F_1|$, $b \leq |F_2|$, $c \leq |F_3|$, $d \leq |F_4|,..., b \leq |F_{k-1}|$, $n \leq |F_k|$. If $p = 1$, we have $F_{1a}, F_{2,b}, F_{3,c}, F_{4,d}..., F_{k-1,b}, F_{k,n}$ as the produced terms. If $p = 2$, then we have a bag of paired attribute values, $F_{1,a}F_{2,b}, F_{3,c}F_{4,d}..., F_{k-1,b}F_{k,n}$ (provided we have even

number of fields). Finally, if we have $p = k$, then we have $F_{1,a}F_{2,b}F_{3,c}F_{4,d}...F_{k-1,b}F_{k,n}$ as a single term produced.

- Step 3) For each encoded term, increment the corresponding counter in the bag of terms or just add the term to the bag of terms if it is not in the bag; the resulting bag of terms can be used to describe the characteristics of a record associated with them.

This encoding process transforms relational datasets into data represented in the vector-space model [7], implemented in DARA [1,2,3]. In this representation, the data can be conveniently clustered [8,9,11,12,13] as a means of aggregating them.

In short, Dynamic Aggregation of Relational Attributes (DARA) algorithm treats records in relational database as a bag of patterns and clusters these records based on the created patterns that they have. The DARA algorithm simply assigns each record in the target table with the cluster number. Each cluster then can describe more information by looking at the most frequent patterns that describe it. In the next section, we describe four methods of discretizing continuous numbers in a relational database, a step that is needed before the dynamic aggregation procedure described in this section can be executed.

3   Types of Discretization

The motivation for the discretization of continuous features is based on the need to obtain higher accuracy rates, although this operation may affect the speed of any learning procedure that may subsequently use it. This paper investigates how the results of discretization affect the results of induction when the *one-to-many* associations between tables are taken into account. In this experiment, we used three methods of discretization: *Equal Height, Equal Weight* and *Entropy-Instance-based discretization* implemented in the DARA algorithm. We discretize all attributes with continuous values before transforming them into DARA's format which can also be used with any traditional data mining tools, such as C4.5.

3.1   Equal Height Discretization

The *Equal Height* interval binning discretizes data so that each bin will have approximately the same number of samples. It involves sorting the observed values together with the record ID. If $|R|$ refers to the size of the records and the $V[|R|]$ refers to the size of the array that stores the sorted values, then the boundaries can be constructed as $V[(|R|/k) \times i]$ where $i = 1, ..., k-1$. The result is a collection of k bins of roughly equal size. This algorithm is class-blind and does not take into consideration the structure of the database, especially the *multi-instance* problem.

3.2   Equal Weight Discretization

The *Equal weight* interval binning considers not only the distribution of numeric values present, but also the groups they appear in. This method involves an idea proposed by Van Laer et al. [14]. It is observed that larger groups have a bigger influence on the choice of boundaries because they have more contributing numeric values. In *equal weight* interval binning, numeric values are weighted

$wt(v) = 1/|class_v|$, where wt is the weight function and v is the value being considered and $|class_v|$ is the size of the class that v belongs to. Instead of producing bins of equal size, we compute boundaries to obtain bins of equal weight. The algorithm starts by computing the size of each class, then it moves through the sorted arrays of values, keeping a running sum of weights wt. Whenever wt reaches a target boundary (|number of classes|/bins), the current numeric value is added as one of the boundaries, and the process is repeated k times (k is the number of bins).

3.3 Entropy-Instance-Based Discretization

Finally, the *Entropy-Instance-Based* interval binning considers the distribution of numeric values present, the groups they appear in, and also the *multiple-instance* problem. A lot of significant research in entropy-based discretization has been carried out, and an early comparison of entropy-based methods for discretization of continuous features and multi-interval discretization methods can be found in [15,16]. In this paper, we modify the *entropy-based* multi-interval discretization method introduced by Fayyad and Irani [17]. Algorithms such as C4.5 try to find a binary cut for each attribute and use a minimal entropy heuristic for discretization continuous attributes. The algorithm uses the class information entropy to select binary boundaries for discretization. Given a set of instances S, a feature A, and a partition boundary T, the class information entropy is

$$E(A,T,S) = \frac{|S_1|}{|S|} Ent(S_1) + \frac{|S_2|}{|S|} Ent(S_2), \tag{1}$$

where S_1 and S_2 correspond to the samples in S satisfying the condition $A < T$ and $A \geq T$, respectively. The entropy function Ent for a given set is calculated based on the class distribution of the samples in the set. For example, given C classes, the class entropy of a subset S is

$$Ent(S) = -\sum_{i=1}^{C} p(C_i, S) \log_2(p(C_i, S)) \tag{2}$$

where $p(C_i, S)$ is the probability of the i-th class in the subset S. This method can be applied recursively to both partitions induced by T until some stopping condition is achieved, thus creating multiple intervals of feature A. So, for k bins, the class information entropy for multi-interval entropy-based discretization is

$$I(A,T,S,k) = \frac{\sum_{b=1}^{k} |S_b| \cdot Ent(S_b)}{|S|} \tag{3}$$

In the *entropy-instance-based* discretization that we peopose, besides the class information entropy, another measure that uses individual information entropy is added to select multi-interval boundaries for discretization. Given n individuals (from target table), the individual information entropy of a subset S is

$$IndEnt(S) = -\sum_{i=1}^{I} p(I_i, S) \log_2(p(I_i, S)) \tag{4}$$

where $p(I_i, S)$ is the probability of the i-th individual in the subset S. The total individual information entropy for all partitions is

$$Ind(A,T,S,k) = \frac{\sum_{b=1}^{k}|S_b| \cdot IndEnt(S_b)}{|S|} \tag{5}$$

For example, suppose we have the following sample of dataset shown in Fig. 2.

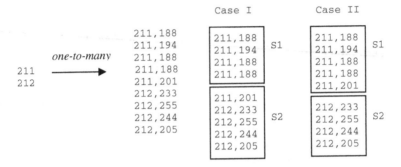

Fig. 2. Case I and Case II: Partitioning of *one-to-many* datasets into two sets

If the partition ranges are (Case I - [S_1:188-200, S_2:201-255]), the individual information entropy of an individual in S_1 is 0.176, computed as -(0.8*log$_2$(0.8)) and S_2 is 0.322, computed as -(0.2*log$_2$(0.2)+1*log$_2$(1))). The smaller the value of the individual information entropy for all partitions, the better is the quality of the partition, taking into consideration the multiple-instance problem. As a result, by minimizing the function $Ind\_I(A,T,S,k)$, Eq. (6), that consists of two base functions $I(A,T,S,k)$, Eq. (3), and $Ind(A,T,S,k)$, Eq. (5), we are discretizing the attribute's values based on the class and individual information entropy.

$$Ind\_I(A,T,S,k) = \frac{\sum_{b=1}^{k}|S_b| * Ent(S_b)}{|S|} + \frac{\sum_{b=1}^{k}|S_b| * IndEnt(S_b)}{|S|}$$

$$= \frac{\sum_{b=1}^{k}|S_b| * (Ent(S_b) + IndEnt(S_b))}{|S|} \tag{6}$$

However, one of the main problems with this discretization criterion is that it is relatively expensive. For instance, for 2 bins (k=2), for a continuous attribute, the expression (6) must be evaluated N-1 times for each attribute, where N is the number of attribute values. Therefore, in this experiment, we use a genetic algorithm-based discretization in order to obtain a multi-interval discretization for continuous attributes.

3.4 Genetic Entropy-Instance-Based Discretization

The genetic strategy consists of an initialization step and the iterative generations of the reproduction phase, the crossover phase and mutation phase, as shown in Fig.3.

```
Begin
  1.  initialize population
  2.  compute fitness for each chromosome
  3.  if termination criterion achieved go to 8
  4.  reproduction phase
  5.  crossover phase
  6.  mutate phase
  7.  go to step 2
  8.  Output best and stop
End
```

Fig. 3. Basic steps in GAs

In the initialization step, a set of strings (or *chromosomes*), where each string consists of b-1 continuous values representing the b partitions, is randomly generated. Here, b is the number of discretization bins and all continuous values must be in the range of the minimum and maximum values of the attribute values. This set of strings of continuous values is the *population* of the genetic algorirhm. For instance, given minimum and maximum values of 1.5 and 20.5 for a continuous field, the number of bins b is 6, the string of sequence continuous values of (2.5,5.5,9.3,12.6,15.5,20.5), is randomly generated within the range [1.5, 20.5]. From this string of continuous values we have six bins with the following range of values: ([1.5, 2.5], [2.6, 5.5], [5.6, 9.3], [9.4, 12.6], [12.7, 15.5], [15.6, 20.5]). After the initialization step, the fitness value for each string is computed. The fitness function for genetic *entropy-instance-based* discretization is defined as

$$f = \frac{1}{Ind\_I(A,T,S,k)} , \qquad (7)$$

so that maximization of the fitness function, f, leads to minimization of $Ind\_I(A,T,S,k)$. The reproduction process selects strings from the old population directed by the survival of the fittest concept of natural genetic systems. In the proportional selection strategy adopted in this paper, a string (*chromosome*) is copied into several copies, which is proportional to its fitness in the population, that go into the reproduction phase for further genetic operations. Roulette wheel selection is one common technique that implements the propositional [6] selection strategy. Next, in the crossover phase, two parent chromosomes will exchange information based on the crossover probability. In this paper, a crossover probability p_c of 0.50 is used. For chromosomes of length l, a random integer, called the crossover point, is generated in the range [1, l-1]. The portions of the chromosomes located to the right of the crossover point are exchanged to produce two offspring for the new population. Finally, in the mutation phase, each chromosome undergoes mutation with a fixed probability p_m of 0.10 is used in this paper. Mutation process is done by generating a number m in the range [0,1]. If the value at a gene position is v, after mutation it becomes $v \pm m * v$, when $v \neq 0$ or $v \pm m * 2$, when $v = 0$. The '+' or '-' sign occurs equal probability.

The processes of fitness computation, reproduction, crossover and mutation are executed for a maximum of iterations. The best set of continuous values seen up to the last generation provides the solution to the multi-interval discretization. The next section provides the results of implementation of the GA-entropy-instance-based discretization (using (7) as fitness function shown in Table 1), along with its comparison with the performance of the entropy-based algorithm (using (3) as the fitness function), instance-based algorithm (using (5) as the fitness function), EqualHeight, and EqualWeight for the Mutagenesis datasets [4].

4   Experimental Evaluations

In our experimental study, we implement the discretization method, described in Section 3 in the DARA algorithm [1,2,3], in conjunction with the C4.5 classifier (J48 in WEKA) [10], as an induction algorithm that is run on the DARA's discretized and transformed data representation. We then evaluate the effectiveness of each discretization method with respect to C4.5 [5]. The setting for all discretization methods are described in Table 1. We chose three varieties of a well-known datasets, the Mutagenesis [4] relational database.

Table 1. Setting of Discretization methods

Algorithm	Fitness Function
Entropy-Based (EB)	$I(A,T,S,k)$
Entropy-Instance-Based (EIB)	$I(A,T,S,k) + Ind(A,T,S,k)$

The data in mutagenesis domain [4] describes 188 molecules falling in two classes, *mutagenic* (active) and *non-mutagenic* (inactive) and 125 of these molecules are mutagenic. The description consists of the atoms and bonds that make up the compound. Thus, a molecule is described by listing its atoms *atom(AtomID, Element, Type, Charge)* and the bonds *bond(Atom1, Atom2, BondType)* between atoms. For the experiments, we have used only three different sets of background knowledge. They will be referred to as experiment B1, B2 and B3.

- B1: The atoms in the molecule are given, as well as the bonds between them; the type of each bond is given as well as the element and type of each atom.
- B2: In addition to data in experiment B1, continuous values about the charge of atoms are added
- B3: In addition to data in experiment B2, two continuous values describing each molecule are added, which are the log of the compound octanol/water partition coefficient (*logP*), and energy of the compounds lowest unoccupied molecular orbital ($^{\mathcal{E}}LUMO$).

Table 2 gives a detailed overview of the accuracy estimates from *leave-one-out cross validation* performance of C4.5 for different number of bins, *b*, tested for B1, B2 and B3.

The predictive accuracy for EqualHeight and EqualWeight is lower on datasets B1 and B2, when the number of bins is smaller, compared to entropy-based and entropy-instance-based performance accuracy. However, the accuracy of entropy-based and entropy-instance-based discretization is lower when the number of bins is smaller on dataset B3, although the accuracy for all four discretization methods is more less the same on average.

Table 2. Performance (%) of *leave-one-out cross validation* of C4.5 on dataset B1 with different methods of discretization

Datasets	b	EH	EWE	EB	EIB
B1	4	74.3	74.3	78.4	80.6
	6	75.6	74.9	75.6	77.8
	8	74.2	74.2	77.8	75.0
	Average	*74.7*	*74.5*	*77.3*	*77.8*
B2	4	72.9	73.3	76.9	75.0
	6	70.8	70.6	76.3	78.1
	8	71.9	74.2	75.3	71.1
	Average	*71.9*	*72.7*	*76.2*	*74.9*
B3	4	83.1	82.8	79.9	79.8
	6	81.3	81.1	76.3	81.2
	8	81.2	81.2	83.1	84.5
	Average	*81.9*	*81.7*	*79.7*	*81.8*

The result of entropy-based and entropy-instance-based discretization on B1, B2 and B3 are virtually identical, which should come as no surprise, as both are using class information. However, the latter performs better (five out of nine tests) compared to entropy-based, as entropy-instance-based discretizes attribute values based on class information and the individual purity. Both entropy-based and entropy-instance-based discretizations outperform EqualHeight and EqualWeight discretizations. Although EqualWeight uses the class information as one of the criterion in determining the cut points for the partitions, it fails to compete with entropy-based and entropy-instance-based. This is due to the fact that both the entropy-based and entropy-instance-based algorithm use a genetic algorithm in order to produce optimized cut points or partitions. Optimization using a genetic algorithm has greatly improved the performance accuracy for entropy-based and entropy-instance-based discretization. As a result, EqualWeight can be ignored as it does not have any advantages over the other discretization methods.

Finally, we also compared our results to other previously published ones, shown in Table 3. The data summarization approach, performed by discretizing and aggregating multiple instances in relational database using DARA, proved particularly successful on large datasets. Data summarization can be performed separately from the target relation, and it makes DARA more scalable and flexible in characterizing a specific item stored in the target relation that has a *one-to-many* association to other non-target relations.

Table 3. Result on Mutagenesis, B1,B2,B3

Algorithm	Mutagenesis		
	B1	B2	B3
PROGOL [4]	76%	81%	83%
FOIL [18]	61%	61%	83%
Tilde	75%	79%	85%
DARA + C4.5	81%	78%	85%

5   Conclusions

Our experiments reveal that all discretization methods in DARA algorithm help us achieve higher percentage of accuracy. The entropy-instance-based and entropy-based discretization methods are recommended for discretization of attribute values in multi-relational datasets, in which multiple instances can be used to improve the discretization process, as it has been shown here. However, when the number of bins is large, the computation of a genetic algorithm-based entropy-based and entropy-instance-based discretization will be very expensive.

In addition to that, the experimental results in the previous section demonstrate that our approach to discretization and aggregation of relational attributes, implemented in DARA, is at least competitive with existing multi-relational techniques, such as Progol and Tilde. Our approach has one major difference with these techniques, which may be the source of the good performance, namely the use of the aggregates to summarize a block of data, containing multiple instances of a single object, without the requirement of any domain knowledge.

In FOL (e.g. Progol, Tilde), the characterization is based on the occurrence of one or more records in the group that maximizes certain properties. On the other hand, DARA algorithm takes all records into consideration as each record has some influences on the value of the aggregation. As a result, FOL and DARA produce two different sets of feature-spaces, though there is still some overlap. We have presented a method called dynamic aggregation of relational attributes (DARA) with entropy-instance-based discretization to propositionalise a multi-relational database, such that the resulting view can be analysed by existing propositional methods. The DARA method has shown a good performance on three well-known datasets in term of performance accuracy.

References

1. Alfred, R., Kazakov, D.: Weighted Pattern-Based Transformation Approach to Relational Data Mining. In: Proc of ICAIET 2006, Kota Kinabalu, Sabah, Malaysia (November 2006)
2. Alfred, R., Kazakov, D.: Data Summarization Approach to Relational Domain Learning Based on Frequent Pattern to Support the Development of Decision Making. In: Li, X., Zaïane, O.R., Li, Z. (eds.) ADMA 2006. LNCS (LNAI), vol. 4093, pp. 889–898. Springer, Heidelberg (2006)
3. Alfred, R., Kazakov, D.: Pattern-Based Transformation Approach to Relational Domain Learning Using DARA. In: the Proc DMIN 2006, USA, pp. 296–302 (2006)
4. Srinivasan, A., Muggleton, S.H., Sternberg, M.J.E., King, R.D.: Theories for mutagenicity: A study in first-order and feature-based induction. Artificial Intelligence 85 (1996)
5. Quinlan, J.R.: C4.5: Programs for Machine Learning. Morgan Kaufmann, Los Altos, California
6. Kramer, S., Lavrač, N., Flach, P.: Propositionalization approaches to relational data mining. In: Dzeroski, S., Lavrač, N. (eds.) Relational Data mining, Springer, Heidelberg (2001)
7. Salton, G., Michael, J.: Introduction to Modern Information Retrieval. McGraw-Hill, Inc., New York (1986)

8. Bezdek, J.C.: Some new indexes of cluster validiy. IEEE Transaction System, Man, Cybern. B 28, 301–315 (1998)
9. Boley, D.: Principal direction divisive partitioning. Data Mining and Knowledge Discovery 2(4), 325–344 (1998)
10. Witten, I., Frank, E.: Data Mining: Practical Machine Learning Tools and Techniques with Java Implementations. Morgan Kaufman (1999)
11. Agrawal, R., Gehrke, J., Gunopulos, D., Raghavan, P.: Automatic subspace clustering of high dimensional data for data mining applications. In: Proceedings of the 1998 ACM SIGMOD International Conference on Management of Data, pp. 94–105. ACM Press, New York (1998)
12. Hofmann, T., Buhnmann, J.M.: Active data clustering. In: Advance in Neural Information Processing System (1998)
13. Hartigan, J.A.: Clustering Algorithms. Wiley, New York (1975)
14. Van Laer, W., De Raedt, L., Deroski, S.: On multi-class problems and discretization in inductive logic programming. In: Raś, Z.W., Skowron, A. (eds.) ISMIS 1997. LNCS, vol. 1325, Springer, Heidelberg (1997)
15. Kohavi, R., Sahami, M.: Error-based and entropy-based discretisation of continuous features. In: Proceedings of the Second International Conference on Knowledge Discovery and Data Mining, AAAI Press (1996)
16. Perner, P., Trautzsch, S.: Multi-interval discretization methods for decision tree learning. In: Advances in Pattern Recognition, Joint IAPR International Workshops SSPR '98 and SPR '98, pp. 475–482 (1998)
17. Fayyad, U.M., Irani, K.B.: Multi-interval discretization of continuous valued attributes for classification learning. In: Proceedings of the Thirteenth International Joint Conference on Artificial Intelligence, pp. 1022–1027 (1993)
18. Srinivasan, A., Muggleton, S., King, R.: Comparing the use of background knowledge by inductive logic programming systems. In: Proceedings of the 5th International Workshop on Inductive Logic Programming (1995)

Adaptive k-Nearest-Neighbor Classification Using a Dynamic Number of Nearest Neighbors*

Stefanos Ougiaroglou[1], Alexandros Nanopoulos[1], Apostolos N. Papadopoulos[1], Yannis Manolopoulos[1], and Tatjana Welzer-Druzovec[2]

[1] Department of Informatics, Aristotle University, Thessaloniki 54124, Greece
{stoug,ananopou,papadopo,manolopo}@csd.auth.gr
[2] Faculty of Electrical Eng. and Computer Science, University of Maribor, Slovenia
welzer@uni-mb.si

Abstract. Classification based on k-nearest neighbors (kNN classification) is one of the most widely used classification methods. The number k of nearest neighbors used for achieving a high accuracy in classification is given in advance and is highly dependent on the data set used. If the size of data set is large, the sequential or binary search of NNs is inapplicable due to the increased computational costs. Therefore, indexing schemes are frequently used to speed-up the classification process. If the required number of nearest neighbors is high, the use of an index may not be adequate to achieve high performance. In this paper, we demonstrate that the execution of the nearest neighbor search algorithm can be interrupted if some criteria are satisfied. This way, a decision can be made without the computation of all k nearest neighbors of a new object. Three different heuristics are studied towards enhancing the nearest neighbor algorithm with an early-break capability. These heuristics aim at: (i) reducing computation and I/O costs as much as possible, and (ii) maintaining classification accuracy at a high level. Experimental results based on real-life data sets illustrate the applicability of the proposed method in achieving better performance than existing methods.

Keywords: kNN classification, multidimensional data, performance.

1 Introduction

Classification is the data mining task [10] which constructs a model, denoted as *classifier*, for the mapping of data to a set of predefined and non-overlapping classes. The performance of a classifier can be judged according to criteria such as its accuracy, scalability, robustness, and interpretability. A key factor that influences research on classification in the data mining community (and differentiates it from classical techniques from other fields) is the emphasis on scalability, that is, the classifier must work on large data volumes, without the need for experts

---

* Work partially supported by the 2004-2006 Greek-Slovenian bilateral research program, funded by the General Secretariat of Research and Technology, Ministry of Development, Greece.

Y. Ioannidis, B. Novikov, and B. Rachev (Eds.): ADBIS 2007, LNCS 4690, pp. 66–82, 2007.
© Springer-Verlag Berlin Heidelberg 2007

to extract appropriate samples for modeling. This fact poses the requirement for closer coupling of classification techniques with database techniques. In this paper, we are interested in developing novel classification algorithms that are accurate *and* scalable, which moreover can be easily integrated to existing database systems.

Existing classifiers are divided into two categories [12], *eager* and *lazy*. In contrast to an eager classifier (e.g., decision tree), a lazy classifier [1] builds no general model until a new sample arrives. A k-nearest-neighbor (kNN) classifier [7] is a typical example of the latter category. It works by searching the training set for the k nearest neighbors of the new sample and assigns to it the most common class among its k nearest neighbors. In general, a kNN classifier has satisfactory noise-rejection properties. Other advantages of a kNN classifier are: (i) it is analytically tractable, (ii) for $k = 1$ and unlimited samples the error rate is never worse than twice the Bayes' rate, (iii) it is simple to implement, and (iv) it can be easily integrated into database systems and exploit access methods that the latter provide in the form of indexes.

Due to the aforementioned characteristics, kNN classifiers are very popular and enjoy many applications. With a naive implementation, however, the kNN classification algorithm needs to compute all distances between training data and a test datum, and requires additional computation to get k nearest neighbors. This impacts the scalability of the algorithm in a negative manner. For this reason, recent research [5] has proposed the use of high-dimensional access methods and techniques for fast computation of similarity joins, which are available in existing database systems, to reduce the cost of searching from linear to logarithmic. Nevertheless, the cost of searching the k nearest neighbors, even with a specialized access method, still increases significantly by increasing k.

For a given test datum, depending on which training data comprise its neighborhood, we may need a small or a large k value to determine its class. In other words, in some cases, a small k value may suffice for the classification, whereas in other cases we may need to examine larger neighborhoods. Therefore, the appropriate k value may vary significantly. This introduces a trade-off: By posing a global and adequately high value for k, we attain good accuracy, but the computational cost of nearest neighbor searching increases for large k values. Higher computational cost reduces the scalability in large data sets. In contrast, by keeping a small k value, we get low computational cost, but this may impact accuracy in a negative manner. What is, thus, required is an algorithm that will combine good accuracy and low computational cost, by locally adapting the required value of k. In this work, we propose a novel framework for a kNN classification algorithm that fulfills the aforementioned property. We also examine techniques that help us for finding the appropriate k value in each case. Our contributions are summarized as follows:

- We propose a novel classification algorithm based on a non-fixed number of nearest neighbors, which is less time consuming than the known kNN classification, without sacrificing accuracy.

- Three heuristics are proposed that aim at the early-break of the kNN classification algorithm. This way, significant savings in computational time and I/O can be achieved.
- We apply the proposed classification scheme to large data sets, where indexing is required. A number of performance evaluation tests are conducted towards investigating the computational time, the I/O time and the accuracy achieved by the proposed scheme.

The rest of our work is organized as follows. The next section briefly describes related work in the area and summarizes our contributions. Section 3 studies in detail the proposed early-break heuristics, and presents the modified kNN classification algorithm. Performance evaluation results based on two real-life data sets are given in Section 4. Finally, Section 5 concludes our work and briefly discusses future work in the area.

2   Related Work

Due to its simplicity and good performance, kNN classification has been studied thoroughly [7]. Several variations have been developed [2], like the distance-weighted kNN, which puts emphasis on nearest neighbors, and the locally-weighted averaging, which uses kernel width to control the size of neighborhood that has large effect. All these approaches propose adaptive schemes to improve the accuracy of kNN classification in the case where not all attributes are similar in their relevance to the classification task. In our research, we are interested in improving the scalability of kNN classification.

Also, kNN classification has been combined with other methods and, instead of predicting a class with simple voting, prediction is done by another machine learner (e.g., neural-network) [3]. Such techniques can be considered complementary to our work. For this reason, to keep comparison clear, we did not examine such approaches.

Böhm and Krebs [5] proposed an algorithm to compute the k-nearest neighbor join using the multipage index (MuX), a specialized index structure for the similarity join. Their algorithm can be applied to the problem of kNN classification and can increase its scalability. However, it is based on a fixed number of k, which (as described in Introduction) if it is not tuned appropriately, it can negatively impact the performance of classification.

3   Adaptive Classification

3.1   The Basic Incremental kNN Algorithm

An algorithm for incremental computation of nearest neighbors using the R-tree family [9,4] has been proposed in [11]. The most important property of this method is that the nearest neighbors are determined in their order of their distance from the query object. This enables the discovery of the $(k + 1)$-th nearest neighbor if we have already determined the previous k, in contrast to the algorithm proposed in [13] (and enhanced in [6]) which requires a fixed value of k.

The incremental nearest neighbors search algorithm maintains a priority queue. Queue entries are Minimum Bounding Rectangles (MBRs) and they are prioritized with respect to their distance from the query point. An object will be examined when it reaches the top of the queue. The algorithm begins by inserting the root elements of the R-tree in the priority queue. Then, it selects the first entry and inserts its children. This procedure is repeated until the first data object reaches the top of the queue. This object is the first nearest neighbor. Figure 1 depicts the Incr-kNN algorithm with some modifications, towards adapting the algorithm for classification purposes. Therefore, each object of the test set is a query point and each object of the training set, contains an additional attribute which indicates the class where the object belongs to. The R-tree is built using the objects of the training set.

The aim of our work is to perform classification by using a smaller number of nearest neighbors than k, if this is possible. This will reduce computational costs and I/O time. However, we do not want to harm accuracy (at least not significantly). Such an early-break scheme can be applied since Incr-kNN determines the nearest neighbors in increasing distance order from the query point. The modifications performed to the original incremental kNN algorithm are summarized as follows:

- We have modified line 3 of the algorithm. The algorithm accepts a maximum value of k and is executed until either k nearest neighbors are found (no early break) or the heuristics criteria are satisfied (early break). Specifically, we added the condition $NNCounter \leq k$ in while statement (line 3).
- We have added the lines 10,11,12 and 13. At this point, the algorithm retrieves a nearest neighbor and checks for early break. Namely, the while loop of the algorithm breaks if the criteria defined by the heuristic that we use are satisfied. So, the new item is classified using $NNCounter$ nearest neighbors, where $NNCounter < k$. Lines 11, 12 and 13 are replaced according to the selected heuristic.
- We have added the lines 25, 26, 27 and 28. These lines perform classification taking into account k nearest neighbors. This code is executed only when the heuristic is not capable of performing an early-break.

3.2   Early-Break Heuristics

In this section, we present three heuristics that interrupt the computation of nearest neighbors when some criteria are satisfied. The classification performance (accuracy and execution cost) depends on the adjustments of the various heuristics parameters. The parameter $MinnNN$ is common to all heuristics and defines the minimum number of nearest neighbors which must be used for classification. After the retrieval of $MinnNN$ nearest neighbors, the check for early-break is performed. The reason for the use of $MinNN$ is that a classification decision is

Algorithm Incr-kNN (QueryPoint q, Integer k)
1. $PriorityQueue$.enqueue(roots children)
2. $NNCounter = 0$
3.   **while** $PriorityQueue$ is not empty **and** $NNCounter \leq k$ **do**
4. $element = PriorityQueue$.dequeue()
5.   **if** $element$ is an object or its MBR **then**
6.   **if** element is the MBR of $Object$ **and** $PriorityQueue$ is not empty
            and objectDist(q, $Object$) > $PriorityQueue$.top **then**
7. $PriorityQueue$.enqueue($Object$, ObjectDist(q, $Object$))
8.   **else**
9.   Report element as the next nearest object (save the class of the object)
10. $NNCounter$++
11.   **if** early-break conditions are satisfied **then**
12.   Classify the new object q in the class where the most nearest neighbors
                 belong to and break the while loop. q is classified using
                 $NNCounter$ nearest neighbors
13.   **endif**
14.   **endif**
15.   **else if** $element$ is a leaf node **then**
16.   **for each** entry (Object, MBR) in $element$ **do**
17. $PriorityQueue$.enqueue ($Object$, dist(q, $Object$))
18.   **endfor**
19.   **else** /*non-leaf node*/
20.   **for each entry** e in $element$ **do**
21. $PriorityQueue$.enqueue(e, dist(q, e))
22.   **endfor**
23.   **endif**
24. **end while**
25. **if** no early-break has been performed **then** // use k nearest neighbors
26.   Find the major class (class where the most nearest neighbors belong to)
27.   Classify the new object q to the major class
28. **endif**

Fig. 1. Outline of Incr-kNN algorithm with early break capability

preferable when it is based on a minimum number of nearest neighbors, otherwise accuracy will be probably poor. These criteria depend on which proposed
heuristic is used. The code of each heuristic replaces lines 11 and 12 of the
Incr-kNN algorithm depicted in Figure 1.

Simple Heuristic (SH). The first proposed heuristic is very simple. According
to this simple heuristic, the early-break is performed when the percentage of
nearest neighbors that vote the major class is greater than a predefined threshold.
We call this threshold $PMaj$.

For example, suppose that we have a data set where the best accuracy is
achieved using 100 nearest neighbors. Also, suppose that we define that $PMaj$
= 0.9 and $MinNN$=7. The Incr-kNN is interrupted when 90% of NNs vote a
specific class. If this percentage is achieved when the algorithm examines the
tenth NN (9 out of 10 NNs vote a specific class), then we avoid the cost of

searching the rest 90 nearest neighbors. Using the Incr-kNN algorithm, we ensure that the first ten neighbors which have been examined, are the nearest.

Furthermore, if the simple heuristic fails to interrupt the algorithm because $PMaj$ is not achieved, it will retry an early-break after finding the next $TStep$ nearest neighbors.

Independent Class Heuristic (ICH). The second early-break heuristic is the Independent Class Heuristic (ICH). This heuristic does not use the $PMaj$ parameter. The early-break of Incr-kNN is based on the superiority of the major class. Superiority is determined by the difference between the sum of votes of the major class and the sum of votes of all the other classes. The parameter $IndFactor$ (Independency Factor) defines the superiority level of the major class that must be met in order to perform an early-break. More formally, in order to apply an early-break, the following condition must be satisfied:

$$SVMC > IndFactor \cdot \left(\sum_{i=1}^{n} SVC_i - SVMC \right) \tag{1}$$

where $SVMC$ is the sum of votes of major class, n is the number of classes and SVC_i is the sum of votes of class i.

For example, suppose that our data set contains objects of five classes, we have set $IndFactor$ to 1 and the algorithm has determined 100 NNs. Incr-kNN will be interrupted if 51 NNs vote a specific class and the rest 49 NNs vote the other classes. If the value of $IndFactor$ is set to 2, then the early-break is performed when the major class has more than 66 votes.

Studying the Independent Class Heuristic, we conclude that the value of the $IndFactor$ parameter should be adjusted by taking into account the number of classes and the class distribution of data set. In the case of a normal distribution, we accept the following rule: when the number of classes is low, $IndFactor$ should be set to a high value. On the other hand, when there are many classes, $IndFactor$ should be set to a lower value.

In ICH, parameter $TStep$ is used in the same way as in the SH heuristic. Specifically, when there is a failure in interruption, the early-break check is again activated after determining the next $TStep$ nearest neighbors.

M-Times Major Class Heuristic (MMCH). The last heuristic that we present is termed M-Times Major Class Heuristic (MMCH). The basic idea is to stop the Incr-kNN when M consecutive nearest neighbors, which vote the major class, are found. In other words, the while-loop of Incr-kNN algorithm terminates when the following sequence of nearest neighbors appears:

$$NN_{x+1}, NN_{x+2}, ..., NN_{x+M} \in MajorClass \tag{2}$$

However, this sequence is not enough to force an early-break. In addition, the $PMaj$ parameter is used in the same way as in the SH heuristic. Therefore, MMCH heuristic breaks the while loop when the percentage of nearest neighbors that vote the major class is greater than $PMaj$ and there is a sequence of M

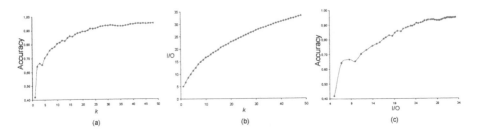

Fig. 2. Accuracy vs k, I/O vs k and accuracy vs I/O for PBC data set

nearest neighbors that belong to the major class. We note that MMCH does not require the $TStep$ parameter.

4   Performance Evaluation

In this section, we present the experimental results on two real-life data sets. All experiments have been conducted on an AMD Athlon 3000+ machine (2000 MHz), with 512 MB of main memory, running Windows XP Pro. The R*-tree, the incremental kNN algorithm, and the classification heuristics have been implemented in C++.

4.1   Data Sets

The first data set is the Pages Blocks Classification (PBC) data set and contains 5,473 items. Each item is described by 10 attributes and one class attribute. We have used the first five attributes of the data set and the class attribute. We reduced the number of attributes considering that the most dimensions we use, the worst performance the family of R-trees has. Each item of the data set belongs to one of the five classes. Furthermore, we have divided the data set into two subsets. The first subset contains 4,322 items used for training and the second contains the rest 1,150 items used for testing purposes.

The traditional kNN classification method achieves the best possible accuracy when $k = 9$. However, this value was very low and so the proposed heuristics can not reveal their full potential. Therefore, we have added noise in the data set in order to make the use of a higher k value necessary. Particularly, for each item of the training set, we modified the value of the class attribute with probability 0.7 (the most noise is added, the highest value of k is needed to achieve the best classification accuracy). This fact forced the algorithm to use a higher k value. This way, we constructed a data set where the best k value is 48. This means that the highest accuracy value is achieved when 48 nearest neighbors contribute to the voting process. This is illustrated in Figure 2(a), which depicts the accuracy value accomplished by modifying k between 1 and 48.

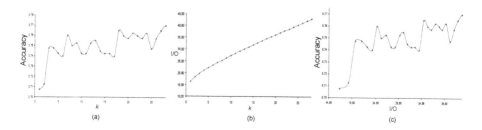

Fig. 3. Accuracy vs k, I/O vs k and accuracy vs I/O for LRI data set

As expected, the higher the value of k, the higher the number of I/O operations. This phenomenon is illustrated by Figure 2(b). We note that if we had used a lower value for k ($k < 48$), we would have avoided a significant number of I/Os and therefore the search procedure would be less time-consuming. For example, if we set $k = 30$, then we avoid 5.84 I/Os for the classification of one item of the test set without significant impact on accuracy. Figure 2(c) combines the two previous graphs. Particularly, this figure shows how many I/Os the algorithm requires in order to accomplish a specific accuracy value.

The second data set is the Letter Image Recognition Data set [8], which contains 20,000 items. We have used 15,000 items for training and 5,000 items for testing. Each item is described by 17 attributes (one of them is the class attribute) and represents an image of a capital letter of the English alphabet. Therefore, the data set has 26 classes (one for each letter). The data set objective is to identify the capital letter represented by the items of the test set using a classification method.

As in the case of the PBC data set, we have reduced the number of dimensions (attributes). In this case, dimensionality reduction has been performed by using Principal Component Analysis (PCA) on the 67% of the original data. The final number of dimensions have been set to 5 (plus the class attribute).

Figure 3 illustrates the data set behavior. Specifically, Figure 3(a) shows the accuracy achieved for k ranging between 1 and 28. We notice that the best accuracy (77%) is achieved when $k = 28$. Figure 3(b) presents the impact of k on the number of I/Os. Almost 43 I/O operations are required to classify an item of the test set for $k = 28$. Finally, Figure 3(c) combines the two previous graphs illustrating the relation between accuracy and the number of I/O operations. By observing these figures, we conclude that if we had used a lower k value, then we could have achieved better execution time by keeping accuracy at high levels.

4.2   Determining Parameters Values

Each heuristic uses a number of parameters. These parameters must be adjusted so that the best performance is achieved (or the best balance between accuracy and execution time). In this section, we present a series of experiments which demonstrates the behavior of the heuristics for different values of the parameters.

We keep the best values (e.g. the parameters that manage to balance execution time and accuracy) for each heuristic and use these values in a subsequent section where heuristics are compared.

As we demonstrate, a heuristic shows its best performance for a range of parameter values. Therefore, for a new data set, these parameters can be adjusted by applying the classification process to a sample instead of using the whole data set.

Pages Blocks Classification Data Set. Initially, we are going to analyze the $MinNN$ parameter. It is a parameter that all heuristics use. Recall that $MinNN$ is the minimum number of NNs that should be used for classification. After determining these NNs, the heuristics are activated. Figure 4 show how the heuristics performance (accuracy and I/O) is affected by modifying the value of $MinNN$. The values of the other parameters have as follows: $TStep = 4$, $IndFactor = 1$, $PMaj = 0.6$, $MTimes = 4$, $k = 48$.

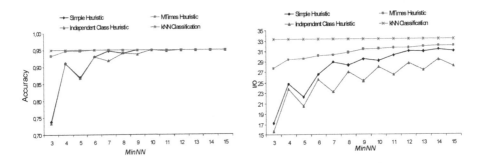

Fig. 4. Impact of $MinNN$ for PBC data set

By observing the results it is evident that accuracy is least affected when the MMCH heuristic is used. Therefore, for this heuristic, we set $MinNN = 4$, which is the value that provides the best balance between I/O and accuracy. In contrast, the accuracy of the other two heuristics is significantly affected by the increase of $MinNN$. We decide to define MinNN = 11 for the Independency Class Heuristic and $MinNN = 7$ for the Simple Heuristic. Our decision is justified by the accuracy and I/O measurements provided for these parameters values.

We continue our experiments by finding the best value for $IndFactor$. Recall that this parameter is used only in ICH heuristic. We modify $IndepFactor$ from 0.4 to 4 and calculate the accuracy achieved. Figure 5 illustrates that the best balance between accuracy and I/O is achieved when $IndFactor = 1$. The values of the other parameters are as follows: $MinNN = 11$, $TStep = 4$, $k = 48$.

Next we study the impact of the parameter $MTimes$, which is used only by the MMCH heuristic. As it is depicted in Figure 6 for $MTimes = 3$, the accuracy level will be high enough and the number of I/O operations is relatively low. So we keep this value as the best possible for this parameter. The values of the other parameters have as follows: $MinNN = 4$, $PMaj = 0.6$, $k = 48$.

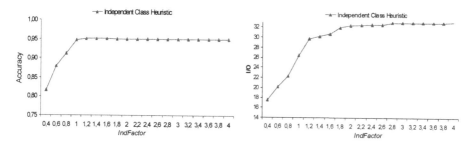

Fig. 5. Impact of $IndFactor$ on the performance of ICH heuristic for PBC data set

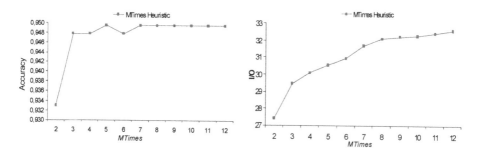

Fig. 6. Impact of $MTimes$ on the performance of MMCH heuristic for PBC data set

Next, we study the impact of $TStep$ parameter on the performance of SC and ICH, since MMCH does not use this parameter. Figure 7 depicts the results. The values of the rest of the parameters have as follows: $MinNN = 6$, $IndFactor = 1$, $PMaj = 0.6$, $MTimes = 3$, $k = 48$. Both SH and ICH heuristics achieve the best accuracy when $TStep = 4$. In fact, SH achieves the same accuracy for $TStep = 3$ or $TStep = 4$, but less I/Os are required when $TStep = 4$. Since MMCH and kNN classification are not affected by $TStep$, their graphs are parallel to the $TStep$ axis. Finally, it is worth to note that although MMCH needs significantly less I/Os than kNN classification, the accuracy that MMCH achieves is the same as that of kNN classification.

Letter Image Recognition Data Set. Next, we repeat the same experiments using the LIR data set. The impact of $MinNN$ is given in Figure 8. It is evident that all heuristics achieve higher accuracy than kNN classification. By studying Figure 8 we determine that the best $MinNN$ values for the three heuristics are: $MinNN = 7$ for SH, $MinNN = 12$ for ICH and $MinNN = 4$ for MMCH. These values achieve the best balance between accuracy and I/O processes. The values of the other parameters have as follows: $TStep = 2$, $IndFactor = 1$, $PMaj = 0.6$, $MTimes = 3$, $k = 28$.

Figure 9 depicts the results for the impact of $IndFactor$. We see that $IndFactor = 1$ is the best value since the ICH heuristic achieve the best possible accuracy

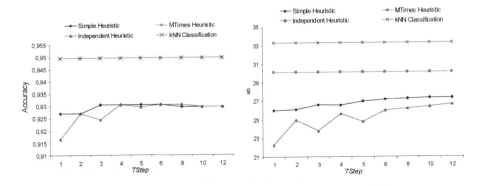

Fig. 7. Impact of $TStep$ parameter for PBC data set

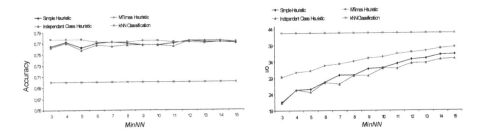

Fig. 8. Impact of $MinNN$ for LIR data set

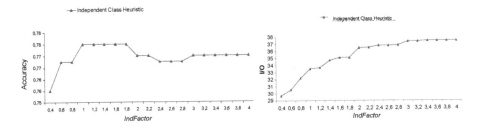

Fig. 9. Impact of $IndFactor$ on the performance of ICH heuristic for LIR data set

value and at the same time saves almost ten I/O operations per query. The values of the other parameters are as follows: $MinNN = 12$, $TStep = 5$, $k = 28$.

Next, we consider parameter $MTimes$, which is used only by the MMCH heuristic. We define $MTimes = 3$, which results in 13 I/O savings for each query and achieves the best possible accuracy value. The results are illustrated in Figure 10. The values of the other parameters have as follows: $MinNN = 4$, $PMaj = 0.6$, $k = 28$.

Finally, in Figure 11 we give the impact of $TStep$. We set $TStep = 4$ for SH and $TStep = 5$ for ICH, since these values give adequate accuracy and execution

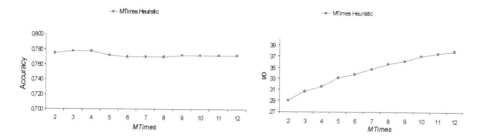

Fig. 10. Impact of $MTimes$ on the performance of MMCH heuristic for LIR data set

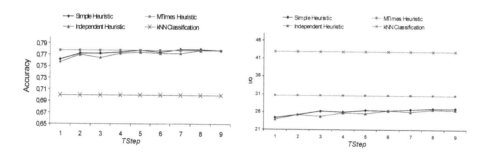

Fig. 11. Impact of $TStep$ parameter for LIR data set

time. The values of the rest of the parameters have as follows: $MinNN = 4$, $IndFactor = 1$, $PMaj = 0.6$, $MTimes = 3$, $k = 28$.

4.3   Comparison of Heuristics

In this section we study how the heuristics compare to each other and to the traditional kNN classification, by setting the parameters to the best values for each heuristic, as they have been determined in the previous section.

Pages Blocks Classification Data Set. Figure 12 depicts the performance results vs $PMaj$. When $PMaj = 0.6$, accuracy is about the same for all heuristics and very close to that achieved by traditional kNN classification. However, our heuristics require significant less I/O for achieving this accuracy. When this accuracy value is accomplished there is no reason to find more nearest neighbors and therefore valuable computational savings are achieved.

According to our results, ICH achieves an accuracy value of 0.947 (kNN's accuracy is 0.9495) while it saves about 6.88 I/Os for the classification of one item. Particularly, when we use ICH, we find 31.29 nearest neighbors on average instead of 48. Similarly, when $PMaj = 0.6$, SH achieves accuracy equal to 0.948 (very close to that of kNN) by performing an early-break when 38.5 NNs on average have been found (almost 10 less than kNN requires). Therefore, SH saves about 4.385 I/Os per each item of the test set. Finally, the same accuracy

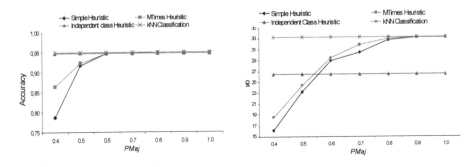

Fig. 12. Accuracy and number of I/Os

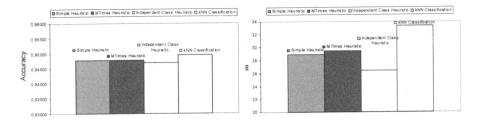

Fig. 13. Accuracy and number of I/Os for $PMaj = 0.6$

value (0.948) is achieved by MMCH. However, this heuristic saves less number of I/Os than SH. Specifically, MMCH spends 3.875 I/O less than kNN because it finds 39.767 NNs on average.

Figures 13 and 14 summarize the results of this experiment using bar charts which have been produced by setting $PMaj = 0.6$ and maintain the same values for the other parameters. Figure 13 shows that the accuracy achieved by the heuristics is very close to that of kNN, whereas the number of required I/Os is significant less than that of kNN. Figure 14 presents the number of nearest neighbors retrieved on average by each method.

We can not directly answer the question "which heuristic is the best". The answer depends on which measure is more critical (accuracy or execution time). If a compromise must be made, Figure 12 will help. We notice that ICH shows the best performance because it achieves an accuracy that is very close to the accuracy of kNN posing the minimum number of I/Os. To declare a winner between SH and MMCH we notice that when $PMaj > 0.6$ the heuristics achieves almost the same accuracy, but MMCH poses more I/Os than SH.

As we have already mentioned, we try to find the parameters values that provide a compromise between accuracy and execution time. However, if one of these measures is more important than the other, the parameters can be adjusted to reflect this preference. Suppose that execution time is more critical than accuracy (when for example a quick-and-dirty scheme should be applied due to application requirements). If we set $PMaj = 0.5$, then 23.245 I/Os per

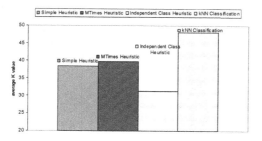

Fig. 14. Required number of nearest neighbors for $PMaj = 0.6$

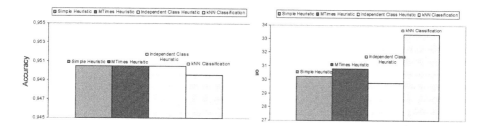

Fig. 15. Accuracy and number of I/Os ($MinNN = 11$)

query will be required (instead of 33.32 required by kNN) by finding 25.72 NNs instead of 48. However, this means that accuracy will be significantly smaller than that of kNN (we saw that when $PMaj = 0.6$, the difference between the accuracy of SH and kNN is minor). Similar results are obtained for MMCH when $PMaj = 0.5$. On the other hand, if accuracy is more critical than time, we can adjust the heuristics parameters towards taking this criticality into account. In this case, it is possible that the heuristics achieve better accuracy than kNN, with execution time overhead. In any case, early-break heuristics always require less execution time than kNN.

By considering the impact of $MinNN$ shown in Figure 4, it is evident that MMCH can achieve a slightly better accuracy than kNN. Specifically, if we set $MinNN = 8$, $PMaj = 0.6$ and $MTimes = 3$, then MMCH achieves an accuracy value equal to 0.950435, whereas 30.793 I/Os are required (while kNN requires 33.3174 I/Os per query and achieves an accuracy of 0.94956). The early-break heuristic is able to avoid 2.5244 I/O per query, whereas at the same time achieves better accuracy when k is adjusted to ensure the best accuracy value. Although the number of saved I/Os may seem small, note that this savings are performed per classified item. Since the PBC test set contains 1,150 items, we realize that the overall number of saved I/Os is 2903, which is significant.

Similar considerations apply to the other two heuristics. For example, ICH can outperform the accuracy of kNN when $IndyFactor = 1.2$, $TStep = 4$, and $MinNN = 11$ (see Figure 5). However, because of the increase of $IndFactor$

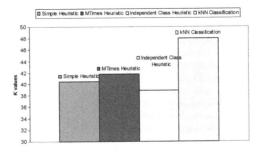

Fig. 16. Required number of nearest neighbors ($MinNN = 11$)

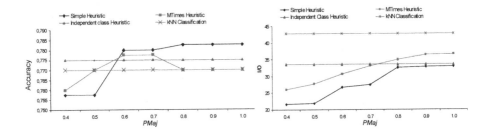

Fig. 17. Accuracy and number of I/Os vs $PMaj$

from 1 to 1.2, the I/O requirements are increased from 26.44 to 29.75. Finally, if we set $MinNN = 11$ instead of 7, SH also outperforms the accuracy of kNN (see Figure 4). These results are illustrated in Figures 15 and 16.

Letter Image Recognition Data Set. We close this section by presenting the comparison results using the LIR data set. Similar conclusions can be drawn as in the case of the PBC data set. The results are given in Figures 17, 18 and 19.

We note that when we $PMaj > 0.5$ (see Figure 17), the three heuristics accomplish better accuracy than kNN and manage to reduce execution time. More specifically, SH achieves an accuracy of 0.7825 with 32.53 I/Os on average for $PMaj = 0.8$, whereas kNN an accuracy of 0.77 spending 42.81 I/Os (see Figure 18). Also, if we set $PMaj = 0.6$, SH achieves an accuracy of 0.78 spending 26.7775 I/Os on average (see Figure 19).

ICH, which is not affected by the $PMaj$ parameter, achieves an accuracy of 0.775 spending 33.52745 I/Os on average. Finally, MMCH achieves the best balance between accuracy and execution time we set $PMaj = 0.6$. Particularly, the heuristic accomplishes an accuracy of 0.7775 and spends 30.6525 I/Os on average (see Figure 18). Comparing the three heuristics using the Letter Image Recognition data set, we conclude that the simple heuristic has the best performance since it achieves the best possible accuracy spending the least possible number of I/Os.

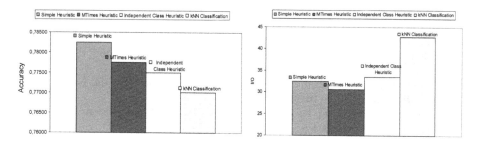

Fig. 18. Accuracy and number of I/Os for $PMaj = 0.8$

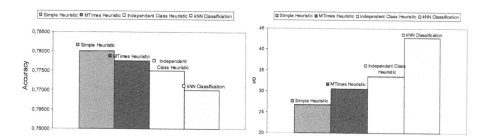

Fig. 19. Accuracy and number of I/Os for $PMaj = 0.6$

5 Conclusions

In this paper, an adaptive kNN classification algorithm has been proposed, which does not require a fixed value for the required number of nearest neighbors. This is achieved by incorporating an early-break heuristic into the incremental k-nearest neighbor algorithm. Three early-break heuristics have been proposed and studied, which use different conditions to enforce an early-break. Performance evaluation results based on two real-life data sets have shown that significant performance improvement may be achieved, whereas at the same time accuracy is not reduced significantly (in some cases accuracy is even better than that of kNN classification). We plan to extend our work towards: (i) incorporating more early-break heuristics and (ii) studying incremental kNN classification by using subsets of dimensions instead of the whole dimensionality.

References

1. Aha, D.W.: Editorial. Artificial Intelligence Review (Special Issue on Lazy Learning) 11(1-5), 1–6 (1997)
2. Atkeson, C., Moore, A., Schaal, S.: Locally weighted learning. Artificial Intelligence Review 11(1-5), 11–73 (1997)
3. Atkeson, C., Schaal, S.: Memory-based neural networks for robot learning. Neurocomputing 9, 243–269 (1995)

4. Beckmann, N., Kriegel, H.-P., Schneider, R., Seeger, B.: The r*-tree: An efficient and robust access method for points and rectangles. In: Proceedings of the ACM SIGMOD Conference, pp. 590–601. ACM Press, New York (1990)
5. Boehm, C., Krebs, F.: The k-nearest neighbour join: Turbo charging the kdd process. Knowledge and Information Systems 6(6), 728–749 (2004)
6. Cheung, K.L., Fu, A.: Enhanced nearest neighbour search on the r-tree. ACM SIGMOD Record 27(3), 16–21 (1998)
7. Dasarathy, B.V.: Nearest Neighbor Norms: NN Pattern Classification Techniques. IEEE Computer Society Press, Los Alamitos (1991)
8. Frey, P.W., Slate, D.J.: Letter recognition using holland-style adaptive classifiers. Machine Learning 6(2), 161–182 (1991)
9. Guttman, A.: R-trees: A dynamic index structure for special searching. In: Proceedings of the ACM SIGMOD Conference, pp. 47–57. ACM Press, New York (1984)
10. Han, J., Kamber, M.: Data Mining: Concepts and Techniques. Morgan Kaufmann, San Francisco (2000)
11. Hjaltason, G.R., Samet, H.: Distance browsing in spatial databases. ACM Transactions on Database Systems 24(2), 265–318 (1999)
12. James, M.: Classification Algorithms. John Wiley & Sons, Chichester (1985)
13. Rousopoulos, N., Kelley, S., Vincent, F.: Nearest neigbor queries. In: Proceedings of the ACM SIGMOD Conference, pp. 71–79. ACM Press, New York (1995)

Database Implementation of a Model-Free Classifier*

Konstantinos Morfonios

Department of Informatics and Telecommunications, University of Athens
kmorfo@di.uoa.gr

Abstract. Most methods proposed so far for classification of high-dimensional data are memory-based and obtain a model of the data classes through training before actually performing any classification. As a result, these methods are ineffective on (a) very large datasets stored in databases or data warehouses, (b) data whose partitioning into classes cannot be captured by global models and is sensitive to local characteristics, and (c) data that arrives continuously to the system with pre-classified and unclassified instances mutually interleaved and whose successful classification is sensitive to using the most complete and/or most up-to-date information. In this paper, we propose LOCUS, a scalable model-free classifier that overcomes these problems. LOCUS is based on ideas from pattern recognition and is shown to converge to the optimal Bayes classifier as the size of the datasets involved increases. Moreover, LOCUS is data-scalable and can be implemented using standard SQL over arbitrary database tables. To the best of our knowledge, LOCUS is the first classifier that combines all the characteristics above. We demonstrate the effectiveness of LOCUS through experiments over both real-world and synthetic datasets, comparing it against memory-based decision trees. The results indicate an overall superiority of LOCUS over decision trees on both classification accuracy and data sizes that it can handle.

Keywords: Lazy Classification, Scalable Classification, Disk-Based Classification, Optimal Bayes.

1   Introduction

Consider a collection of "labeled" objects available, often called a *training set*. The objects are described by a number of attributes, called *features*, and are modeled as *feature vectors*. The object labels are used to categorize each object into one of several predefined *classes*. Assuming that the class of every object can be expressed as a function of the object's attributes, features are often called *predictor attributes*, while the corresponding class is the *dependent attribute*. Classification is the task of labeling new objects, whose class is unknown, using the a-priori labeling in the training set as a basis. Some example applications of common classification tasks

---

* The project is co-financed within Op. Education by the ESF (European Social Fund) and National Resources.

Y. Ioannidis, B. Novikov, and B. Rachev (Eds.): ADBIS 2007, LNCS 4690, pp. 83–97, 2007.
© Springer-Verlag Berlin Heidelberg 2007

include automated medical diagnosis, target group identification, email filtering, character or speech recognition, and fraud detection.

Interestingly, traditional classification algorithms mainly focus on providing high accuracy even at the cost of iterative traversals of the training data, neglecting the potential access costs of such operations. Clearly, this is acceptable only for small datasets that fit in main memory: the majority of the literature involves datasets with only a few hundreds of instances at most.

Nevertheless, increasing use of computers and database management systems (DBMSs) in modern business environments, decreasing storage costs, and other similar trends have generated a wealth of data stored in databases and data warehouses. *Data mining* over such databases can discover useful information that can help their owners predict future events based on past observations. As a special case of data mining, classification presents new challenges when applied to training sets that are orders of magnitude larger than the ones typically considered. Revisiting traditional classification algorithms to make them applicable in today's large-scale environment is highly desirable.

Existing classifiers can be placed into two main categories: *eager* and *lazy*. Eager classifiers use an off-line training phase during which they build a model of the training set. During classification they use this model to label unknown objects. Popular eager classifiers include Decision Trees [18] and Support Vector Machines [5]. On the contrary, lazy classifiers need no training as they do not rely on a global model. For every unknown object, they search the training set to find the objects that are most similar to the unknown one (under various similarity measures on the objects' features, e.g., Euclidian distance). Then, they classify the given object based on the classes of these most similar objects. The most widely known lazy classifier is Nearest Neighbors [12].

Comparing the aforementioned general classification schemes, we can see that eager classifiers pay a considerable cost for (off-line) training but provide faster answers during decision, exploiting their use of a global model. On the contrary, lazy methods spend no time in training at the expense of increased costs during decision, due to on-line searching. Besides trivial training costs, another great advantage of lazy classifiers is that they exhibit greater accuracy on complex datasets when a global model is too hard to find, since they exploit only local information. Furthermore, they need no incremental maintenance when the training set is expanded with new known instances. This is extremely important in applications that involve a continuous flow of new training data and classification tasks that are sensitive to the most up-to-date information. For example, automatically deciding whether to buy or sell a given share in a stock-market application seems to be more effective if based on a window consisting of the most recent transactions, rather than on a global model built on older observations. Clearly, keeping such a window up to date is much easier in lazy schemes.

Our study of related work has revealed that most disk-based classification methods proposed so far mainly focus on the development of eager classifiers. To overcome this drawback, in this paper, we propose LOCUS (**L**azy **O**ptimal **C**lassification of **U**nlimited **S**calability), an effective, efficient, and disk-based lazy classifier. LOCUS is data-scalable and can be implemented using standard SQL over arbitrary database tables. It overcomes the efficiency problems of existing lazy methods providing very fast on-line answers, while it still enjoys all the advantages of laziness described above. To the best of our knowledge, LOCUS is essentially the first disk-based lazy classifier with the following properties:

- It exhibits good classification accuracy, which improves as training sets become larger. This can be justified theoretically based on its convergence to the optimal Bayes classifier, which minimizes the classification error probability. The same is also verified experimentally in comparison to Decision Trees, a very popular and accurate existing classifier.
- It is database-friendly and, to the best of our knowledge, the only lazy method that uses a small and constant number of highly selective range queries[1] in order to classify unknown objects. Such queries actually need to access a very small part of the underlying database and have been well studied in the database literature. They can be expressed in standard SQL and existing query optimizers guarantee their fast response times with the use of traditional indices, e.g., B^+-Trees.

The rest of this paper is organized as follows: In Section 2, we review related work and, in Section 3, we describe LOCUS and show its optimality with respect to classification error probability. In Section 4, we present the results of our experimental evaluation using both real-world and synthetic datasets and, finally, we conclude in Section 5 and present our directions for future work.

2   Related Work

The problem of classification over small datasets that fit in main memory has been thoroughly studied in the past and there are a very large number of related papers. The most influential methods have been well described in existing textbooks related to pattern recognition [21] and machine learning [15]. Although LOCUS is a disk-based approach built on fundamental ideas originally conceived in these areas, surveying existing memory-based techniques in detail exceeds our purpose.

The need for scalable and database-friendly data mining techniques over extremely large datasets has given thrust to revisiting traditional methods under new constraints, in an attempt to transform them into scalable solutions. General ideas towards this end can be found in related surveys [6,17].

Interestingly, a large number of these efforts have focused on Decision Trees, mainly due to their ability of learning faster than other eager classifiers and their accuracy, which has been found comparable or superior to other classification models [8]. Moreover, every path in a Decision Tree can be easily converted into an SQL statement that can be used to access databases efficiently [1]. Popular

---

[1] The number of queries depends on the training set. In all cases we have seen in practice one or two queries are enough.

methods in this category include SLIQ [13], SPRINT [19], RainForest [9], Med-Gen [11], and BOAT [8]. Note that these methods do not actually propose a new classification model. Instead, they provide a disk-based implementation of a popular one, i.e., Decision Trees. The rationale behind LOCUS is similar. It provides a disk-based implementation for a rather familiar classifier based on counting training vectors in a fixed neighborhood centered at an incoming unknown vector. The main differences between LOCUS and the other techniques is that LOCUS is lazy and converges to an optimal solution with respect to classification error probability as the training dataset becomes larger.

Another popular approach that can be combined with known classifiers to reduce resource requirements is sampling and dataset size reduction. For example, CB-SVM [23] first applies BIRCH [24], a disk-based clustering algorithm, and then trains a classifier based on support vector machines using the centroids of the identified clusters instead of the original data. Such methods induce potentially expensive pre-processing that can be thought of as another form of training. Moreover, they leave a taste of defeat as they depend on lossy techniques. LOCUS considers every training vector as valuable. Both theoretical and experimental results show that its accuracy improves as datasets become larger, rendering it superior to sampling.

As we have already mentioned, lazy classifiers do not build a general model in a training stage, but access the training data on-line for every decision. Relying on local rather than global rules, they promise higher accuracy in inherently complex situations. Popular lazy classifiers include Nearest Neighbors [12], IB1-IB5 [3], and LazyDT [7]. Clearly, evaluating a similarity measure between an unknown vector and every training vector for ranking the latter accordingly is impractical in very large datasets. To overcome this drawback, several approaches have been proposed, based on specialized multidimensional indices for accelerating the search without scanning the entire training set [12]. (These methods mainly focus on Nearest Neighbors.) However, it has been recently shown [20] that accessing the data using such techniques can be even worse than a sequential scan through the entire dataset under very broad conditions. This is due to the fact that the distance difference between the nearest and the furthest neighbor is often so small that it turns methods based on Nearest Neighbors inaccurate [4]. Based on these conclusions, we consider LOCUS as essentially the first disk-based lazy classifier that guarantees both speed and accuracy over any large dataset.

In order to achieve this, LOCUS uses an SQL Interface Protocol (SIP) [10] and is based on highly selective range queries leaving the burden of data access to the DBMS. Using SQL for implementation of data mining methods has been proposed elsewhere [10] as a general hint. Querying the data on the server where it is originally stored (a) provides efficiency, (b) saves time from expensive export operations, (c) provides increased security, since data is available only to authorized users, and (d) enables scientists from different fields, potentially unwilling to learn the usage of new software, to use familiar interfaces on top of a database system. SQL interfaces have also been applied on construction of Decision Trees

[10] as well as in DBPredictor [14]. The latter is a lazy method that builds a custom model consisting of an IF-THEN rule for every unknown vector. To do this, it queries the training database iteratively, until some stopping criteria are met. Executing an arbitrary number of queries per instance for building a local model generates concerns about the performance of this method. Furthermore, the model constructed is rather ad-hoc and its accuracy is not based on a mathematical foundation. Unlike DBPredictor, LOCUS uses a small fixed number of queries per instance. Moreover, it builds no model but is based on raw counting, which is proven to converge to the optimal Bayes classifier.

3 LOCUS Classification

In this section, motivated by the need for a scalable and accurate disk-based lazy classifier, we propose LOCUS and argue that, to the best of our knowledge, it is the first lazy classifier with these properties that converges to optimality with respect to minimizing the classification error probability for large training sets. In principle, given an unlabeled feature vector $\mathbf{x}$, LOCUS counts the number of neighbors per class that reside in a fixed neighborhood around $\mathbf{x}$. Although this idea is not novel (counting neighbors is common in pattern recognition), its naive implementation overlooks that selecting the training vectors that live in the given neighborhood of an arbitrary $\mathbf{x}$ by scanning the entire training set is very expensive in large datasets. Moreover, although reasonable, it seems to be rather ad-hoc. In the following subsections, we provide answers for both: First, we show that our simple counting method can be based on a sound mathematical background [21] and converges to the optimal Bayes classifier. Second, we provide a very efficient disk-based method to implement the otherwise familiar task of counting. This is analogous to the contribution of algorithms like RainForest [9] that provide efficient disk-based algorithms for the construction of well-known Decision Trees in the area of eager classifiers.

3.1 Intuition of LOCUS

Assume a classification task in a D-dimensional feature space. For the sake of simplicity, let all features be numerical with discrete domains (LOCUS can also work with continuous and categorical features, as shown later). If C_i denotes the number of distinct values of the i-th feature ($i \in [1, D]$), then there are $C = C_1 \times C_2 \times \cdots \times C_D$ possible different feature vectors. Suppose that our training set is very dense and large, so that it contains an instance of every possible vector. Then, intuitively, when an unknown vector $\mathbf{x}$ comes, it seems reasonable to classify it according to the class of the training vector $\mathbf{y}$ that matches $\mathbf{x}$ exactly (i.e., $\mathbf{y} = \mathbf{x}$).

Unfortunately, this ideal scenario is very unrealistic, mainly due to two reasons: the number of features is usually large and most features have large domains. Both factors result in an increase of C. For example, if $D = 10$ and $C_i = 10 \ \forall \ i \in [1, D]$, then $C = 10^{10}$. Hence, even in this rather simple case an ideal training set should consist of 10 billion feature vectors, which seems impractical.

In practice, the available training set is usually sparse, i.e., it contains a very small subset of all possible feature vectors. Hence, finding an exact match to $\mathbf{x}$ has very low probability. To overcome this drawback, a reasonable solution is to loosen the condition of "exact match". Since finding a training vector $\mathbf{y}$ such that $\mathbf{y} = \mathbf{x}$ is usually infeasible, we can alternatively search for training vectors in a narrow neighborhood Y centered at $\mathbf{x}$. Formally, we can say that $\mathbf{y} \in Y$ iff $y_i \in [x_i\text{-}\delta_i, x_i+\delta_i] \; \forall \; i \in [1, D]$, where y_i (x_i) is the value of $\mathbf{y}$ $(\mathbf{x})$ for the i-th feature and δ_is denote our tolerance regarding the extent of the neighborhood Y around $\mathbf{x}$. If δ_i values are small enough to ensure a small deviation from the ideal scenario described above and large enough to make Y non-empty with high probability, then classifying $\mathbf{x}$ according to the majority class in Y seems intuitively reliable.

In the 2-dimensional example of Fig. 1, we would classify the unknown vector $\mathbf{x} = <x_1, x_2>$ as "+", as indicated by the majority of training points that fall within a small area centered at $\mathbf{x}$.

Clearly, loosening the "exact match" condition as described above is possible for all numerical features (either discrete or continuous). Furthermore, it is possible for categorical features, if their values can be ordered. In such cases the range $[x_i\text{-}\delta_i, x_i+\delta_i]$ must be substituted by a range of contiguous values around x_i according to the defined order. On the other hand, such loosening is not possible in non-ordered categorical features, in which case, we leave the equality condition unmodified. In general, this is of minor importance for two reasons: (a) Usually, there is a mixture of numerical and categorical features and loosening the equality condition for those that include ordering is just enough. (b) Categorical attributes have usually small domains, which makes the satisfaction of an exact match condition highly possible.

In the following subsection, we flesh out a known argument from pattern recognition [21] that proves optimality for the classifier described above, showing that it converges to the optimal Bayes classifier as training datasets become larger.

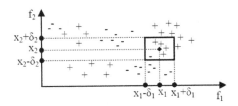

Fig. 1. Classification example of vector $\mathbf{x} = <x_1, x_2>$

3.2 Optimality

Assume that $\mathbf{x}$ is an unlabelled feature vector and that there are M possible classes $\omega_1, \cdots, \omega_M$ to which it can be classified. Then our task is to classify $\mathbf{x}$ into *the most probable* of these classes. Naturally, the term "most probable" brings in mind the conditional probabilities $P(\omega_i \mid \mathbf{x})$, $i \in [1, M]$. Hence, it seems reasonable to classify $\mathbf{x}$ as ω_i if $P(\omega_i \mid \mathbf{x}) > P(\omega_j \mid \mathbf{x}) \ \forall \ j \in [1, M]$ such that $j \neq i$. This rule is known as the *Bayes classification rule*. Actually, this reasonable rule, which seems rather empirical, turns out to minimize the classification error probability [21]. Hence, the Bayes classifier, which performs classification according to the Bayes classification rule, is optimal with respect to minimizing the classification error probability.

So, our original classification problem has now been transformed into that of comparing the conditional probabilities $P(\omega_i \mid \mathbf{x})$, $i \in [1, M]$. In order to fulfill this task, let us recall the Bayes rule from the probability theory basics:

$$P(\omega_i|\mathbf{x}) = \frac{p(\mathbf{x}|\omega_i)P(\omega_i)}{p(\mathbf{x})} \tag{1}$$

In this formula $P(\omega_i)$ is the a priori probability of class ω_i, $p(\mathbf{x} \mid \omega_i)$ is the class-conditional probability density function[2], and $p(\mathbf{x})$ is the pdf of $\mathbf{x}$.

For the sake of simplicity, let us focus on the 2-class case (M=2). Then, using formula (1) transforms the Bayes classification rule as follows: classify $\mathbf{x}$ to ω_1 if the following holds

$$P(\omega_1|\mathbf{x}) > P(\omega_2|\mathbf{x}) \Rightarrow$$

$$\frac{p(\mathbf{x}|\omega_1)P(\omega_1)}{p(\mathbf{x})} > \frac{p(\mathbf{x}|\omega_2)P(\omega_2)}{p(\mathbf{x})} \Rightarrow$$

$$\frac{p(\mathbf{x}|\omega_1)}{p(\mathbf{x}|\omega_2)} > \frac{P(\omega_2)}{P(\omega_1)} \tag{2}$$

Let N denote the total number of training vectors (N_1 of which belong to ω_1 and N_2 to ω_2), V the volume of some neighborhood centered at $\mathbf{x}$, and n_1 (n_2) the number of training vectors that belong to ω_1 (ω_2) and reside in the given neighborhood. Then the elements of formula (2) can be estimated as follows [21]:

$$P(\omega_i) \approx \frac{N_i}{N} \text{ and } p(\mathbf{x}|\omega_i) \approx \frac{1}{V} \times \frac{n_i}{N_i} \ (i \in [1, 2])$$

These estimators converge to the real values when $N_i \to \infty$, provided that $V \to 0$, $n_i \to \infty$, and $\frac{n_i}{N_i} \to 0$ ($i \in [1, 2]$) at the same time.

In other words, these conditions indicate that using the aforementioned estimators is more reliable when (a) the number of training vectors is very large, (b) the neighborhood around $\mathbf{x}$ is rather small, and (c) there is a large number of training vectors within the borders of the neighborhood, which is though much smaller than the total number of training vectors.

Replacing the elements of formula (2) with their estimators gives:

$$\frac{\frac{1}{V} \times \frac{n_1}{N_1}}{\frac{1}{V} \times \frac{n_2}{N_2}} > \frac{\frac{N_2}{N}}{\frac{N_1}{N}} \Rightarrow n_1 > n_2$$

---

[2] We assume that $\mathbf{x}$ can take any value in the D-dimensional feature space. In the case that feature vectors can take only discrete values, pdfs become probabilities.

The last inequality implies classifying $\mathbf{x}$ to ω_1 if $n_1 > n_2$. This simplified criterion is an estimator of the Bayes classification rule and can be generalized for the multi-class case as follows: classify $\mathbf{x}$ to ω_i if $n_i > n_j \ \forall \ j \in [1, M]$ such that $j \neq i$.

This result proves that the intuitive classifier described in the previous subsection, which simply relies on counting training vectors that fall inside a given neighborhood around $\mathbf{x}$, converges to the optimal Bayes classifier, which minimizes the classification error probability.

Let us revisit the pdf estimator:

$$p(\mathbf{x}|\omega_i) \approx \frac{1}{V} \times \frac{n_i}{N_i}$$

This can be written as

$$p(\mathbf{x}|\omega_i) \approx \frac{1}{V} \times \frac{\sum_{j=1}^{N_i} \phi(\mathbf{x}_j)}{N_i} \tag{3}$$

where $\phi(\mathbf{x}_j)$ is a kernel function that returns 1 if $\mathbf{x}_j$ resides in the given neighborhood, or 0 otherwise. In the literature [21], there are smoother kernel functions that better estimate continuous pdfs. However, in our case count seems to be enough, since it is easy to implement and our goal is not to find the absolute values of pdfs, but to identify the class that maximizes the pdf in the given neighborhood.

3.3   Disk-Based Implementation

LOCUS is based on the criterion described above fixing a small neighborhood around an incoming feature vector $\mathbf{x}$ and counting the number of its neighbors for every class in a lazy fashion. The majority class wins. Note that this strategy differs from that of the K-Nearest Neighbors, where the number of neighbors is fixed instead of the volume of the neighborhood. As we have shown, this intuitive alternative converges to an optimal solution. Furthermore, lazily counting seems to be very attractive, due to its simplicity. However, a naive implementation can be very expensive in terms of memory requirements and I/O costs. Recall that, formula (3) includes an invocation of the kernel function ϕ for every training vector. This implies a complete scan of the training set for every incoming unknown object, which is unacceptable for very large datasets, on which we focus.

A straightforward solution would be to apply sampling in order to keep only a small fragment of the original data that fits in main memory. However, as we have shown, LOCUS converges to optimality when the underlying dataset is as large as possible. Hence, sampling is out of question.

An alternative solution is to implement LOCUS over a DBMS using standard SQL. This approach is popular in data mining and it has been shown effective in different mining tasks [10]. Note that counting neighbors in a small area can be easily transformed into a highly selective range query that accesses directly the feature vectors (stored as tuples) that reside in the given ranges. Such queries have been thoroughly studied in the database literature and existing query optimizers guarantee high efficiency with the use of traditional indexing techniques. This property actually turns LOCUS into an efficient disk-based approach that is database-friendly.

Back to the example in Fig. 1, assume that known objects are stored in a relation R with the following schema $R(f_1, f_2, class)$. Then the number of objects of each class that fall within the given volume can be found by the following query:

SELECT class, count(*)
FROM R
WHERE $f_1 \geq x_1 - \delta_1$ AND $f_1 \leq x_1 + \delta_1$ AND $f_2 \geq x_2 - \delta_2$ AND $f_2 \leq x_2 + \delta_2$
GROUP BY class

Since the volume V is chosen small so that $n_i/N_i \rightarrow 0$, the number of tuples accessed by the query above is actually orders of magnitude smaller than the total number of tuples in R. This property makes LOCUS scalable, enabling it work over very large datasets, which minimizes the classification error probability, as shown in the previous subsection.

If in our previous example feature f_2 was categorical with no ordering defined for its values, the corresponding query would be:

SELECT class, count(*)
FROM R
WHERE $f_1 \geq x_1 - \delta_1$ AND $f_1 \leq x_1 + \delta_1$ AND $f_2 = x_2$
GROUP BY class

As we have already explained, the existence of categorical features does not generate problems.

3.4   Selection of Neighborhood Volume

As shown in Section 3.2, a proper value for the neighborhood volume V depends on the characteristics of the underlying training set. Clearly, V must be smaller when the dataset is denser and vice versa. A dataset becomes denser as (a) the size N of the training set increases, (b) the number of relevant features D decreases, and (c) the number of different values of every feature C_i decreases.

Choosing a proper value for the neighborhood volume V can be automated by using cross-validation for minimizing the classification error. Generally, if V is too small the result set of the queries used by LOCUS is empty and classification accuracy is marginal. The execution time of such queries is very fast, since they actually access a limited number of tuples, if any, making initial experimentation with small values of V very cheap. As V increases, the probability that a result set is empty decreases and this can be easily observed, since in this case, LOCUS starts returning useful results and classification accuracy increases. Finally, when V becomes too large, a considerable proportion of feature vectors lies in the corresponding neighborhood, which invalidates the preconditions of optimality and results once more in a decrease of classification accuracy overall. Hence, we propose selecting V with cross-validation, i.e., by identifying a value that strikes a balance between the two trends and minimizes classification error.

4   Experimental Evaluation

To evaluate the efficiency of the proposed techniques, we have compared LOCUS with Decision Trees (DTs) [18], as they have been popular and effective, and have been widely used for benchmarking in the past. The actual implementation of DTs we have used is J48, which is a variation of C4.5 [18] offered in weka [22], a standard, open-source suite of machine learning algorithms. We have also run some initial experiments with Nearest-Neighbors, implemented in weka as well. Its scalability has been found poor, even for moderate-sized datasets, and its accuracy comparable to that of DTs. Hence, we have excluded it from further investigation. We have implemented LOCUS in C++ and we have used an open-source DBMS for query processing. We have run our experiments on a Pentium 4 (2.8 GHz) PC with 512 MB memory under Windows XP. Below, we present the results of our experimental evaluation of the algorithms of interest over appropriate subsets of the features of both synthetic and real-world datasets.

Synthetic datasets: Due to lack of publicly available very large real-world datasets we have used ten functions, first proposed elsewhere [2], that have been widely used in the past for generating synthetic data proper for evaluating disk-based classification algorithms. The resulting datasets consist of nine predictor attributes (6 numerical, 3 categorical) of various domain sizes and two classes. Please refer to the bibliography [2] for more information.

We have generated datasets of various sizes and devoted a reasonable size of 200 MB of memory to memory-based DTs, which have managed to run over small and medium datasets (of 5×10^3 and 5×10^4 tuples) but not over larger ones (of 5×10^5 tuples or more). We have further tested LOCUS with datasets two orders of magnitude beyond this limit (of 5×10^5 and 5×10^6 tuples). All test sets have consisted of 10^3 tuples. Below, we illustrate the most indicative results.

Fig. 2 shows the error rates generated by LOCUS and DTs, respectively, for medium datasets ($N=5 \times 10^4$ tuples) generated by the ten functions mentioned above. We see that LOCUS wins in four cases (2, 7, 8, 9), is equal with DTs in three (1, 3, 10), and loses in three (4, 5, 6). Its greatest success over DTs occurs in function 2, which is complicated enough to prevent DTs from building a global model. LOCUS overcomes this based on local information only.

Expectedly, the accuracy of LOCUS improves converging to optimality as datasets become larger. We show this in Fig. 3, which presents the error rates generated by LOCUS for all ten functions when the number of training tuples varied from 5×10^3 to 5×10^6. Clearly, error rates tend to decrease and finally LOCUS reaches the accuracy of DTs for all cases it lost in Fig. 2. These results

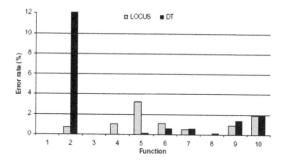

Fig. 2. Error rate using synthetic data (N=5×10⁴)

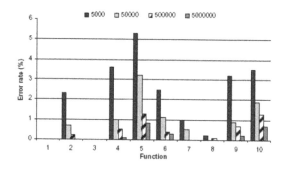

Fig. 3. Error rate using synthetic data wrt dataset size

show the potentiality of LOCUS, mainly in very large datasets, for all of which error rates have dropped under 1%.

As an example of how the prerequisites (mentioned in Section 3.2) for convergence to Bayes optimality hold when the size of the training dataset grows, Fig. 4 shows the size of the neighborhood volume V used by LOCUS as a function of the number of tuples stored in the database. The specific volumes have been found with cross-validation over datasets generated with function 5 (behavior is similar for all ten functions). The values indicated are the normalized volumes (i.e., the fractions of the volumes over the volume corresponding to the largest training set) and clearly show that V tends to zero as the size of the dataset increases.

Fig. 5 illustrates the scalability of LOCUS with respect to the number of tuples stored in the database (x-axis is logarithmic). Each point represents the average time for making a decision. Averages refer to all ten functions. Performing classification in approximately half a second over datasets consisting of 5 × 10⁶ tuples is very promising. Note that in these experiments we have used a single B⁺-Tree for each dataset, indexing the corresponding attribute with the largest domain. Optimistically, since the classification algorithm can be applied directly over the database relation R that holds the original data, it is highly likely that R is already indexed for other purposes, implying that effective reuse

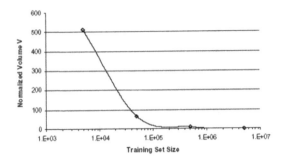

Fig. 4. Normalized volume V used by LOCUS on synthetic data wrt dataset size

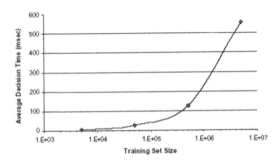

Fig. 5. Time scalability of LOCUS on synthetic data wrt dataset size

Dataset	F	N	M
Patient	8	90	3
Glass	9	214	7
Liver	6	345	2
BreastCancer	9	699	2
Diabetes	8	768	2
Letters	16	2×10^4	26
CovType 5000	10	5×10^3	7
CovType 50000	10	5×10^4	7
CovType 500000	10	5×10^5	7

Fig. 6. Properties of the real datasets we have used

of existing resources may just be enough. On the other hand, multidimensional index structures like R-trees and techniques like pre-sorting stored data or materializing views could be used to generate better results. The effect of these techniques in this case is identical to that when applied to any (simple) queries expressed in standard SQL, so it is orthogonal to and beyond the scope of this paper.

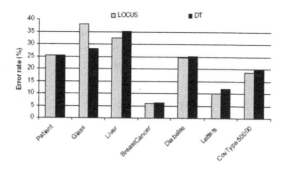

Fig. 7. Error rate using real datasets

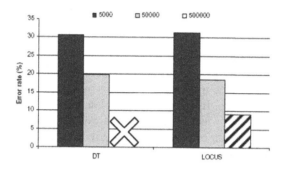

Fig. 8. Error rate using CovType wrt dataset size

Real datasets: As we have already mentioned, LOCUS behaves best under large datasets; hence, testing its behavior over real data, which are rather small, is really challenging. In our study, we have chosen commonly used publicly available real-world datasets [16]. Fig. 6 shows the properties of the real datasets we have used (number of features F, number of tuples N, and number of classes M, respectively). Datasets appear ordered according to the average number of tuples per class they contain. We expect better results as this number increases. In small datasets (up to Diabetes), we have treated 2/3 of all tuples as known and 1/3 as unknown (separation has been random), which is common. In Letters and CovType we have used 1,000 unknown tuples instead. All results are averages of five different experiments per dataset.

Fig. 7 shows the error rates generated by LOCUS and DTs respectively. Error rates are in general high over most of these datasets, which exhibit inherent difficulty in being used for prediction tasks. LOCUS performs worst over Glass, which is too sparse, consisting of only 214 tuples for 7 classes. Nevertheless, surprisingly, it outperforms DTs in most cases overall, even over very small datasets. As datasets become larger (mainly in Letters and CovType) its superiority increases. This is also clear in Fig. 8, which shows the accuracy improvements over CovType with respect to the training set size. The symbol "X" denotes that DTs failed to deal with the case of 5×10^5 tuples.

Overall, we have shown that LOCUS is scalable, its accuracy is comparable to that of eager DTs in small datasets and becomes superior when datasets become larger. Hence, we have provided strong evidence that it is a promising classifier, mainly suited for datasets with large and constantly growing sizes.

5   Conclusions and Future Work

In this paper, we proposed LOCUS, an accurate and efficient disk-based lazy classifier that is data-scalable and can be implemented using standard SQL. We have shown that in most cases it exhibits high classification accuracy, which improves as training sets become larger, based on its convergence to the optimal Bayes. Overall, the results are very promising with respect to the potential of LOCUS as the basis for classification, mainly over large or inherently complex datasets.

Note that, in this thread of our work, we have focused on classification scalability with respect to the number of known vectors N and have deliberately neglected scalability with respect to the number of dimensions D. The latter problem is tightly related to methods for feature selection, which are orthogonal to our work presented here. In the future, we plan to investigate the applicability of similar techniques for feature selection as well. Furthermore, we intend to implement a parallel version of LOCUS and study its effectiveness on regression problems (possibly replacing the "count" aggregate function with "average").

References

1. Agrawal, R., Ghosh, S.P., Imielinski, T., Iyer, B.R., Swami, A.N.: An Interval Classifier for Database Mining Applications. In: VLDB 1992 (1992)
2. Agrawal, R., Imielinski, T., Swami, A.N.: Database Mining: A Performance Perspective. IEEE Trans. Knowl. Data Eng. 5(6), 914–925 (1993)
3. Aha, D.W., Kibler, D.F., Albert, M.K.: Instance-Based Learning Algorithms. Machine Learning 6, 37–66 (1991)
4. Beyer, K.S., Goldstein, J., Ramakrishnan, R., Shaft, U.: When Is "Nearest Neighbor" Meaningful? In: Beeri, C., Bruneman, P. (eds.) ICDT 1999. LNCS, vol. 1540, Springer, Heidelberg (1998)
5. Burges, C.J.C.: A Tutorial on Support Vector Machines for Pattern Recognition. Data Min. Knowl. Discov. 2(2), 121–167 (1998)
6. Chen, M.S., Han, J., Yu, P.S.: Data Mining: An Overview from a Database Perspective. IEEE Trans. Knowl. Data Eng. 8(6), 866–883 (1996)
7. Friedman, J.H., Kohavi, R., Yun, Y.: Lazy Decision Trees. In: AAAI/IAAI, vol. 1, pp. 717–724 (1996)
8. Gehrke, J., Ganti, V., Ramakrishnan, R., Loh, W.Y.: BOAT-Optimistic Decision Tree Construction. In: SIGMOD 1999 (1999)
9. Gehrke, J., Ramakrishnan, R., Ganti, V.: RainForest - A Framework for Fast Decision Tree Construction of Large Datasets. In: VLDB 1998 (1998)
10. John, G.H., Lent, B.: SIPping from the Data Firehose. In: KDD 1997 (1997)
11. Kamber, M., Winstone, L., Gon, W., Han, J.: Generalization and Decision Tree Induction: Efficient Classification in Data Mining. In: RIDE 1997 (1997)

12. Katayama, N., Satoh, S.: The SR-tree: An Index Structure for High-Dimensional Nearest Neighbor Queries. In: SIGMOD 1997 (1997)
13. Mehta, M., Agrawal, R., Rissanen, J.: SLIQ: A Fast Scalable Classifier for Data Mining. In: Apers, P.M.G., Bouzeghoub, M., Gardarin, G. (eds.) EDBT 1996. LNCS, vol. 1057, Springer, Heidelberg (1996)
14. Melli, G.: A Lazy Model-Based Algorithm for On-Line Classification. In: Zhong, N., Zhou, L. (eds.) Methodologies for Knowledge Discovery and Data Mining. LNCS (LNAI), vol. 1574, Springer, Heidelberg (1999)
15. Mitchel, T.: Machine Learning. McGraw-Hill, New York (1997)
16. Newman, D.J., Hettich, S., Blake, C.L., Merz, C.J.: UCI Repository of machine learning databases, http://www.ics.uci.edu/~mlearn/MLRepository.html
17. Provost, F.J., Kolluri, V.: A Survey of Methods for Scaling Up Inductive Algorithms. Data Min. Knowl. Discov. 3(2), 131–169 (1999)
18. Quinlan, J.R.: Induction of Decision Trees. Machine Learning 1(1), 81–106 (1986)
19. Shafer, J.C., Agrawal, R., Mehta, M.: SPRINT: A Scalable Parallel Classifier for Data Mining. In: VLDB 1996 (1996)
20. Shaft, U., Ramakrishnan, R.: When Is Nearest Neighbors Indexable? In: Eiter, T., Libkin, L. (eds.) ICDT 2005. LNCS, vol. 3363, Springer, Heidelberg (2004)
21. Theodoridis, S., Koutroumbas, K.: Pattern Recognition, 3rd edn. Academic Press, London (2005)
22. Witten, I.H., Frank, E.: Data Mining: Practical machine learning tools and techniques, 2nd edn. Morgan Kaufmann, San Francisco (2005)
23. Yu, H., Yang, J., Han, J.: Classifying large data sets using SVMs with hierarchical clusters. In: KDD 2003 (2003)
24. Zhang, T., Ramakrishnan, R., Livny, M.: BIRCH: An Efficient Data Clustering Method for Very Large Databases. In: SIGMOD 1996 (1996)

Update Support for Database Views Via Cooperation

Stephen J. Hegner[1] and Peggy Schmidt[2]

[1] Umeå University, Department of Computing Science
SE-901 87 Umeå, Sweden
hegner@cs.umu.se
http://www.cs.umu.se/~hegner
[2] Christian-Albrechts-University Kiel, Department of Computer Science
Olshausenstraße 40, D-24098 Kiel, Germany
pesc@is.informatik.uni-kiel.de
http://www.is.informatik.uni-kiel.de/~pesc

Abstract. Support for updates to views of database schemata is typically very limited; only those changes which can be represented entirely within the view, or changes which involve only generic changes outside of the view, are permitted. In this work, a different point of view towards the view-update problem is taken. If a proposed update cannot be performed within the view, then rather than rejecting it outright, the cooperation of other views is sought, so that in their combined environments the desired changes can be realized. This approach has not only the advantage that a wider range of updates are supported than is possible with more traditional approaches, but also that updates which require the combined access privileges of several users are supported.

Keywords: update, view.

1 Introduction

Support for updates to database views has long been recognized as a difficult problem. An update which is specified on a view provides only partial information on the change of state of the main schema; the complementary information necessary to define a complete *translation* of that update to the main schema must be determined in other ways. Over the years, a number of approaches have been developed for such translations. In the *constant-complement strategy*, first defined in [1] and later refined in [2] and [3], the fundamental idea is that the translation must leave unaltered all aspects of the main database which are not visible from the view; formally, the so-called *complementary view* is held constant. Theoretically, it is the cleanest approach, in that it defines precisely those translations which are free from so-called *update anomalies* [2, 1.1] which involve changes to the database which are not entirely visible within the view itself. Unfortunately, the family of anomaly-free updates is relatively limited, and for this reason the constant-complement strategy has

Y. Ioannidis, B. Novikov, and B. Rachev (Eds.): ADBIS 2007, LNCS 4690, pp. 98–113, 2007.
© Springer-Verlag Berlin Heidelberg 2007

been viewed as inadequate by some investigators [4], and so numerous more liberal approaches have been forwarded, including both direct approaches [5] and those which relax some, but not all, of the constraints of the constant-complement strategy [6]. All of these more liberal approaches involve, in one way or another, updates to the main schema which are not visible within the view itself.

Even if one accepts that view update strategies which are more liberal than the constant-complement approach are necessary and appropriate, there is a significant further issue which must be taken into consideration — access rights. It is a fundamental design principle of modern database systems that users have access rights, and that all forms of access, both read and write, must respect the authorization of those rights. With the constant-complement update strategy, in which only those parts of the main schema which are visible in the view may be altered in an update, this issue poses no additional problems beyond those of specifying properly the access rights on each view. However, with a more liberal update approach, changes to the main schema may be mandated which are not visible within the view. This implies that the user of the view must have write access privileges beyond that view, which is often unrealistic. Thus, even if one is willing to accept some update anomalies, view update support beyond the constant-complement strategy is to a large extent unacceptable because of the serious problems surrounding access rights which it implies.

To address these concerns, a quite different approach to supporting view updates is proposed in this paper. When an update u to view Γ cannot be supported by the constant-complement strategy, the *cooperation* of other views is enlisted. If translation of u implies that an update to the main schema must be made which is not visible within Γ, then these additional changes must be embodied in the cooperating views. If the user of Γ who desires to effect u does not have the necessary access privileges on the cooperating views, then the cooperation of suitable users of these views must also be enlisted, in order to effect the update "in unison". This, in turn, provides information on the workflow pattern which is necessary to realize the update.

For such a theory of cooperation to take form, it is necessary to be able to regard a database schema as a collection of interconnected views. The fundamental ideas of such representations of database schemata are found in *component-based modelling*, as forwarded by Thalheim [7] [8] [9]. Roughly speaking, a component is an encapsulated database schema, together with *channels* which allow it to be connected to other components. A database schema is then modelled as an interconnection of such components. The work of Thalheim is, in the first instance, oriented towards conceptual modelling and the design of database schemata using the *higher-order entity-relationship model (HERM)* [10]. In [11], the ideas of component-based modelling have been recast and formalized in a way which makes them more amenable to view-update problems. It is this latter work which is used, in large part, as the basis of this paper.

2   The Core Concepts by Example

To present the ideas underlying cooperative updates in a complete and unambiguous fashion, a certain amount of formalism is unavoidable. However, it is possible to illustrate many of the key ideas with a minimum of formalism; such an illustration, via a running example, is the goal of this section. First, the main ideas of database components will be illustrated via an example, with that same example then used as the basis for the illustration of a cooperative update.

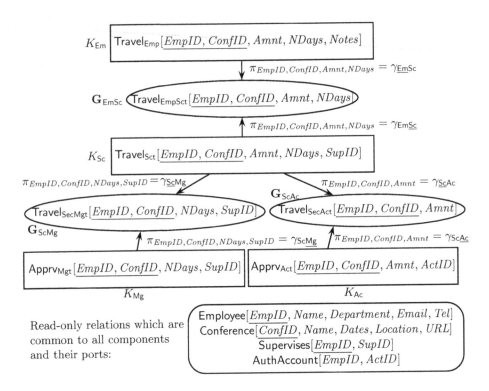

Fig. 1. Graphical depiction of the components of the running example

Discussion 2.1 (An informal overview of database components). For a much more thorough and systematic presentation of the ideas underlying the database components of this paper, the reader is referred to [11]. A *component* is an ordered pair $C = (\mathsf{Schema}(C), \mathsf{Ports}(C))$ in which $\mathsf{Schema}(C)$ is a database schema and $\mathsf{Ports}(C)$ is a finite set of nonzero views of $\mathsf{Schema}(C)$, called the *ports* of C. The running relational example is depicted in Fig. 1; there are four components, the *employee component* K_{Em}, the *secretariat component* K_{Sc}, the *management component* K_{Mg}, and the *accounting component* K_{Ac}. The relation unique to the schema of a given component is shown enclosed in a rectangle. In addition, there is a set of relations which are common to all components; these are shown in a box with rounded corners at the bottom of the figure. For example, the schema of K_{Em} consists of the relations $Travel_{Emp}$, Employee, Conference, Supervises, and AuthAccount. The primary key of each relation is underlined; in addition, the following inclusion dependencies are assumed to hold:

$Travel_{Emp}[EmpID] \subseteq Employee[EmpID]$,
$Travel_{Sct}[EmpID, SupID] \subseteq Supervises[EmpID, SupID]$,
$Apprv_{Mgt}[EmpID, SupID] \subseteq Supervises[EmpID, SupID]$,
$Apprv_{Act}[EmpID, ActID] \subseteq AuthAccount[EmpID, ActID]$,
$Supervises[EmpID] \subseteq Employee[EmpID]$,
$Supervises[SupID] \subseteq Employee[EmpID]$,
$AuthAccount[EmpID] \subseteq Employee[EmpID]$.

The ports of the components are also represented in Fig. 1. The schema of each view (qua port) Γ_x is represented within an ellipse, labelled with the name of that schema. The associated view mapping γ_x for each port is shown next to the arrow which runs from the component schema to the port schema. For example, the component K_{Em} has only one port, which is denoted by $\Gamma_{\underline{EmSc}} = (\mathbf{G}_{EmSc}, \gamma_{\underline{EmSc}})$, with γ_{EmSc} the projection $\pi_{EmpID, ConfID, Amnt, NDays}$: Schema$(K_{Em}) \rightarrow \mathbf{G}_{EmSc}$. In other words, Ports$\langle K_{Em} \rangle = \{\Gamma_{EmSc}\}$. The component K_{Sc}, on the other hand, has three ports, $\Gamma_{\underline{EmSc}} = (\mathbf{G}_{EmSc}, \gamma_{\underline{EmSc}})$, $\Gamma_{\underline{ScMg}} = (\mathbf{G}_{ScMg}, \gamma_{\underline{ScMg}})$, and $\Gamma_{\underline{ScAc}} = (\mathbf{G}_{ScAc}, \gamma_{\underline{ScAc}})$, with the port definitions as given in the figure. Similarly, the component K_{Mg} has one port, $\Gamma_{\underline{ScMg}} = (\mathbf{G}_{ScMg}, \gamma_{\underline{ScMg}})$, and the component K_{Ac} has one port, $\Gamma_{\underline{ScAc}} = (\mathbf{G}_{ScAc}, \gamma_{\underline{ScAc}})$. The underlying *interconnection family*, which describes which ports are connected to which, is $\{\{\Gamma_{\underline{EmSc}}, \Gamma_{\underline{EmSc}}\}, \{\Gamma_{\underline{ScMg}}, \Gamma_{\underline{ScMg}}\}, \{\Gamma_{\underline{ScAc}}, \Gamma_{\underline{ScAc}}\}\}$. Each member of this family is called a *star interconnection*. The names of ports may be arbitrary, although for convenience in this example a special naming convention has been used. The view $\Gamma_{\underline{xy}} = (\mathbf{G}_{xy}, \gamma_{\underline{xy}})$ is a port of the component K_x, and is connected to the port $\Gamma_{x\underline{y}} = (\mathbf{G}_{xy}, \gamma_{x\underline{y}})$ of the component K_y.

The ports of connected components must have *identical* (and not just isomorphic) schemata. It becomes clear why this is necessary when defining the state of a combined interconnection. Let $M = (M_{Em}, M_{Sc}, M_{Mg}, M_{Ac}) \in$ LDB(Schema(K_{Em})) $\times$ LDB(Schema(K_{Sc})) $\times$ LDB(Schema(K_{Mg})) $\times$ LDB(Schema(K_{Ac})), with LDB$(-)$ denoting the set of legal databases. For M to be a legal state of the interconnected component, the local states must agree on all ports. More precisely, it is necessary that the following hold: $\gamma_{\underline{EmSc}}(M_{Em}) = \gamma_{EmSc}(M_{Sc})$, $\gamma_{\underline{ScMg}}(M_{Sc}) = \gamma_{ScMg}(M_{Mg})$, and $\gamma_{\underline{ScAc}}(M_{Sc}) = \gamma_{ScAc}(M_{Ac})$.

Some difficulties can arise if the underlying hypergraph is cyclic [11, 3.2]; i.e., if there are cycles in the connection. While these can often be overcome, the details become significantly more complex. Therefore, in this paper, it will always be assumed that the interconnections are acyclic.

Example 2.2 (An example of cooperative update). Suppose that Lena is an employee, and that she wishes to travel to a conference. The successful approval of such a request is represented by the insertion of an appropriate tuple in the relation $\mathsf{Travel_{Emp}}$. She has insertion privileges for the relation $\mathsf{Travel_{Emp}}$ for tuples with her *EmpID* (which is assumed to be Lena, for simplicity); however, these privileges are qualified by the additional requirement that the global state of the interconnected components be consistent. Thus, any insertion into $\mathsf{Travel_{Emp}}$ must be matched by a corresponding tuple in $\mathsf{Travel_{Sct}}$ — that is, a tuple whose projection onto $\mathsf{Travel_{EmpSct}}$ matches that of the tuple inserted into $\mathsf{Travel_{Emp}}$. This tuple must in turn be matched by corresponding tuples in $\mathsf{Apprv_{Mgt}}$ and $\mathsf{Apprv_{Act}}$. Thus, to accomplish this update, Lena requires the cooperation someone authorized to update the component K_{Sc}, which in turn requires the cooperation of those authorized to update K_{Mg} and K_{Ac}.

The process of cooperative update proceeds along a project-lift cycle. The update request of Lena is *projected* (see Definition 3.2) to the ports of K_{Em}; the connected components (in this case just K_{Sc}) then *lift* (see Definition 3.3) these projections to their schemata. The process then continues, with K_{Sc} projecting its proposed update to K_{Mg} and K_{Sc}. These projections and liftings cannot modify the state of the database immediately, as they are only proposed updates until all parties have agreed. Rather, a more systematic process for managing the negotiation process is necessary. The process is controlled by a nondeterministic automaton, which maintains key information about the negotiation and determines precisely which actions may be carried out by components (and their associated actors) at a given time. It also manages the actual update of the database when a successful negotiation has been completed. This automaton is described formally in Definition 3.4. In this section, it will be described more informally, and consequently somewhat incompletely, by example.

The central data structures of this automaton are two sets of registers for managing proposed updates. For each component C there is a *pending-update register* $\mathsf{PendingUpdate}(C)$ which records proposed updates which are initiated by that component, but not yet part of the permanent database. In addition, for each component C and each port Γ of C, there is a *port-status registers* $\mathsf{PortStatus}(C, \Gamma)$ which is used to record projections of updates received by neighboring components. In addition, for each component C, the register $\mathsf{CurrentState}(C)$ records the actual database state for that component. The automaton also has a Status, which indicates the phase of the update process in which the machine lies. In *Idle*, it is waiting for an initial update request from one of the components. In *Active*, it is processing such a request by communication and refinement; the bulk of the processing occurs in this state. The value *Accepted* indicates that all components have agreed on a suitable set of updates. In the *Final* phase, one of these updates is selected, again by propagating proposals amongst the components. Finally, the value of the variable $\mathsf{Initiator}$ identifies the component which initiated the update request. A state of this update automaton is given by a value for each of its state variables. These variables, together with their admissible and initial values, are shown in Table 1.

In Table 2, a sequence of twelve steps which constitutes a successful realization of a travel request from Lena is shown. An attempt has been made to represent all essential information, albeit in compact format. Notation not already described,

Table 1. The state variables of the update automaton

Name	Range of values	Initial Value
Status	$\in \{$*Idle, Active, Accepted, Final* $\}$	*Idle*
Initiator	$\in X \cup \{$NULL$\}$	NULL
For each component C:		
CurrentState(C)	$\in$ LDB(Schema$\langle C \rangle$)	$(M_C)_0$
PendingUpdate(C)	$\in$ RDUpdates(Schema$\langle C \rangle$) $\cup \{$NULL$\}$	NULL
For each component C and $\Gamma \in$ Ports$\langle C \rangle$:		
PortStatus(C, Γ)	$\in$ RDUpdates(Schema$\langle \Gamma \rangle$) $\cup \{$NULL$\}$	NULL

such as the step of the update automaton which is executed, will be described as the example proceeds. The formal descriptions of these steps may be found in Definition 3.4.

To understand how this request is processed, it is first necessary to expand upon the request itself, which involves alternatives. Suppose that Lena wishes to travel either to ADBIS or else to DEXA. For ADBIS, she requires a minimum of €800; for DEXA €1000. For ADBIS she needs to travel for at least five days; for DEXA only three. Although she is flexible, she also has preferences. A trip to ADBIS is to be preferred to a trip to DEXA, and within a given conference, more money and time is always to be preferred. To express these alternatives, a *ranked directional update* (see Definition 3.1) is employed. It is *directional* in the sense that it is either an insertion or a deletion (for technical reasons, only insertions and deletions are supported in the current framework), and it is *ranked* in the sense that there is a partial order which expresses preferences on the updates. The update request, call it $\mathbf{u}_0$, would then consist of all tuples of the form Travel$_{\mathsf{Emp}}$[Lena, ADBIS, e_A, d_A, η] together with those of the form Travel$_{\mathsf{Emp}}$[Lena, DEXA, e_D, d_D, η], with $800 \leq e_A \leq 2000$, $5 \leq d_A \leq 10$, $1000 \leq e_D \leq 2000$, $3 \leq d_D \leq 10$. Here η denotes a null value for the *Notes* field. There is also a technical requirement that a set of ranked updates be finite; hence the upper bounds on the values for time and money. The set of all possible ranked directed updates on a schema $\mathbf{D}$ is denoted RDUpdates($\mathbf{D}$). At the end of a successful update negotiation, all parties will agree to support some subset of the elements of $\mathbf{u}_0$, with Lena then choosing one of them.

In processing this request, the automaton is initially in the state *Idle*, awaiting an update request from some component. This is illustrated in step 0 of Table 2. Lena initiates her travel request by executing INITIATEUPDATE($K_{\mathsf{Em}}, \mathbf{u}_0$), which communicates the projection Proj($\mathbf{u}_0$,) of Travel$_{\mathsf{Emp}}$ onto the common port relation Travel$_{\mathsf{SecMgt}}$, as shown in step 1 of the table. Because this is only an update request, and not an actual update, it is placed in the appropriate port status register, in this case the status register for Γ_{EmSc}; the database state remains unchanged. A value for a state variable of the form PortStatus(C, Γ) represents

Table 2. The state evolution of Example 2.2

K_{Em}		K_{Sc}				K_{Mg}		K_{Ac}	
Pending Update	Port Status	Pending Update	Port Status	Port Status	Port Status	Pending Update	Port Status	Pending Update	Port Status
K_{Em}	Γ_{EmSc}	K_{Sc}	Γ_{EmSc}	Γ_{ScMg}	Γ_{ScAc}	K_{Mg}	Γ_{ScMg}	K_{Ac}	Γ_{ScAc}

0. Initial database state $= (M_{Em}, M_{Sc}, M_{Mg}, M_{Ac})$ Status: *Idle*   Initiator: NULL

NULL	NULL	NULL	NULL	NULL	NULL	NULL	NULL	NULL	NULL

1. $\textsc{InitiateUpdate}(K_{Em}, \mathbf{u}_0)$ Status: *Active*   Initiator: K_{Em} $\mathbf{u}_0' = \mathrm{Proj}(\mathbf{u}_0, \Gamma_{EmSc})$

$\mathbf{u}_0$	NULL	NULL	$\mathbf{u}_0'$	NULL	NULL	NULL	NULL	NULL	NULL

2. $\textsc{PromoteInitialUpdate}(K_{Sc})$ Status: *Active*   Initiator: K_{Em}

$\mathbf{u}_1 \in \mathrm{NERestr}(\mathrm{MinLift}(\mathbf{u}_0', \Gamma_{\underline{EmSc}}, M))$ $\mathbf{u}_1' = \mathrm{Proj}(\mathbf{u}_1, \Gamma_{\underline{ScMg}})$ $\mathbf{u}_1'' = \mathrm{Proj}(\mathbf{u}_1, \Gamma_{\underline{ScAc}})$

$\mathbf{u}_0$	NULL	$\mathbf{u}_1$	NULL	NULL	NULL	NULL	$\mathbf{u}_1'$	NULL	$\mathbf{u}_1''$

3. $\textsc{PromoteInitialUpdate}(K_{Mg})$ Status: *Active*   Initiator: K_{Em}

$\mathbf{u}_2 \in \mathrm{NERestr}(\mathrm{MinLift}(\mathbf{u}_1', \Gamma_{\underline{ScMg}}, M))$ $\mathbf{u}_2' = \mathrm{Proj}(\mathbf{u}_2, \Gamma_{\underline{ScMg}})$

$\mathbf{u}_0$	NULL	$\mathbf{u}_1$	NULL	$\mathbf{u}_2'$	NULL	$\mathbf{u}_2$	NULL	NULL	$\mathbf{u}_1''$

4. $\textsc{PromoteInitialUpdate}(K_{Ac})$ Status: *Active*   Initiator: K_{Em}

$\mathbf{u}_3 \in \mathrm{NERestr}(\mathrm{MinLift}(\mathbf{u}_1'', \Gamma_{\underline{ScAc}}, M))$ $\mathbf{u}_3' = \mathrm{Proj}(\mathbf{u}_3, \Gamma_{\underline{ScMg}})$

$\mathbf{u}_0$	NULL	$\mathbf{u}_1$	NULL	$\mathbf{u}_2'$	$\mathbf{u}_3'$	$\mathbf{u}_2$	NULL	$\mathbf{u}_3$	NULL

5. $\textsc{RefineUpdate}(K_{Sc})$ Status: *Active*   Initiator: K_{Em}

$\mathbf{u}_4 \in \mathrm{NERestr}(\mathrm{Refine}(\mathbf{u}_1, ((\Gamma_{\underline{ScMg}}, \mathbf{u}_2'), (\Gamma_{\underline{ScAc}}, \mathbf{u}_3'))))$ $\mathbf{u}_4' = \mathrm{Proj}(\mathbf{u}_4, \Gamma_{\underline{EmSc}})$

$\mathbf{u}_0$	$\mathbf{u}_4'$	$\mathbf{u}_4$	NULL	NULL	NULL	$\mathbf{u}_2$	NULL	$\mathbf{u}_3$	NULL

6. $\textsc{RefineUpdate}(K_{Em})$ Status: *Active*   Initiator: K_{Em}

$\mathbf{u}_5 = \mathrm{Refine}(\mathbf{u}_1, ((\Gamma_{\underline{EmSc}}, \mathbf{u}_4')))$

$\mathbf{u}_5$	NULL	$\mathbf{u}_4$	NULL	NULL	NULL	$\mathbf{u}_2$	NULL	$\mathbf{u}_3$	NULL

7. $\textsc{AcceptUpdate}$ Status: *Accepted*   Initiator: K_{Em}

$\mathbf{u}_5$	NULL	$\mathbf{u}_4$	NULL	NULL	NULL	$\mathbf{u}_2$	NULL	$\mathbf{u}_3$	NULL

8. $\textsc{SelectFinalUpdate}(K_{Em}, u_5)$ Status: *Final*   Initiator: K_{Em}

$u_5 = (M_{Em}, M_{Em}') \in \mathrm{Updates}(\mathbf{u}_5)$ $u_5' = \mathrm{Proj}(u_5, \Gamma_{\underline{EmSc}})$

u_5	NULL	u_4	u_5'	NULL	NULL	u_2	NULL	u_3	NULL

9. $\textsc{RefineFinalUpdate}(K_{Sc})$ Status: *Final*   Initiator: K_{Em}

$u_4 = (M_{Sc}, M_{Sc}') \in \mathrm{NERestr}(\mathrm{Refine}(\mathbf{u}_4, (u_5', \Gamma_{\underline{EmSc}})))$ $u_4' = \mathrm{Proj}(u_4, \Gamma_{\underline{ScMg}})$

$u_4'' = \mathrm{Proj}(u_4, \Gamma_{\underline{ScAc}})$

u_5	NULL	u_4	NULL	NULL	NULL	u_2	u_4'	u_3	u_4''

10. $\textsc{RefineFinalUpdate}(K_{Mg})$ Status: *Final*   Initiator: K_{Em}

$u_2 = (M_{Mg}, M_{Mg}') \in \mathrm{NERestr}(\mathrm{Refine}(\mathbf{u}_2, (u_4', \Gamma_{\underline{ScMg}})))$

u_5	NULL	u_4	NULL	NULL	NULL	u_2	NULL	u_3	u_4''

11. $\textsc{RefineFinalUpdate}(K_{Ac})$ Status: *Final*   Initiator: K_{Em}

$u_3 = (M_{Ac}, M_{Ac}') \in \mathrm{NERestr}(\mathrm{Refine}(\mathbf{u}_3, (u_4'', \Gamma_{\underline{ScAc}})))$

u_5	NULL	u_4	NULL	NULL	NULL	u_2	NULL	u_3	NULL

12. $\textsc{CommitUpdate}$ Status: *Idle*   Initiator: NULL

New database state $= (M_{Em}', M_{Sc}', M_{Mg}', M_{Ac}')$

NULL	NULL	NULL	NULL	NULL	NULL	NULL	NULL	NULL	NULL

an *unprocessed* update request to component C; that is, a request which has yet to be lifted to that component. Once the lifting takes place, $\mathsf{PortStatus}(C, \Gamma)$ is reset to NULL. Note further that the request is not placed in the status register for Γ_{EmSc}, since the initiating component already knows about it.

The secretariat component must agree to the selected update by proposing a corresponding update to the $\mathsf{Travel_{Sct}}$ relation, so that the projections of $\mathsf{Travel_{Emp}}$ and $\mathsf{Travel_{Sct}}$ agree on $\mathsf{Travel_{EmpSct}}$. Formally, this is accomplished via a *lifting* of $\mathbf{u}_0$; that is, a ranked directed update $\mathbf{u}_1$ on $\mathsf{Schema}(K_{\mathsf{Sc}})$ which projects to $\mathbf{u}_0$ under γ_{EmSc}. In principle, this could be any such lifting, but since the secretariat is assumed to be largely an administrative arm in this example, it is reasonable to assume that the lifting retains all possibilities requested by the employee. This lifting must in turn be passed along to the other components to which K_{Sc} is connected; namely the management component K_{Mg} and the accounting component K_{Ac}. It is not passed back to K_{Em} at this point, since it is assumed that $\mathsf{Proj}(\mathbf{u}_1, \Gamma_{\mathsf{EmSc}}) = \mathsf{Proj}(\mathbf{u}_0, \Gamma_{\mathsf{EmSc}})$. A lifted update is passed back to the sending component only when the lifting alters the projection on to their interconnection. Thus, if the secretariat had made restrictions to the proposed update (by limiting the number of days, say), then the lifting would need to be passed back to K_{Em} as well. In any case, the lifting which is selected by the secretariat is passed along to K_{Mg} and K_{Act} as $\mathbf{u}_1'$ and $\mathbf{u}_1''$, respectively, as represented in step 2 of Table 2: $\textsc{PromoteInitialUpdate}(K_{\mathsf{Sc}})$.

Management responds by lifting $\mathbf{u}_1'$ to an update $\mathbf{u}_2$ on the entire component K_{Mg}, as illustrated in step 3 of Table 2: $\textsc{PromoteInitialUpdate}(K_{\mathsf{Mg}})$. Suppose, for example, that $\mathbf{u}_2$ approves travel to ADBIS for seven days, but denies travel to DEXA completely. This decision must then be passed back to K_{Sc}; this is represented as $\mathbf{u}_2'$ in the table. On the other hand, if K_{Mg} decides to allow all possible travel possibilities which are represented in $\mathbf{u}_1'$, the lifting is not passed back to K_{Sc}. This point will be discussed in more detail later.

Similarly, K_{Ac} must respond to $\mathbf{u}_1''$, which is a ranked update regarding travel funds, but not the number of days. Accounting must decide upon an appropriate lifting $\mathbf{u}_3$. For example, in the lifting $\mathbf{u}_3$ of $\mathbf{u}_1''$, it may be decided that €1500 can be allocated for travel to DEXA, but only €900 for travel to ADBIS. In step 4 of Table 2, this component reports its lifting decision back to K_{Sc} via $\mathbf{u}_3'$.

In step 5, $\textsc{RefineUpdate}(K_{\mathsf{Sc}})$, the decisions $\mathbf{u}_2'$ and $\mathbf{u}_3'$ of K_{Mg} and K_{Ac}, which were reported to K_{Sc} in steps 3 and 4, are lifted to $\mathbf{u}_4$ and then reported back to K_{Em} via $\mathbf{u}_4'$. In this example, the employee Lena would discover that she may travel only to ADBIS and not to DEXA, with a maximum funding allocation of €900 and for at most seven days. She acknowledges this with $\textsc{RefineUpdate}(K_{\mathsf{Em}})$ in step 6, producing $\mathbf{u}_5$. As she is the originator of the update request it is highly unlikely that $\mathbf{u}_5$ would be anything but the maximal lifting of $\mathbf{u}_4$. As such, further update is passed back to K_{Sc} during this step.

The system next observes that no component has any pending updates in its port-status registers, and so marks the cooperative update process as successful via the $\textsc{AcceptUpdate}$ action in step 7, which includes a transition to *Accepted* status. Unlike the previous steps, this action is taken entirely by the system, and it is the only possibility from the state reached after step 6.

The process is not yet complete, however, as Lena must select a particular update, and then that update must be lifted to the entire network of components. She may select any member of $\mathbf{u}_5$; however, in this case, there is a maximal entry in what remains of her initial ranked update: travel to ADBIS for seven days with €900. Given her indicated preferences, this would likely be her choice. A second round of confirmation via project-lift requires each of the other components to select a specific update to match her choice. These choices (for example, the *ActID* which pays for the trip) will be invisible to Lena. The details are contained in steps 8-11 of Table 2 via a SELECTFINALUPDATE($K_{\mathsf{Em}},$) step, followed by three REFINEFINALUPDATE(C) commands, one for each of the three other components, and are similar in nature to the previous negotiation. Keep in mind that boldface letters (e.g., $\mathbf{u}$) represent ranked updates, while italics letters (e.g. u) represent simple updates. After these, the final step is the system-initiated action COMMITUPDATE, in which the agreed-upon update is committed to the database.

There are a few further points worth mentioning. First of all, the order in which the steps were executed is not fixed. For example, steps 3 and 4 can clearly be interchanged with no difference in subsequent ones. However, even greater variation is possible. Step 5, REFINEUPDATE(K_{Sc}), could be performed before PROMOTEINITIALUPDATE(K_{Ac}) of step 4. In that case, upon completion of REFINEUPDATE(K_{Ac}), a second execution of REFINEUPDATE(K_{Sc}) would be necessary. The final result would nonetheless be the same. More generally, the final result is independent of the order in which decisions are made. See Observation 3.5 for a further discussion.

Suppose that in PROMOTEINITIALUPDATE(K_{Ac}) of step 4, all incoming updates in $\mathbf{u}_1''$ are supported by $\mathbf{u}_3$, i.e., $\mathsf{Proj}(\mathbf{u}_3, \Gamma_{\mathsf{ScAc}}) = \mathbf{u}_1''$. In that case, no revised request of the form $\mathbf{u}_3'$ is transmitted back to K_{Sc} via the port-status register. It is thus natural to ask how K_{Sc} "knows" that no such request will appear. The answer is that it does not matter. The entire process is nondeterministic, and K_{Sc} can execute REFINEUPDATE(K_{Sc}) based upon the input $\mathbf{u}_2'$ from K_{Mg} alone. If an update request $\mathbf{u}_3'$ comes from K_{Ac} later, a second REFINEUPDATE(K_{Sc}) based upon it would allow precisely the same final result as would the process described in Table 2. Again, see Observation 3.5 below.

It is also worthy of note that some decisions may lead to a rejection. For example, management might decide to allow travel only to ADBIS, while the accountant might find that there are funds for travel to DEXA but not ADBIS. In that case, the refinement step 5 would fail, and the only possible step to continue would be a rejection. Additionally, any component can decide to reject any proposed update on its ports at any time before the *Accepted* state is reached simply by executing a REJECTUPDATE(C), even if there is a possible update. For example, a supervisor might decide to disallow a trip.

3   A Formal Model of Cooperative Updates

In this section, some of the technical details regarding updates and update families are elaborated, and then a more complete description of the behavior of the update automaton is given. In Definition 3.1–Definition 3.3 below, let $\mathbf{D}$ be a database schema, and let $\Gamma = (\mathbf{V}, \gamma)$ be a view of $\mathbf{D}$; that is, $\gamma : \mathbf{D} \to \mathbf{V}$ is a database morphism whose underlying mapping $\overset{\circ}{\gamma} : \mathsf{LDB}(\mathbf{D}) \to \mathsf{LDB}(\mathbf{V})$ is surjective. Consult [11] for details.

Definition 3.1 (Updates and update families). Following [2, Sec. 3], an *update* on $\mathbf{D}$ is a pair $u = (M_1, M_2) \in \mathsf{LDB}(\mathbf{D}) \times \mathsf{LDB}(\mathbf{D})$. This update is called an *insertion* if $M_1 \leq_{\mathbf{D}} M_2$, and a *deletion* if $M_2 \leq_{\mathbf{D}} M_1$. A *directional update* is one which is either an insertion or else a deletion. A *ranked update* on $\mathbf{D}$ is a triple $\mathbf{u} = (M, S, \leq_{\mathbf{u}})$ in which $M \in \mathsf{LDB}(\mathbf{D})$, S is a finite subset of $\mathsf{LDB}(\mathbf{D})$, and $\leq_{\mathbf{u}}$ is a preorder (i.e., a reflexive and transitive relation [12, 1.2]) on S, called the *preference ordering*. The set of *updates of* $\mathbf{u}$ is $\mathsf{Updates}(\mathbf{u}) = \{(M, M') \mid M' \in S\}$. $\mathbf{u}$ is *deterministic* if $\mathsf{Updates}(\mathbf{u})$ contains exactly one pair, and *empty* if $\mathsf{Updates}(\mathbf{u}) = \emptyset$. In general, ranked updates are denoted by boldface letters (e.g., $\mathbf{u}$), while ordinary updates will be denoted by italic letters (e.g., u). The ranked update $\mathbf{u}$ is an *insertion* (resp. a *deletion*) if every $(M, M') \in \mathsf{Updates}(\mathbf{u})$ is an insertion (resp. a deletion), and $\mathbf{u}$ is a *ranked directional update* if it is either an insertion or else a deletion. Every (ordinary) update can be regarded as a ranked update in the obvious way; to (M_1, M_2) corresponds the ranked update $(M_1, \{M_2\}, \leq_{\mathbf{D}|\{M_1,M_2\}})$. The set of all ranked directional updates on $\mathbf{D}$ is denoted $\mathsf{RDUpdates}(\mathbf{D})$. For $S' \subseteq S$, the *restriction* of u to S' is $\mathsf{Restr}(\mathbf{u}, S') = (M_1, S', \leq_{\mathbf{u}_{|S'}})$ with $\leq_{\mathbf{u}_{|S'}}$ the restriction of $\leq_{\mathbf{u}}$ to S'. The set of all *nonempty restrictions* of u is $\mathsf{NERestr}(\mathbf{u}) = \{\mathsf{Restr}(\mathbf{u}, S') \mid (S' \subseteq S) \wedge (S' \neq \emptyset)\}$.

Definition 3.2 (Projection of updates and update families). For $u = (M_1, M_2)$ an update on $\mathbf{D}$, the *projection of* u *to* Γ is the update $(\overset{\circ}{\gamma}(M_1), \overset{\circ}{\gamma}(M_2))$ on $\mathbf{V}$. This update is often denoted $\gamma(u)$. Now let $\mathbf{u} = (M_1, S, \leq_{\mathbf{u}})$ be a ranked update on $\mathbf{D}$. The *projection of* $\mathbf{u}$ *to* Γ, denoted $\mathsf{Proj}(\mathbf{u}, \Gamma)$, is the ranked update $\gamma(\mathbf{u}) = (\overset{\circ}{\gamma}(M_1), \overset{\circ}{\gamma}(S), \leq_{\gamma(\mathbf{u})})$ in which $\overset{\circ}{\gamma}(S) = \{\gamma(M) \mid M \in S\}$ and $N \leq_{\gamma(\mathbf{u})} N'$ iff for every $M, M' \in S$ for which $\overset{\circ}{\gamma}(M) = N$ and $\overset{\circ}{\gamma}(M') = N'$, $M \leq_{\mathbf{u}} M'$. Observe that projection preserves the property of being an insertion (resp. deletion) in both the simple and ranked cases. Another important operation in component-based updating is refinement. Suppose that $\mathbf{u}$ is a proposed ranked update to a component, and that for each of its ports is given a ranked update which is a restriction of $\mathbf{u}$ onto that component. The refinement of $\mathbf{u}$ by those restrictions is the largest restriction of $\mathbf{u}$ which is compatible with all of the ranked updates on the ports. The formal definition is as follows. Let $\{\Gamma_i \mid 1 \leq i \leq n\}$ be a set of views of $\mathbf{D}$, $\mathbf{u}'_i = \mathsf{Restr}(\mathsf{Proj}(\mathbf{u}, \Gamma_i), S)$, and for $1 \leq i \leq n$, let $S_i \subseteq \{\overset{\circ}{\gamma_i}(M) \mid M \in S\}$, with $S' = \{M \in S \mid (\forall i \in \{1, 2, \ldots, n\})(\exists N_i \in S_i)(\overset{\circ}{\gamma_i}(M) = N_i)\}$. The *refinement* of $\mathbf{u}$ by $U = \{(\Gamma_i, \mathbf{u}'_i) \mid 1 \leq i \leq n\}$ is defined to be $\mathsf{Restr}(\mathbf{u}, S')$, and is denoted $\mathsf{Refine}(\mathbf{u}, U)$.

Definition 3.3 (Liftings of updates and update families). The operation which is inverse to projection is lifting, in which an update to a view is "lifted" to the main schema. In contrast to projection, lifting is inherently a nondeterministic operation. Let $u = (N_1, N_2)$ be an update on $\mathbf{V}$, and let $M_1 \in \mathsf{LDB}(\mathbf{D})$ with $\mathring{\gamma}(M_1) = N_1$ as well. A *lifting* of u to $\mathbf{D}$ for M_1 is an update $u' = (M_1, M_2)$ on $\mathbf{D}$ with the property that $\mathring{\gamma}(M_2) = N_2$. If u is an insertion (resp. deletion), the lifting u' is *direction preserving* if u' is also an insertion (resp. deletion). If u is an insertion and u' is direction preserving, u' is *minimal* if for for any lifting (M_1, M_2') of u to $\mathbf{D}$ for M_1 with $M_2' \leq_u M_2$, it must be the case that $M_2' = M_2$, and u' is *least* if for any lifting (M_1, M_2') of u to $\mathbf{D}$ for M_1, $M_2 \leq_u M_2'$. A corresponding definition holds for deletions, with "$\leq_u$" replaced by "$\geq_u$". These ideas extend in a straightforward manner to ranked updates. Let $\mathbf{u} = (N_1, S, \leq_{\mathbf{u}})$ be a ranked update on $\mathbf{V}$, and let $M_1 \in \mathsf{LDB}(\mathbf{D})$ with $\mathring{\gamma}(M_1) = N_1$. A *lifting* of $\mathbf{u}$ to $\mathbf{D}$ for M_1 is a ranked update $\mathbf{u}' = (M_1, S', \leq_{\mathbf{u}'})$ on $\mathbf{D}$ which satisfies the following three properties:

(lift–i) $(\forall M_2 \in S')(\exists N_2 \in S)(\mathring{\gamma}(M_2) = N_2)$.
(lift–ii) $(\forall N_2 \in S)(\exists M_2 \in S')(\mathring{\gamma}(M_2) = N_2)$.
(lift–iii) For $M_2, M_2' \in S'$, $M_2 \leq_{\mathbf{u}'} M_2'$ iff $\mathring{\gamma}(M_2) \leq_{\mathbf{u}} \mathring{\gamma}(M_2')$.

If $\mathbf{u}$ is a ranked directional update, then the lifting $\mathbf{u}'$ is *direction preserving* if it satisfies the obvious conditions —- if $\mathbf{u}$ is an insertion (resp. deletion), then so too is $\mathbf{u}'$. A direction-preserving ranked directional update u is *minimal* (resp. *least*) if each member of $\mathsf{Updates}(\mathbf{u})$ is minimal (resp. least). The set of all minimal liftings of $\mathbf{u}$ for M_1 is denoted $\mathsf{MinLift}(\mathbf{u}, \Gamma, M_1)$.

Definition 3.4 (The update automaton). Details regarding the precise conditions under which the actions of the automaton may be applied are expanded here. For the most part, information which has already been presented in Example 2.2 and Table 1 will not be repeated.

The machine operates nondeterministically. There are eight classes of actions; any member of any of these classes may be selected as the next step, provided that its preconditions are satisfied. The listed actions are then executed in the order given. All of these operations, with the exception of ACCEPTUPDATE and *Commit*, must be initiated by a user of the associated component. Operation is synchronous; that is, only one operation may be executed at a time. Subsequent operations must respect the state generated by the previous operation. Formally, a *computation* of this automaton is a sequence $D = \langle D_1, D_2, \ldots, D_n \rangle$ in which $D_1 = \text{INITIATEUPDATE}(C, \mathbf{u})$ for some $C \in X$ and $\mathbf{u} \in \mathsf{RDUpdates}(\mathsf{Schema}(C))$, and for $1 \leq i \leq n - 1$, D_{i+1} is a legal step to follow D_i according to the rules spelled out below. The computation *defines a single negotiation* if $D_n = \text{COMMITUPDATE}$, while $D_k \neq \text{COMMITUPDATE}$ for any $k < n$. The *length* of the computation is n. In the description which follows, X is taken to be a finite set of components with J an interconnection family for X.

INITIATEUPDATE($C, \mathbf{u}$): This is the first step in the update process, and is initiated by a user of the component C by proposing a ranked update $\mathbf{u}$ to its current state. The appropriate projections of this update are propagated to all ports of components which are connected to C and whose state is altered by the update.

Preconditions:
P_{11}: Status = Idle
Actions:
Q_{11}: Initiator $\leftarrow C$
Q_{12}: Status $\leftarrow$ Active
Q_{13}: PendingUpdate$(C) \leftarrow \mathbf{u}$
Q_{14}: $(\forall \Gamma \in \text{Ports}\langle C \rangle)(\forall \Gamma' \in \text{AdjPorts}\langle C, J \rangle)((\text{Proj}(\mathbf{u}, \Gamma) \neq \text{ identity })$
$\Rightarrow \text{PortStatus}(\text{SrcCpt}(\Gamma'), \Gamma') \leftarrow \text{Proj}(\mathbf{u}, \Gamma)).$

PROMOTEINITIALUPDATE(C): This step is relevant in the situation that a component has received an update request on one of its ports, but it has not yet proposed any corresponding update to its own state. A user of that component selects a lifting of this update request for its own state, and propagates its projections to its ports to all neighboring components.

Preconditions:
P_{21}: Status = Active
P_{22}: PendingUpdate(C) = NULL
P_{23}: $(\exists \Gamma' \in \text{Ports}\langle C \rangle)(\text{PortStatus}(C, \Gamma') \neq \text{NULL})$
/* Since the hypergraph of J is acyclic, Γ' must be unique. */
Actions:
Q_{21}: PendingUpdate$(C) \leftarrow$ Choose $\mathbf{u} \in \text{NERestr}(\text{MinLift}(\mathbf{u}, \Gamma', \text{Schema}\langle C \rangle))$
where $(\Gamma' \in \text{Ports}\langle C \rangle) \wedge (\text{PortStatus}(C, \Gamma') \neq \text{NULL})$
Q_{22}: $(\forall \Gamma \in \text{Ports}\langle C \rangle)(\text{PortStatus}(C, \Gamma) \leftarrow \text{NULL})$
Q_{23}: $(\forall \Gamma \in \text{Ports}\langle C \rangle)(\forall \Gamma'' \in \text{AdjPorts}\langle C, J \rangle)((\text{Proj}(\mathbf{u}, \Gamma) \neq \text{ identity })$
$\Rightarrow \text{PortStatus}(\text{SrcCpt}(\Gamma''), \Gamma'') \leftarrow \text{Proj}(\mathbf{u}, \Gamma)).$

REFINEUPDATE(C): In this step, a user of component C further restricts its current proposal for an update, based upon additional ranked updates received at its ports. Upon successful completion, the new proposed update of the component is consistent with the proposed updates which were on its ports.

Preconditions:
P_{31}: Status = Active
P_{32}: PendingUpdate$(C) \neq$ NULL
P_{33}: Refine(PendingUpdate(C),
$\{(\Gamma, \text{PortStatus}(C, \Gamma)) \mid \text{PortStatus}(C, \Gamma) \neq \text{NULL}\}) \neq \emptyset.$
Actions:
Q_{31}: PendingUpdate$(C) \leftarrow$ Choose $\mathbf{u} \in \text{NERestr}(\text{Refine}(\text{PendingUpdate}(C),$
$\{(\Gamma, \text{PortStatus}(C, \Gamma)) \mid \text{PortStatus}(C, \Gamma) \neq \text{NULL}\}))$
Q_{32}: $(\forall \Gamma \in \text{Ports}\langle C \rangle)(\text{PortStatus}(C, \Gamma) \leftarrow \text{NULL})$
Q_{33}: $(\forall \Gamma \in \text{Ports}\langle C \rangle)(\forall \Gamma' \in \text{AdjPorts}\langle C, J \rangle)((\text{Proj}(\mathbf{u}, \Gamma) \neq \text{ identity })$
$\Rightarrow \text{PortStatus}(\text{SrcCpt}(\Gamma'), \Gamma') \leftarrow \text{Proj}(\mathbf{u}, \Gamma)).$

REJECTUPDATE(C): This step is a crude, all-purpose rejection step. Any component can reject the proposed update and terminate the entire process at any time for any reason. One reason might be that it cannot unify the update proposals on its ports, but it could also be that one of its users wishes to terminate the update for other reasons. Upon such termination, the update automaton is returned to its initial state; all proposed updates are discarded.

Preconditions:
   P_{41}: Status = Active
Actions:
   Q_{41}: Status ← Idle
   Q_{42}: Initiator ← NULL
   Q_{43}: $(\forall C \in X)$(PendingUpdate(C) ← NULL)
   Q_{44}: $(\forall C' \in X)(\forall \Gamma \in \text{Ports}\langle C'\rangle)$(PortStatus($C', \Gamma$) ← NULL)

ACCEPTUPDATE: This step is executed when all components which are involved in the update process agree that the update can be supported. This agreement is indicated by the fact that no port has a pending update; all such port updates have been integrated into component updates. It is initiated automatically; users cannot effect it directly.

Preconditions:
   P_{51}: Status = Active
   P_{52}: $(\forall C \in X)(\forall \Gamma \in \text{Ports}\langle C\rangle)$(PortStatus($C, \Gamma$) = NULL)
Actions:
   Q_{51}: Status ← Accepted

SELECTFINALUPDATE(C, u): Upon acceptance, the components have agreed upon a ranked update. However, the database must be updated to a single new state; thus, a single update must be chosen from the ranked set. The purpose of this step is to initiate that selection process; it is always executed by a user of the component in which the update request initiated.

Preconditions:
   P_{61}: Status = Accepted
   P_{62}: Initiator = C
   P_{63}: $\mathbf{u} \in$ NERestr(PendingUpdate(C)) ∧ $\mathbf{u}$ is deterministic.
Actions:
   Q_{61}: For each $\Gamma \in \text{Ports}\langle C\rangle$, if Proj($\mathbf{u}, \Gamma$) is not the identity update, then
       for all $\Gamma' \in \text{AdjPorts}\langle C, J\rangle$, PortStatus(SrcCpt($\Gamma'$), Γ') ← Proj($\mathbf{u}, \Gamma$)
   Q_{62}: Status ← Final

REFINEFINALUPDATE(C): In this step, a user of component C selects a final deterministic update which is consistent with that chosen by the initiator.

Preconditions:
   P_{71}: Status = Final

Actions:

Q_{71}: PendingUpdate(C) ← some *deterministic* restriction of
Refine(PendingUpdate(C),
$\quad\quad \{(\Gamma, \text{PortStatus}(C, \Gamma)) \mid \text{PortStatus}(C, \Gamma) \neq \text{NULL}\})$

Q_{72}: $(\forall \Gamma \in \text{Ports}\langle C \rangle)(\text{PortStatus}(C, \Gamma) \leftarrow \text{NULL})$

COMMITUPDATE: In this step, the update is committed to the database, and the update automaton is returned to its initial state (albeit with the new database state). This step is executed automatically when its preconditions are satisfied; users cannot initiate it.

Preconditions:

P_{81}: Status = *Final*

P_{82}: $(\forall \Gamma \in \text{Ports}\langle C \rangle)(\text{PortStatus}(C, \Gamma) = \text{NULL})$

Actions:

Q_{81}: $(\forall C \in X)(\text{CurrentState}(C) \leftarrow \text{PendingUpdate}(C))$

Q_{81}: $(\forall C \in X)(\text{PendingUpdate}(C) \leftarrow \text{NULL})$

Q_{83}: Initiator ← NULL

Q_{84}: Status ← *Idle*

Because the model of negotiation which has been presented is very simple, it has some nice theoretical properties. Firstly, infinite computations are not possible; the machine will always halt. Second, although the machine is nondeterministic, it is not necessary to guess correctly to make things work. These ideas are formalized in the following observation.

Observation 3.5 (Computations are well behaved). *In the automaton model given in Definition 3.4, the following conditions hold.*

(a) *For any $C \in X$ and any $\mathbf{u} \in$ RDUpdates(Schema(C)), there is a natural number n (which may depend upon the current state of the database) such that every computation beginning with* INITIATEUPDATE$(C, \mathbf{u})$ *has length at most n.*

(b) *Let $D = \langle D_1, D_2, \ldots, D_{k-1}, D_k, \ldots, D_n \rangle$ be a computation of the update automaton which defines a single negotiation, and let $E = \langle E_1, E_2, \ldots, E_{k-1}, E_k \rangle$ be a computation of that same machine with $E_i = D_i$ for $1 \leq i \leq k-1$. Assume further that E_k is not of the form* REJECTUPDATE(C). *Then there is a choice of lifting associated with E_k and a computation $E' = \langle E_1, E_2, \ldots, E_{k-1}, E_k, \ldots, E_{n'} \rangle$ which also defines a single negotiation, and with the further property that D and E' result in the same update on the database. (The values of n and n' need not be the same).*

PROOF OUTLINE: Part (a) follows from the observation that only one decision for an initial ranked update may be made for each component. After that, the process only refines these initial decisions. Since a ranked update is finite by definition, there can only be a finite number of such refinements, and each such refinement must reduce the number of possibilities in some pending update. Part (b) follows by observing that one may simply choose for the lifting of E_k the final ranked update for that component under the computation D.   □

4   Conclusions and Further Directions

The basic idea of supporting view update by negotiating with other views (*qua* components) has been presented. A formal model of this process has been developed using a nondeterministic automaton as the underlying computational model. Although it has certain limitations, it does provide a formal model of update by component cooperation within that context, thus providing a firm basis for further development of these ideas, a few of which are identified below.

Extension of the basic model: The basic model of cooperative update is limited in several ways. For these reasons, the it should be viewed as a proof-of-concept effort rather than a comprehensive solution; further research will address the following issues. First, the current model of cooperation is monotonic. Once the actors have made initial proposals for the updates which they support, the process of identifying the solution update is solely one of refining those initial proposals. On the other hand, in realistic situations, it is often necessary for the parties to negotiate nonmonotonically, by retracting their initial proposals and then submitting new ones, or by modifying their existing proposals adding new alternatives rather than by just refining existing ones. A key extension for future work is to develop an extended model which supports such nonmonotonic negotiation. A second important direction for future work is the development of a computational formalism which embodies more specific modelling of communication between components. The current automaton-based rendering does not provide the necessary information to the actors to effect an efficient negotiation; the nondeterminism allows very long and inefficient although completely correct solutions. With a more detailed model of communication, much more efficient negotiation will be possible.

Relationship to workflow: The topic of *workflow* involves the systematic modelling of processes which require the coordinated interaction of several actors [13]; closely related ideas are known to have central importance in information systems [14]. Within the context of the development of interactive database systems and cooperative work, long-term transactions and workflow loops are central topics [15] [16] [17]. There is a natural connection between these ideas and those surrounding cooperative update which have been introduced in this paper. Indeed, underlying the process of negotiating a cooperative solution to the view-update problem, as illustrated, for example, in Table 2, is a natural workflow of update tasks, each defined by a step in the execution of the automaton.

An important future direction for this work is to develop the connection between workflow for database transactions and the models of cooperative view update which have been presented in this paper. Indeed, given a database schema defined by components and a requested update on one of these components, it should be possible to use the ideas developed in this paper to define and identify the workflow pattern which is required to effect that update. This, in effect, would provide a theory of *query-based workflow construction*.

Acknowledgment. Much of this research was carried out while the first author was a visitor at the Information Systems Engineering Group at the University of Kiel. He is indebted to Bernhard Thalheim and the members of the group for the invitation and for the many discussions which led to this collaborative work.

References

1. Bancilhon, F., Spyratos, N.: Update semantics of relational views. ACM Trans. Database Systems 6, 557–575 (1981)
2. Hegner, S.J.: An order-based theory of updates for database views. Ann. Math. Art. Intell. 40, 63–125 (2004)
3. Hegner, S.J.: The complexity of embedded axiomatization for a class of closed database views. Ann. Math. Art. Intell. 46, 38–97 (2006)
4. Langerak, R.: View updates in relational databases with an independent scheme. ACM Trans. Database Systems 15, 40–66 (1990)
5. Bentayeb, F., Laurent, D.: View updates translations in relational databases. In: Quirchmayr, G., Bench-Capon, T.J.M., Schweighofer, E. (eds.) DEXA 1998. LNCS, vol. 1460, pp. 322–331. Springer, Heidelberg (1998)
6. Gottlob, G., Paolini, P., Zicari, R.: Properties and update semantics of consistent views. ACM Trans. Database Systems 13, 486–524 (1988)
7. Thalheim, B.: Database component ware. In: Schewe, K.D., Zhou, X. (eds.) Database Technologies, Proceedings of the 14th Australasian Database Conference, ADC 2003, Adelaide, South Australia, February 2003, pp. 13–26. Australian Computer Society (2003)
8. Schmidt, P., Thalheim, B.: Component-based modeling of huge databases. In: Benczúr, A.A., Demetrovics, J., Gottlob, G. (eds.) ADBIS 2004. LNCS, vol. 3255, pp. 113–128. Springer, Heidelberg (2004)
9. Thalheim, B.: Component development and construction for database design. Data Knowl. Eng. 54, 77–95 (2005)
10. Thalheim, B.: Entity-Relationship Modeling. Springer, Heidelberg (2000)
11. Hegner, S.J.: A model of database components and their interconnection based upon communicating views. In: Jakkola, H., Kiyoki, Y., Tokuda, T. (eds.) Information Modelling and Knowledge Systems XXIV. Frontiers in Artificial Intelligence and Applications, IOS Press, 2007 (in press)
12. Davey, B.A., Priestly, H.A.: Introduction to Lattices and Order, 2nd edn. Cambridge University Press, Cambridge (2002)
13. van der Aalst, W., van Hee, K.: Workflow Management: Models, Methods, and Systems. MIT Press, Cambridge (2002)
14. Flores, F., Graves, M., Hartfield, B., Winograd, T.: Computer systems and the design of organizational interaction. ACM Trans. Inf. Syst. 6, 153–172 (1988)
15. Alonso, G., Agrawal, D., Abbadi, A.E., Kamath, M., Günthör, R., Mohan, C.: Advanced transaction models in workflow contexts. In: Su, S.Y.W. (ed.) Proceedings of the Twelfth International Conference on Data Engineering, February 26 - March 1, 1996, pp. 574–581. IEEE Computer Society, New Orleans, Louisiana (1996)
16. Rusinkiewicz, M., Sheth, A.P.: Specification and execution of transactional workflows. In: Kim, W. (ed.) Modern Database Systems, pp. 592–620. ACM Press, Addison-Wesley (1995)
17. Hidders, J., Dumas, M., van der Aalst, W.M.P., ter Hofstede, A.H.M., Verelst, J.: When are two workflows the same? In: Atkinson, M.D., Dehne, F.K.H.A. (eds.) CATS. CRPIT, vol. 41, pp. 3–11. Australian Computer Society (2005)

An Agile Process for the Creation of Conceptual Models from Content Descriptions

Sebastian Bossung[1], Hans-Werner Sehring[2],
Henner Carl[1], and Joachim W. Schmidt[2]

Hamburg University of Technology, Germany
Sustainable Content Logistics Centre, Hamburg, Germany
{sebastian.bossung,hw.sehring,h.carl,j.w.schmidt}@tuhh.de

Abstract. It is widely accepted practice to build domain models as a conceptual basis for software systems. Normally, the conceptual schema cannot be supplied by domain experts but is constructed by modelling experts. However, this is infeasible in many cases, e.g., if the system is to be generated ad hoc from a conceptual schema.

This paper presents an iterative process that helps domain experts to create a conceptual schema without the need for a modelling expert. The process starts from a set of sample instances provided by the domain expert in a very simple form. The domain expert is assisted in consolidating the samples such that a coherent schema can be inferred from them. Feedback is given by generating a prototype system which is based on the schema and populated with the provided samples.

The process combines the following three aspects in a novel way: (1) it is based on a large amount of samples supplied by the domain expert, (2) it gives feedback by agile generation of a prototype system, and (3) it does not require a modelling expert nor does it assume modelling knowledge with the domain expert.

1 Introduction

It is common practice to begin the development of software systems with an analysis of the application domain, which results in a conceptual schema. It describes the relevant entities in the domain without regard to implementation issues and serves as a basis for later steps in the software development process. Traditionally, the transformation of the conceptual schema into application program code or data schemata was done manually. In recent years, model-driven development (MDD) approaches, e.g., the Model Driven Architecture (MDA) approach, has aimed to automatise this transformation to a large degree. Nevertheless, the underlying conceptual schema must exist before the development of the software system and must remain unchanged during its lifetime. In the following we will use the term *schema* as a synonym of conceptual schema as we are not interested in other types of schemata.

In previous work we have presented the Conceptual Content Management (CCM) approach, which generates content management systems from a conceptual schema. In this respect, it is similar to MDD. However, unlike most MDD

Y. Ioannidis, B. Novikov, and B. Rachev (Eds.): ADBIS 2007, LNCS 4690, pp. 114–129, 2007.

approaches, CCM does not assume the schema to remain unchanged during the lifetime of the system. Instead, schema evolution (for the whole user community of the system) and schema personalisation (for individual users) are explicitly supported by CCM. Users of the system, i.e., experts in the application domain, use our conceptual modelling language to state the initial schema as well as evolutional and personalisational changes to it. System generation in reaction to schema changes works well in CCM and includes the transformation of relevant instances.

The creation of an appropriate conceptual schema of the application domain is a challenge in many development approaches, including MDD and CCM. Both assume that the domain expert is able to express the model with sufficiently formal means. It is generally acknowledged that the help of a modelling expert is necessary as domain experts in many fields cannot be assumed to be familiar with the formalisms used in a conceptual schema (e.g., inheritance between classes). While the consultation of a modelling expert is an acceptable approach in traditional software development, modelling experts cannot always be available at short notice during schema evolution or personalisation steps in a CCM system. This paper explores possibilities to allow domain experts to create a conceptual schema without the aid of a modelling expert. To this end, we introduce the *Asset Schema Inference Process* (ASIP) that incorporates work from several areas: sample-based schema inference, agile and iterative schema refinement, prototype systems as well as collaborative schema creation processes. The idea of creating the conceptual schema from samples is based the observation that domain experts often find it troublesome to give an abstracted account of their application domain. Most are, however, able to "tell their story" by providing examples of the entities they work with.

We continue in section 2 with an overview of the modelling techniques used by the ASIP to express the input samples and to describe the resulting schema. The input samples are provided in a graph representation that makes heavy use of multimedial content. Emphasis was put on simplicity—such that domain experts can intuitively understand the samples—with just enough structure to make schema inference possible. The created schema is expressed in Asset classes (section 2.2) which provide a dualistic model of an entity with a multimedial as well as a conceptual representation. In section 2.3 we introduce Conceptual Content Management Systems (CCMSs) and their architecture. CCMSs can be generated from Asset schemata.

Our process is introduced in section 3. It consists of four phases: (1) sample acquisition, (2) schema inference, (3) feedback and (4) system creation. Section 4 takes a closer look at two alternative implementations of the phases 2 and 3. The first approach we propose is based on traditional schema inference techniques and user feedback questions. Alternatively, domain experts can influence the schema by supervising a cluster-based algorithm.

Before we conclude with a summary and outlook in section 6, we discuss the ASIP in section 5 with some reflections about the quality of the generated schema and a comparison with related approaches.

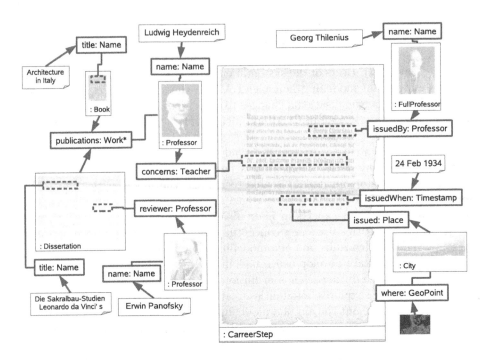

Fig. 1. Some sample entity definitions. Types on plain content are omitted.

2   Conceptual Content Management

In a series of interdisciplinary projects, in particular in cooperation with scientists from the humanities, the need for open dynamic conceptual content management (CCM) has been recognised. In this section we review the approach, which relies on experienced domain experts who also have modelling skills. Such domain experts create a conceptual schema based on previously acquired knowledge. CCM modelling allows conceptual schemata to evolve dynamically, even when CCMSs derived from such a schema are already in use.

2.1   Sample Entity Definition

In many disciplines, especially those which are not grounded on mathematical models, a first step towards the understanding of a problem lies in the collection of problem instances. In particular in history and art history the (past) phenomena are recognised by some kind of trace they left, e.g., a piece of art. Knowledge on such phenomena is collected through samples of entity descriptions.

Figure 1 illustrates a sample collected by a domain expert and explored in a free-form notation for entity descriptions. The figure is loosely based on concept graphs [30].

Such descriptions use multimedia content that constitutes the perception of an entity. Content is paired with explanations of the depicted entity in order to represent it conceptually. Explanations are shown as boxes in figure 1. Each box contains a name (in front of the colon) and a type (after the colon). Types in this example just name a subject for the explanation and do not prescribe a structure. Explanation types can be of single or multiple cardinality (the latter is denoted by an asterisk following the type; e.g., Work*). Explanations are provided by multimedia content again—as shown by the arrows—and can in turn have explanations attached.

Based on insights into many such samples a domain expert will come up with a model of the respective domain. Note that current CCM projects follow a "model first" approach in which domain experts have to invent an initial schema—however complete it may be. Through open modelling (see section 2.2) domain experts are free to refine the schema anytime they like to.

2.2   Open and Dynamic Domain Modelling

For entity descriptions like those discussed in the previous section the *Conceptual Content Management* approach [26,28] has been developed. Its principles are: (1) entities are described by pairs of content and a conceptual model and (2) modelling is open and dynamic.

Entity descriptions are maintained as indivisible pairs of content and a conceptual model, called *Assets*. Entities are visualised by the content which can take any multimedial form and is addressed by a content handle. A conceptual model is provided by different attributes associated with the entity, as indicated by the lines between parts of the content and descriptive attributes in figure 1.

Open modelling is provided through the *Asset definition language* [26]. It allows domain experts to state a conceptual schema of a domain by defining *Asset classes*. The aspect of dynamics is achieved by automatic system evolution and will be addressed in the subsequent section.

For example, assume a domain expert with samples similar to those in figure 1 in mind. Such an expert might create a model like the following:

```
model Historiography
from Time import Timestamp
from Topology import Place
class Professor {
 content image
 concept characteristic name :String
        relationship publications:Work* }
class Document {
 content scan
 concept characteristic title :String
        relationship    concerns  :Professor
        relationship    reviewer  :Professor*
        relationship    issuing   :Issuing }
class Issuing { concept relationship issued     :Place
                        relationship issuedBy   :Professor
                        relationship issuedWhen:Timestamp}
```

The schema is called Historiography. In the example it is assumed that there are conceptual schemata for the base domains of time (Time) and space (Topology). Classes are defined that reflect the entities of the domain under consideration and allow the creation of the required Asset instances.

Asset class definitions declare content handles as well as a conceptual model. The conceptual model part consists of attribute definitions, where each attribute is either a characteristic that is inherent to an entity, a relationship that refers to other asset instances which in turn describe independent entities or a constraint that restricts attribute values and references.

Such a schema reflects modelling decisions that a domain expert made based on previously acquired expertise. For example, even though there is evidence of books (see figure 1), in the example the decision was made to not model books by Asset classes, but only Documents in general. There might be only few relevant books, so there is no need for a class "Book". Another example is the class Issuing which is not obvious from the samples. However, the domain expert knows the domain concept of issuing and models it accordingly. Object domains are chosen for characteristic attributes. Here, names and titles are represented by strings.

Since schemata are determined by personal decisions to a high degree, other experts might disagree with a given schema. To this end, the open modelling and dynamic system evolution allow the personalisation of schemata. For example, one domain expert might want to work with a model that is redefined in the following way:

```
model MyHistoriography
from Historiography import Document, Professor
class Document {
 concept relationship reviewer unused }
class Dissertation refines Document {
 concept relationship reviewer :Professor* }
```

In this personalisation example the class Document is imported from the existing Historiography model and redefined by providing an alternative definition under the same name. The relationship reviewer is moved from Document to a new subclass Dissertation. An attribute is removed from a class by the keyword unused instead of a type constraint. Note that removing an attribute does not pose a problem to the type system, since Document is not refined, but instead its definition is replaced with a new one. Other typical redefinitions include the introduction of new attributes or changing an attribute's kind from characteristic to relationship or vice versa.

2.3 CCM Technology

The property of dynamics is established by CCM technology. When a domain expert applies an open schema redefinition, an existing CCMS that has been generated from the original schema has to be adapted. Since open model changes are permitted at runtime, a CCMS might already be populated with Asset instances. Furthermore, communication with domain experts who do not apply a schema change—or who apply a different one—has to be maintained.

These requirements are met by an interaction of a model compiler for the Asset definition language and a CCMS architecture that enables the required system evolution. Details on the technological aspects can be found in [28,27].

3   The Asset Schema Inference Process

Figure 1 gives an example of entity description samples that contribute to a domain expert's knowledge. For the discussion of the Asset Schema Inference Process (ASIP) we review figure 1 as a simple approach to providing free-form entity descriptions. For the remainder of the paper we refer to concrete explanation needs for entity descriptions as *abstractions*. Abstractions are given a name and are also assigned a *semantic type*. Semantic types are also assigned to multimedia content.

The semantic types are arranged in a single-inheritance hierarchy and have no structural definitions. A similar notion is adopted in [19] by differentiating between classification (semantic types) and typing (structural types). The simplicity of our semantic types is deliberate to enable domain experts to work with these types without the help of a modelling expert. We assume here that the hierarchy of semantic types has been created in a group effort and changes much more slowly than the structural definitions in the conceptual schema. Practical experience suggests that a group of domain experts is able to create a hierarchy of semantic types using a controlled process after some initial training.

The graphical notation of figure 1 is backed by a more formal language in the implementation of the ASIP. For this language a type system is defined to check well-typedness of descriptions according to semantic types [4]. However, this need not be of concern to domain experts. It should be stressed that figure 1 does not depict an Asset instance according to a CCM Asset class definition.

In this section we present the basic structure of the ASIP. Two alternative implementations for the core steps are outlined in section 4.

3.1   An Example for Asset Model Inference

We give an example of a schema construction for the sample in figure 1. It contains several entities from the application domain which are combined into a single sample: The document in the centre, three persons, two publications as well as names and geographic places. This example will focus on the persons. First, three classes are created, one for each person: two classes are structurally identical (just one name attribute and a content), another also has an attribute publications. Structural comparison leads to the merging to the two identical classes. The new joint class can automatically be named Professor as this is the most specific common semantic type of all backing samples. Next, the class that also handles publications is analysed. No mergable classes are found (in this sample; there would usually more samples), but an inheritance relationship is

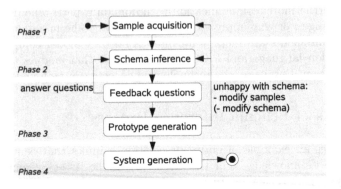

Fig. 2. Overview of the ASIP

detected with the class Professor and automatically introduced. The new class should also be named Professor but due to a name clash it is renamed Professor-2. In this example, the system would prompt the user for feedback, whether the Professor-2 class should really be kept, as there is only one sample backing it. The generated part of the model looks like this:

```
model InferredHistoriography
class Professor {
   content c :java.awt.Image
   concept characteristic name :Name }
class Professor-2 refines Professor {
   concept relationship publications :Work* }
```

The referred classes are created by recursion on the structure of entities.

3.2   Phases of the Model Inference Process

The ASIP is partitioned into four phases which are depicted in figure 2. Input artefacts to the ASIP are the samples in phase 1 and user feedback during phase 3. The results are the created schema and the CCMS which can be generated from it. The ASIP works in the following four phases:

1. *Acquisition:* Samples can be acquired by creation of new entity descriptions. Alternatively, the domain expert can choose to import some Asset instances from an existing CCMS. These instances are then converted into the free-form model presented in section 2.1. The domain expert will usually modify these to reflect the new requirements, for example by adding an abstraction.
2. *Schema generation:* In this phase the provided samples are analysed and a corresponding schema is created. This is done automatically to the extent possible. If ambiguities arise, feedback questions are collected to pose to the domain expert in the next phase. This second phase is thus performed fully automatic. We have two alternative implementations for schema generation which we discuss in section 4.

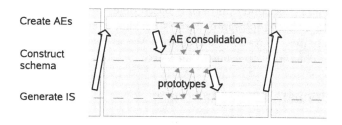

Fig. 3. Schema construction stages

3. *Feedback:* The generated schema is presented to the user. Because an—even graphical—presentation of the schema is often not understood (see section 1 and [29]) we use other means to provide feedback. The exact possibilities depend on the particular implementation of the previous phase and are, therefore, given in section 4. Both approaches use the generation of prototypes. Others [22] have done this as well and have found an increase in the motivation to participate [25, ch. 3.3.5].

   If the user is unhappy with the generated schema, feedback is considered and phases 1 and 2 are re-executed. Besides answering questions by the system, the user can also modify the underlying samples (e.g., remove an attribute that is not considered so important after all).

4. *System construction:* A full system is set up, including potential additional modules that are required to adapt the old parts of the system which function according to the previous schema. The samples can be fed into the generated system. System generation is described in [27].

The ASIP can be used in the lifecycle of a CCMS to support schema evolution and personalisation steps as well as the initial construction of the system. Outside the CCM approach, the ASIP can be employed for schema creation. Figure 3 shows the ASIP in the context of the lifecycle of a CCMS. When the need for a schema modification arises (first arrow), samples are created that reflect the new wishes of the domain expert. The samples are analysed in the ASIP and possibly consolidated by the domain expert. Meanwhile, prototype systems are generated to give the domain expert an impression of the new system. Once this is satisfactory, the new system is generated. The cycle starts over when the next need for personalisation arises. We also have investigated the embedding of the ASIP in the whole system creation process but omit this here for brevity [5].

4 Schema Inference

In this section we present two alternatives for the core steps of the ASIP: The first uses schema inference [23] as described in the literature, the second is a cluster-based analysis. The first approach will be referred to as "traditional" schema inference below. Schema inference is well-understood for a variety of applications including object-oriented systems [20], semi-structured data in general [9] and

XML data [34]. The second approach is based on a statistical learning approach that first divides the samples into clusters. Inside the clusters, traditional means are used to create a class for each cluster. We are not aware of any previous application of clustering to schema inference.

4.1 "Traditional" Schema Inference

The traditional schema inference approach implements the process steps 2 and 3 in several parts. First, a naive Asset class, which directly reflects the structure of the sample, is created for each of the input samples. Next, the set of classes is analysed to find identical classes as well as classes that can automatically be put in inheritance relationships (see below). Other modifications to the schema are also considered, but require feedback from the user. This feedback is given in step 3 of the process and the schema is modified according to the answers obtained from the domain expert.

Schema inference algorithms usually work by first constructing a data structure that will accomodate all instances, e.g., [9]. Afterwards, this data structure is often simplified to remove redundancies and sometimes also relaxed in an effort to better meet user expectations. Accordingly, the naive Asset classes in the ASIP are constructed from the samples by creating an attribute for each abstraction in the sample. The content compartment is filled according to the technical type of the content from the sample. Classes are created with the help of an intensional type system for samples [4], that assigns types to samples in terms of Asset classes. This ensures that all valid samples can be reflected in an Asset class.

Next, the schema with naive classes is simplified. During the simplification phase, feedback questions are collected that will be posed to the domain expert later. There are four cases of simplification:

1. *Identical class*: Two classes whose attributes and contents are identical in both names and types are considered identical. These classes are automatically merged.
2. *Inheritance*: If a class introduces additional attributes or contents but is otherwise identical to another class, the two classes are automatically put in an inheritance relationship.
3. *Type match*: Two classes that consist of attributes and contents which are of identical types but have differing names are considered a type match. Fully-automatic inference algorithms often treat such classes as identical. The ASIP prompts the user for input in such cases.
4. *Inheritance orphan*: A class with only few backing samples is called an inheritance orphan. The user is prompted whether the samples should be moved into the superclass.

Simplifications are carried out in the order specified here. After simplification, feedback questions are posed to the user. The schema is modified according to the answers. Further simplifications which are detected in a re-execution of the four simplification cases might become possible as a result.

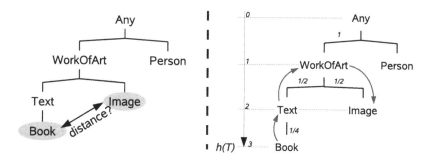

Fig. 4. Distance of semantic types

4.2 Cluster-Based Sample Analysis

Besides traditional schema inference, we have experimented with an approach that first clusters the samples according to input from the user. Asset classes are constructed from the clusters.

Our clustering approach is based on the *k-means* algorithm [11]. The general idea is to divide samples into groups (called *clusters*) and to create an Asset class for each cluster. In contrast to common clustering methods we employ user interaction for the segmentation of samples into clusters. From a statistical point of view, our approach is not data clustering, but a method of semi-supervised learning. The clusters in *k-means* are represented by a cluster-centre which has the characteristics of mean value of the cluster's samples. In our approach the clusters (and cluster-centres) are motivated semantically (based on the semantic types annotated to the samples) and structurally (based on the common attributes of the samples in a cluster). Therefore, an Asset class can be created from each cluster by using the structural information only.

To assign each sample s to a cluster c, a distance measure $d(s, c)$ is introduced as a weighted average of semantic d_{sem} and structural distance d_{struct}.

$$d(s, c) = \alpha \cdot d_{sem}(s, c) + (1 - \alpha)d_{struct}(s, c) \qquad \alpha \in [0 \ldots 1]$$

The semantic distance is based on the number of steps one has to take to travel from one semantic type to another in the hierarchy of semantic types. The general idea is to annotate each edge in the semantic type tree with exponentially decreasing weights. The semantic distance is the sum of weights along the edges of the shortest path in the semantic tree between the semantic type of the sample T_1 and that of the cluster T_C (figure 4):

$$d_{sem}(T_1, T_C) := \begin{cases} \frac{1}{2^{h(T_1)}} & \text{if } T_1 \text{ is direct supertype of } T_C \\ d_{sem}(T_1, T_m) + d_{sem}(T_m, T_C) & \text{if } T_1 \text{ is direct supertype of } T_m \\ & \text{and } T_m \text{ is supertype of } T_C \\ d_{sem}(T_S, T_1) + d_{sem}(T_S, T_C) & \text{if } T_S \text{ is the most specific common supertype of } T_1 \text{ and } T_C \end{cases}$$

Table 1. Costs of edit operations on classes

Operation	Cost	Operation	Cost
add attribute	low	narrow attribute type	very low
remove attribute	very high	change cardinality to many	medium
change attribute name	low	change cardinality to single	very low
broaden attribute type	medium		

The structural distance is based on an edit distance of class attributes. It is constructed in a similar way as the edit distance between character strings [14] where each edit action (introduction or removal of attributes, modification of their types or cardinalities; table 1 gives an overview of the costs) is assigned a penalty. For full definitions of both distances see [5].

Using the edit distance, our algorithm has these four iterative steps:

1. *Classification:* All samples are assigned to the closest cluster based on the distance measure $d(s, c)$.
2. *Optimisation:* Recomputation of the centre of each cluster according to the samples in the cluster. Semantically this is achieved by finding the minimal set of semantic types (according to d_{sem}) that subsume the semantic types of all samples in the cluster. Structural optimisation is done by creating the set of attributes that has the minimal cumulated distance d_{struct} to all samples of the cluster. Full details on the algorithms for finding initial clusters and structural optimisation, along with their implementation, can be found in [5]. Steps 1 and 2 are repeated as long as change is detected in the optimisation.
3. *Inheritance hierarchy creation:* Put the Asset classes that correspond to each cluster in an inheritance hierarchy. This is done by structural matching as in the traditional inference approach. If a change is detected, repeat steps 1 through 3.
4. *Feedback:* Visualise the resulting clusters to the user to allow feedback. There are two ways to show a visualisation of the clusters: directly (with their backing samples) or to generate a prototype system that reflects the current state of the schema. In addition to previously presented feedback options, the domain experts can directly request additional partitionings of clusters that they deem too heterogenous.

Due to its ability to treat many samples at a time, the clustering approach is able to handle the same amount of samples with less user interaction than the traditional approach.

5   Discussion

In this section we first consider the quality of the generated schemata (section 5.1) and briefly compare the ASIP to other related approaches, in particular to modelling approaches (section 5.2) and to ontology construction (section 5.3).

Table 2. Assessment of schema quality according to three dimensions [6]. (DE) = influenced by samples and feedback given by domain expert, * = discussed in text.

Specification		Usage	
graph. legibility:	n/a	completeness:	(DE)
simplicity:	good (DE) *	understandability:	good (DE) *
expressiveness:	(DE)	**Implementation**	
synt. correctness:	by construction	implementability:	high, generated *
sem. correctness:	good (DE) *	maintainability:	high, built-in *

5.1   On the Judgement of Schema Quality

It is generally difficult to quantify schema quality. Several approaches are used [3] including the simplicity of the schema and the tightness of fit to the samples. When creating a schema for an application domain, there is a significant limitation to these measures: the goal is not to find a schema that exactly fits the samples, but to find a schema which describes the application domain well. Frameworks to assess schema quality, therefore, use less quantifiable criteria ([18] reviews such frameworks). We use three criteria from [6]: *specification, usage* and *implementation*, which are broken down into several aspects each (table 2).

Specification is concerned with how well the requirements to and extent of the domain can be understood from the schema. *Simplicity* is impacted by the structural diversity of the samples, caused by samples that describe the same kind of entity but use different abstractions to do so. If both approaches are truly different and equally valid, the creation of two separate classes by the ASIP is beneficial. Otherwise the ASIP has to rely on type matches. If diversity is too great to cause a type match, it is up to the domain expert to reconsider some samples. Failure to do so hurts the specification of the schema. The semantic *correctness* can be negatively impacted by structurally coinciding classes with different meanings. Such errors need to be resolved by the domain expert. They are especially likely in weakly modelled parts of the application domain (for example, descriptions of entities of different kinds by a name only).

Usage includes the *understandability* of the schema for human readers. The choice of appropriate names is an important factor here. Attribute names can be carried over from samples (where they have been introduced by domain experts), class names need to be invented (based on the semantic types of samples, see example in section 2.1). Collisions in names cause generated class names, hurting understandability. The automatic generation of prototype systems decreases the impact of understandability issues as classes never have to be understood at an abstract level by the domain expert.

Finally, implementation measures the effort that is necessary to create a system based on the schema. CCMSs are generated from the schema. Practical experiences indicate that generation takes time comparable to a conventional compiler run on hand-written code (exact timing of course depends on various aspects of the setup). Implementation also contains *maintainability* which measures how well the schema can evolve. Since evolution is explicitly supported in the CCM approach and the ASIP is set up to support the creation of the necessary conceptual schema, the maintainability of our approach can be considered high.

Generally, the quality of the schema depends on the extent and individual quality of the input samples. The crucial parameters are the degree of coverage of the application domain and the (lack of) structural diversity of the samples. This influences the aspects marked "DE" in table 2.

5.2   Modelling Approaches

Contemporary modelling approaches assume the presence of a modelling expert who talks to domain experts, knows the modelling formalism being used and has the skill to abstract from the concrete application domain.

For example, software development usually starts in the requirements analysis phase with the interaction between a domain expert and a modelling expert ("system analyst") [7,16,8]. Languages for requirements analysis have been proposed, but they are not widely accepted in practice [24, D13.5]. A central issue in practice is the difficulty of obtaining sufficiently formal requirements descriptions.

An initial modelling phase can be supported by structuring the interaction between domain and modelling experts [33,17] or by automatically constructing parts of a model [8]. Both approaches start from textual descriptions and help the modelling expert in creating conceptual models from them. Since this way of creating conceptual models is a complex task, the *Klagenfurt Conceptual Predesign* [17] adds an additional step to the usual sequence of modelling phases.

In support of open modelling, iterative conceptual model creation has also been explored [21]. Model creation is performed in a series of small steps leading from an abstract model to a concrete database schema. The decisions necessary to make a model more concrete require the presence of a modelling expert.

Often the involvement of domain experts can be reduced by providing the modelling expert with an ontology of general knowledge [32].

5.3   Ontology Creation

The creation of conceptual models (and hence the ASIP) shares problems with ontology creation for application domains in the respects that the reuse of existing ontologies is complicated (e.g., because of mismatches in the levels of abstractions of the ontologies in question) and that the complexity of modelling languages can prevent domain experts from providing models on their own. Several process for the development of ontologies have been proposed [10].

An approach that is iterative and interactive like the ASIP is the approach to *ontological modelling by Holsapple and Joshi* [12], referred to as *OMHJ* in the following. Starting from an initial model the OMHJ uses iterative model refinement in the second of its four phases. OMHJ does not use machine generated questions like the ASIP, but employs the Delphi Method [15] involving a human moderator who poses the questions. This moderator furthermore integrates additional input from domain experts. OMHJ lays an emphasis on the collaboration of users, an aspect that is currently not considered in the ASIP. The experts in OMHJ are expected to be able to provide initial abstractions on their own whereas the ASIP is based on samples provided by the experts.

A process with a similar scope as the ASIP is the *On-To-Knowledge meta-process* [31] which leads to an ontology of an application domain. In contrast to the ASIP, On-To-Knowledge is not based on samples and requires the intervention of a modelling expert. On-to-Knowledge has two iterative elements: the refinement during development and the evolution of ontology as part of the maintenance of an application. Refinements are based on input supplied by domain experts, where this input is formalised by a modelling expert. Ontology evolution is similar to the open modelling provided by CCM and leads to the re-execution of refinement phase. In contrast to CCM, however, the impact of ontology evolution onto a software system has to be handled manually.

Linguistic analysis [1] starts with sufficiently verbose textual descriptions of an application domain, for example descriptions of use cases [13, section 9.5]. Such textual descriptions are mined for entities and their attributes by a modelling expert. In linguistic analysis the textual descriptions take a role similar to that of the samples in the ASIP. An appropriate coverage of the application domain under consideration is necessary in both approaches. [8] introduces additional modelling steps to add missing model parts. In contrast to the samples used in thee ASIP, texts are informal. Therefore, automatic model extraction is difficult and a modelling expert is necessary. Furthermore, a high amount of "noise" in textual descriptions is a problem for linguistic analysis [2]. ASIP avoids most data cleaning by the more formal input provided by the samples.

6   Summary and Outlook

Practical experience has shown that many domain experts find it difficult to abstractly describe their application domain in a conceptual schema. To enable them to create such schemata nonetheless, we have presented the Asset Model Inference Process (ASIP) which combines three important aspects: (1) no modelling knowledge is required from the domain experts, (2) the schema is inferred from samples and (3) the feedback during the process is given in the form of a prototype system. Besides traditional inference methods we have also presented a cluster-based approach. We have conducted first experiments with the process.

For the future we are interested in several subjects related to the ASIP. The enhancement of distribution and collaboration of users will be interesting. Given the CCMSs' ability to work with multiple models, collaborative schema creation can be enabled inside the system in the manner of the Delphi method [15]. The idea is essentially to create one schema (and thus one set of modules in the CCMS) per iteration. Since samples can be imported into the ASIP, this allows for round-trip schema development. Another interesting aspect is the fine-tuning of the distance measures in the cluster-based inference. Finally, we will further investigate the difference between and interrelation of classification (as in semantic types) and structural typing (as in Asset classes). We expect that some observations of [19] and others are applicable to our scenario. In past projects we have observed that classification is an important part of all our CCM applications, making its native support desirable.

References

1. Abbott, R.J.: Program design by informal english descriptions. Communications of the ACM 26(11), 882–894 (1983)
2. Ahonen, H., Heinonen, O., Klemettinen, M., Verkamo, A.I.: Applying data mining techniques for descriptive phrase extraction in digital document collections. In: Proceedings of the Advances in Digital Libraries Conference, ADL '98, pp. 2–11 (1998)
3. Angluin, D., Smith, C.H.: Inductive inference: Theory and methods. ACM Computing Surveys 15(3), 237–269 (1983)
4. Bossung, S.: Conceptual Content Modeling. PhD thesis, Technische Universität Hamburg-Harburg, 2007 (to be published)
5. Carl, H.: Konzeptuelle Modellierung durch Domänenexperten. Master's thesis, Technische Universität Hamburg-Harburg (2007)
6. Cherfi, S.S.-S., Akoka, J., Comyn-Wattiau, I.: Conceptual modeling quality - from EER to UML schemas evaluation. In: Spaccapietra, S., March, S.T., Kambayashi, Y. (eds.) ER 2002. LNCS, vol. 2503, pp. 414–428. Springer, Heidelberg (2002)
7. Díaz, I., Moreno, L., Fuentes, I., Pastor, O.: Integrating natural language techniques in OO-method. In: Gelbukh, A. (ed.) CICLing 2005. LNCS, vol. 3406, pp. 560–571. Springer, Heidelberg (2005)
8. Fliedl, G., Kop, C., Mayr, H.C.: From textual scenarios to a conceptual schema. Data & Knowledge Engineering 55(1), 20–37 (2004)
9. Goldman, R., Widom, J.: DataGuides: Enabling query formulation and optimization in semistructured databases. In: Proceedings of 23rd International Conference on Very Large Data Bases, pp. 436–445. Morgan Kaufmann, San Francisco (1997)
10. Gómez-Pérez, A., Fernández-López, M., Corcho, O.: Ontological Engineering. Springer, Heidelberg (2004)
11. Hastie, T., Tibshirani, R., Friedman, J.: The elements of Statistical Learning: Data Mining, Inference and Prediction. Springer, Heidelberg (2001)
12. Holsapple, C.W., Joshi, K.D.: A collaborative approach to ontology design. Communications of the ACM 45(2), 42–47 (2002)
13. Larman, C.: Object-oriented Analsyis and Design and Iterative Development, 3rd edn. Prentice-Hall, Englewood Cliffs (2005)
14. Levenshtein, V.I.: Binary codes capable of correcting deletions, insertions and reversals. Soviet Physics Doklady, 707–710 (1966)
15. Lindstone, H.A., Furoff, M.: The Delphi Method: Techniques and Applications. Addison-Wesley, Reading (1975)
16. Maiden, N.A.M., Jones, S., Flynn, M.: Innovative requirements engineering applied to ATM. In: Proceedings Air Traffic Management, Budapest (2003)
17. Mayr, H.C., Kop, C.: Conceptual predesign – bridging the gap between requirements and conceptual design. In: Proceedings of 3rd Int. Conference on Requirements Engineering, pp. 90–100. IEEE Computer Society, Los Alamitos (1998)
18. Moody, D.L.: Theoretical and practical issues in evaluating the quality of conceptual models: current state and future directions. Data & Knowledge Engineering 55(3), 243–276 (2005)

19. Norrie, M.C.: Distinguishing typing and classification in object data models. In: Information Modelling and Knowledge Bases, vol. VI, IOS (1995)
20. Palsberg, J., Schwartzbach, M.I.: Object-oriented type inference. In: Proceedings of the Conference on Object-Oriented Programming, Systems, Languages & Applications, pp. 146–161 (1991)
21. Proper, H.A.E.: Data schema design as a schema evolution process. Data Knowledge Engineering 22(2), 159–189 (1997)
22. Puerta, A., Micheletti, M., Mak, A.: The UI pilot: A model-based tool to guide early interface design. In: Proceedings of the 10th international conference on Intelligent user interfaces, pp. 215–222. ACM Press, New York (2005)
23. Rahm, E., Bernstein, P.A.: A survey of approaches to automatic schema matching. VLDB Journal 10(4), 334–350 (2001)
24. Rechenberg, P., Pomberger, G. (eds.): Informatik Handbuch, 4th edn. Hanser(2006)
25. Rupp, C.: Requirements-Engineering und -Management: Professionelle, iterative Anforderungsanalyse für die Praxis, 3rd edn. Carl Hanser Verlag (2004)
26. Schmidt, J.W., Sehring, H.-W.: Conceptual Content Modeling and Management: The Rationale of an Asset Language. In: Broy, M., Zamulin, A.V. (eds.) PSI 2003. LNCS, vol. 2890, pp. 469–493. Springer, Heidelberg (2004)
27. Sehring, H.-W., Bossung, S., Schmidt, J.W.: Content is capricious: A case for dynamic systems generation. In: Manolopoulos, Y., Pokorný, J., Sellis, T. (eds.) ADBIS 2006. LNCS, vol. 4152, pp. 430–445. Springer, Heidelberg (2006)
28. Sehring, H.-W., Schmidt, J.W.: Beyond Databases: An Asset Language for Conceptual Content Management. In: Benczúr, A.A., Demetrovics, J., Gottlob, G. (eds.) ADBIS 2004. LNCS, vol. 3255, pp. 99–112. Springer, Heidelberg (2004)
29. Shipman III, F.M., Marshall, C.C.: Formality considered harmful: Experiences, emerging themes, and directions on the use of formal representations in interactive systems. Computer Supported Cooperative Work 8(4), 333–352 (1999)
30. Sowa, J.: Knowledge Representation. Brooks/Cole (2000)
31. Staab, S., Studer, R., Schnurr, H.-P., Sure, Y.: Knowledge processes and ontologies. IEEE Intelligent Systems 16(1), 26–34 (2001)
32. Storey, V.C., Goldstein, R.C., Ullrich, H.: Naive semantics to support automated database design. IEEE Transactions on Knowledge and Data Engineering 14(1), 1–12 (2002)
33. Tam, R.C.-M., Maulsby, D., Puerta, A.R.: U-TEL: A tool for eliciting user task models from domain experts. In: Proc. of the 3rd Int. Conference on Intelligent User Interfaces, pp. 77–80. ACM Press, New York (1998)
34. Wong, R.K.: Sankey, J.: On structural inference for XML data. Technical report, University of New South Wales (June 2003)

ODRA: A Next Generation Object-Oriented Environment for Rapid Database Application Development

Michał Lentner and Kazimierz Subieta

Polish-Japanese Institute of Information Technology
ul. Koszykowa 86, 02-008 Warszawa, Poland
{m.lentner,subieta}@pjwstk.edu.pl

Abstract. ODRA (Object Database for Rapid Application development) is an object-oriented application development environment currently being constructed at the Polish-Japanese Institute of Information Technology. The aim of the project is to design a next-generation development tool for future database application programmers. The tool is based on the query language SBQL (Stack-Based Query Language), a new, powerful and high level object-oriented programming language tightly coupled with query capabilities. The SBQL execution environment consists of a virtual machine, a main memory DBMS and an infrastructure supporting distributed computing. The paper presents design goals of ODRA, its fundamental mechanisms and some relationships with other solutions.

1 Introduction

With the growth of non-classical database application, especially in the rapidly growing Internet context, the issue of bulk data processing in distributed and heterogeneous environments is becoming more and more important. Currently, increasing complexity and heterogeneity of this kind of software has led to a situation where programmers are no more able to grasp every concept necessary to produce applications that could efficiently work in such environments. The number of technologies, APIs, DBMSs, languages, tools, servers, etc. which the database programmer should learn and use is extremely huge. This results in enormous software complexity, extensive costs and time of software manufacturing and permanently growing software maintenance overhead. Therefore the research on new, simple, universal and homogeneous ideas of software development tools is currently very essential.

The main goal of the ODRA project is to develop new paradigms of database application development. We are going to reach this goal by increasing the level of abstraction at which the programmer works. To this end we introduce a new, universal, declarative programming language, together with its distributed, database-oriented and object-oriented execution environment. We believe that such an approach provides functionality common to the variety of popular technologies (such as relational/object databases, several types of middleware, general purpose

Y. Ioannidis, B. Novikov, and B. Rachev (Eds.): ADBIS 2007, LNCS 4690, pp. 130–140, 2007.

programming languages and their execution environments) in a single universal, easy to learn, interoperable and effective to use application programming environment.

The principle ideas which we are implementing in order to achieve this goal are the following:

- **Object-oriented design.** Despite the principal role of object-oriented ideas in software modeling and in programming languages, these ideas have not succeeded yet in the field of databases. As we show in this paper, our approach is different from current ways of perceiving object databases, represented mostly by the ODMG standard [4] and database-related Java technologies (e.g. [5, 6]). Instead, we are building our system upon a methodology called the Stack-Based Approach (SBA) to database query and programming languages [13, 14]. This allows us to introduce for database programming all the popular object-oriented mechanisms (like objects, classes, inheritance, polymorphism, encapsulation), as well as some mechanisms previously unknown (like dynamic object roles [1,7] or interfaces based on database views [8, 10]).
- **Powerful query language extended to a programming language.** The most important feature of ODRA is SBQL (Stack-Based Query Language), an object-oriented query and programming language. SBQL differs from programming languages and from well-known query languages, because it is a query language with the full computational power of programming languages. SBQL alone makes it possible to create fully fledged database-oriented applications. The possibility to use the same very-high-level language for most database application development tasks may greatly improve programmers' efficiency, as well as software stability, performance and maintenance potential.
- **Virtual repository as middleware.** In a networked environment it is possible to connect several hosts running ODRA. All systems tied in this manner can share resources in a heterogeneous and dynamically changing, but reliable and secure environment. Our approach to distributed computing is based on object-oriented virtual updatable database views [9]. Views are used as *wrappers* (or mediators) on top of local servers, as a *data integration* facility for global applications, and as *customizers* that adopt global resources to needs of particular client applications. This technology can be perceived as contribution to distributed databases, Enterprise Application Integration (EAI), Grid Computing and Peer-To-Peer networks.

The rest of the paper is organized as follows. In Section 2 we shortly present the main motivations and features of SBQL. In Section 3 we discuss application integration using ODRA. In Section 4 we discuss various deployment scenarios concerning ODRA-based applications. Section 5 concludes.

2 SBQL

The term *impedance mismatch* denotes a well-known infamous problem with mapping data between programming languages (recently Java) and databases. The majority of Java programmers spend between 25% and 40% of their time trying to map objects to relational tables and v/v. In order to reduce the negative influence of the feature, some automatic binding mechanisms between programming language objects and database structures have been suggested. This approach is expressed in the ODMG standard, post-relational DBMS-s and several Java technologies.

Unfortunately, all these solutions have only shown that despite the strong alignment of the database constructs with the data model of the programming languages used to manipulate them, the impedance mismatch persists. It could be reduced at the cost of giving up the support for a higher-level query language, which is unacceptable. Impedance mismatch is a bunch of negative features emerging as a result of too loose coupling between query languages and general-purpose programming languages. The incompatibilities concern syntax, type checking, language semantics and paradigms, levels of abstraction, binding mechanisms, namespaces, scope rules, iteration schemas, data models, ways of dealing with such concepts as null values, persistence, generic programming, etc. The incompatibilities cannot be resolved by any, even apparently reasonable approach to modify functionality or additional utilities of existing programming languages.

All these problems could be completely eliminated by means of a new self-contained query/programming language based on homogeneous concepts. Our idea concerns an imperative object-oriented programming language, in which there is no distinction between expressions and queries. Such expressions/queries should have features common to traditional programming language expressions (literals, names, operators), but also should allow for declarative data processing. Query operators can be freely combined with other language's constructs, including imperative operators, control statements and programming abstractions. The language could be used not only for application programming, but also to query/modify databases stored on disk and in main memory. Our proposal of such a language is named SBQL and it is the core of ODRA.

2.1 Queries as Expressions

SBQL is defined for a very general data store model, based on the principles of object relativism and internal identification. Each object has the following properties: an internal identifier, an external name and some value. There are three kinds of objects:

- simple (<OID, name, atomic value>),
- complex (<OID, name, set of subobjects>),
- reference (<OID, name, target OID>).

There are no dangling pointers: if a referenced object is deleted, the reference objects that point at it, are automatically deleted too. There are no null values - lack of data is not recorded in any way (just like in XML). This basic data model can be used to represent relational and XML data. We can also use it to build more advanced data structures which are properties of more complex object-oriented data models (supporting procedures, classes, modules, etc).

SBQL treats queries in the same way as traditional programming languages deal with expressions (we therefore use the terms query and expression interchangeably). Basic queries are literals and names. More complex queries are constructed by connecting literals and names with operators. Thus, every query consists of several subqueries and there are no limitations concerning query nesting. Because of the new query nesting paradigm, we avoid the *select-from-where* sugar, which is typical for SQL-like query languages. In SBQL expressions/queries are written in the style of programming languages, e.g. 3+1; $(x+y)*z$; *Employee* **where** *salary* > $(x+y)$; etc.

In SBQL we have at least six query result kinds: atomic values, references, structures, bags, sequences and binders. They can be combined and nested in fully orthogonal way (limited only by type constraints). Unlike ODMG, we do not introduce explicitly collections (bags, sequences) in the data store; collections are substituted by many objects with the same name (like in XML). Structures are lists of fields (possibly unnamed) which are together treated as a single value. Binders are pairs <*name, result*>, written as *name*(*result*), where *result* is any query result. Query results are not objects - they have no OIDS and may have no names. Query results and procedure parameters belong to the same domain, hence it is possible to pass a query as a parameter. SBQL queries never return objects, but references to them (OIDs). Due to this feature not only *call-by-value*, but also *call-by-reference* and other parameter passing styles are possible.

It is possible to create a procedure which returns a collection of values. Such a procedure reminds a database view, but it can encapsulate complex processing rather than a single query (like in SQL). However, in SBQL procedures and (updateable) views are different abstractions; the second one has no precedents in known query and programming languages. Names occurring in queries are bound using the environment stack, which is a structure common to most programming languages. Its sections are filled in with binders. Stack sections appear not only as results of procedure calls, but also due to a specific group of query operators, called *non-algebraic*. Among them there are such operators as: . (dot), *where, join, order by, forall, forany,* etc. All non-algebraic operators are macroscopic, i.e. work on collections of data.

Query operators called *algebraic* do not use the environment stack. Some of them (e.g. *avg, union*) are macroscopic, some other (e.g. +, -, *) are not. Algebraic and non-algebraic operators, together with the environment stack and the query result stack make up typical functionality expected from a query language designed to deal with structured and semi-structured data.

Some SBQL operators provide functionality which is absent or very limited in popular query languages. Among them are transitive closures and fixed-point equations. Together with procedures (which can be recursive), they constitute three styles of recursive programming in SBQL. All SBQL operators are fully orthogonal with imperative constructs of the language.

SBQL queries can be optimized using methods known from programming languages and databases. The SBQL optimizer supports well known techniques, such as rewriting, cost-based optimization, utilization of indices and features of distributive operators (like shifting selections before joins). Several powerful strategies unknown in other query languages (like shifting independent subqueries before non-algebraic operators) have been developed [11, 13]. The process of optimization usually occurs on the client-side during the program compilation process. The traditional server-side optimization is also possible. Since the SBQL compiler provides static counterparts of runtime mechanisms (database schema, static environment stack and static result stack), it is possible to use these counterparts to simulate the whole program evaluation process during compile time. This makes it possible strong and semi-strong type checking of SBQL queries/programs [12] and compile-time optimizations.

Thanks to the data independence, global declarations are not tied with programs. Instead, they are parts of a database schema and can be designed, administered and maintained independently of programs. If necessary, a client-side SBQL compiler automatically downloads the metabase (which contains the database schema, database statistics, and other data) from the server to accomplish strong type checking and query optimization.

2.2 Advanced SBQL Features

SBQL supports popular imperative programming language constructs and mechanisms. There are well known control structures (if, loop, etc.), as well as procedures, classes, interfaces, modules, and other programming and database abstractions. All are fully orthogonal with SBQL expressions. Most SBQL abstractions are first-class citizens, which means they can be created, modified, deleted and analyzed at runtime. Declarations of variables in SBQL may determine many objects of the same name and type. Thus, the concept of variable declaration is similar to table creation in relational databases.

Apart from variable name and type, variable declarations determine cardinalities, usually [0..*], [1..*], [0..1] and [1..1]. A cardinality constraint assigned to a variable is the way in which SBQL treats collections. It is possible to specify whether variable name binding should return ordered (sequences) or unordered (bags) collections. Arrays are supported as ordered collections with fixed length.

In classical object-oriented languages (e.g. Java) types are represented in a relatively straightforward manner and suitable type equivalence algorithms are not hard to specify and implement. However, type systems designed for query languages have to face such problems as irregularities of data structures, repeating data (collections with various cardinality constraints), ellipses, automatic coercions, associations among objects, etc. These peculiarities of query languages make existing approaches to types too limited. SBQL provides a semi-strong type system [12] with a relatively small set of types and with structural type conformance. SBQL is perhaps the only advanced query language with the capability of static type checking. Static type checking in SBQL is based on the mechanisms used also during static optimization (described above): the static environment stack, the static query result stack, and the metabase (which contains variable declarations). Note that traditional query processing assumes that queries are embedded in a host language as strings, which makes static type checking impossible.

SBQL programs are encapsulated within modules, which constrain access from/to their internals through import and export lists. Modules are considered complex objects which may contain other objects. Because modules are first-class citizens, their content may change during run-time. Classes are complex objects too consisting of procedures and perhaps other objects. Procedures stored inside classes (aka *methods*) differ from regular procedures only by their execution environment, which additionally contains internals of a currently processed object.

Procedure parameters and results can be bulk values. All procedure parameters (even complex ones) can be passed by values and by references. Due to the stack-based semantics procedures can be recursively called with no limitations and with no special declarations.

SBQL supports two forms of inheritance: class inheritance (static) and object inheritance (dynamic). The former is typical of popular programming languages. Unlike Java, multiple inheritance is allowed. The second inheritance form is also known as dynamic object roles [1, 7]. This concept assumes that throughout its lifetime an object can gain and lose multiple roles. For example, the object Person can have simultaneously such roles as Employee, Student, Customer, etc. Roles may be dynamically inserted or deleted into/from an object. An object representing a role inherits all features of its super-objects. This method models real-life scenarios better than multiple inheritance. For instance, persons can be students, employees or customers only for some time of their entire life. The mechanism of dynamic object roles solves many problems and contradictions, in particular, with multiple inheritance, tangled aspects, historical objects, etc.

Many SBQL concepts have their roots in databases rather than in programming languages. One of them is the mechanism of updatable views [8, 10]. Such views not only present the content of a database in different ways, but also allow to perform update operations on virtual data in a completely transparent way. In SBQL a view definition contains a procedure generating so-called *seeds* of virtual objects. The definition also bears specification of procedures that are to be performed on stored objects in response to update operations addressing virtual objects generated by the view. The procedures overload generic operations (create, retrieve, update, insert, delete) performed on virtual objects. The view definer is responsible for implementing every operation that must be performed after a virtual object is updated. There are no restrictions concerning which operations on virtual objects are allowed and which are not (unlike view updateability criteria known from other proposals). Because a view definition is a complex object, it may contain other objects, in particular, nested view definitions, ordinary procedures, variables (for stateful views), etc.

Updatable views in ODRA have many applications, ranging from traditional (virtual mapping of data stored in the database) to complete novelty (mediators or integrators). In particular, since SBQL can handle semi-structured data, SBQL views can be used as an extremely powerful transformation engine (instead of XSLT). The power of SBQL views is the power of a universal programming language, concerning both the mapping of stored objects into virtual ones and the mapping of a updates of virtual objects into updates of stored ones.

Another case is the concept of the interface, which is also expressed as a view. Because an interface is a first-class citizen, apart from tasks common to traditional interfaces, it can also serve as an element of the security subsystem. A more privileged user has access to more data inside an object, while another user sees its internals through a separate, limited interface.

2.3 SBQL Runtime Environment

An SBQL program is not directly compiled into a machine code. It is necessary to have some intermediate forms of programs and a virtual execution environment which executes them. The first form is a syntactic tree, which is the subject of optimizations and type checking. This form is transformed into a bytecode (different from a Java bytecode, for several important reasons). The SBQL execution environment consists of a virtual machine (VM) acting on a bytecode. The VM functionality provides services typical for hardware (virtual instruction set, virtual memory, etc.) and operating systems (loading, security, scheduling, etc.). Once compiled, a bytecode can be run on every system for which ODRA has been ported. We plan that SBQL programs can also move from one computer to another during runtime (e.g. from a busy computer to an idle one).

The DBMS part controls the data store and provides such mechanisms as transaction support, indexing, persistence, etc. It is a main memory DBMS, in which persistence is based on modern operating systems' capabilities, such as memory mapped files. ODRA assumes the orthogonal persistence model [2].

3 Application Integration Using ODRA

The distributed nature of contemporary information systems requires highly specialized software facilitating communication and interoperability between applications in a networked environment. Such software is usually referred to as middleware and is used for application integration. ODRA supports information-oriented and service-oriented application integration. The integration can be achieved through several techniques known from research on distributed/federated databases. The key feature of ODRA-based middleware is the concept of transparency. Due to transparency many complex technical details of the distributed data/service environment need not to be taken into account in the application code. ODRA supports such transparency forms as transparency of updating made from the side of a global client, transparency of distribution and heterogeneity, transparency of data fragmentation, transparency of data/service redundancies and replications, transparency of indexing, etc.

These forms of transparency have not been solved to a satisfactory degree by current technologies. For example, Web Services support only transparency of location and transparency of implementation. Transparency is achieved in ODRA through the concept of a virtual repository (Fig. 1). The repository seamlessly integrates distributed resources and provides a global view on the whole system, allowing one to utilize distributed software resources (e.g. databases, services, applications) and hardware (processor speed, disk space, network, etc.). It is responsible for the global administration and security infrastructure, global transaction processing, communication mechanisms, ontology and metadata management. The repository also facilitates access to data by several redundant data structures (global indexes, global caches, replicas), and protects data against random system failures.

The user of the repository sees data exposed by the systems integrated by means of the virtual repository through the global integration view. The main role of the integration view is to hide complexities of mechanisms involved in access to local data sources. The view implements CRUD behavior which can be augmented with logic responsible for dealing with horizontal and vertical fragmentation, replication, network failures, etc. Thanks to the declarative nature of SBQL, these complex

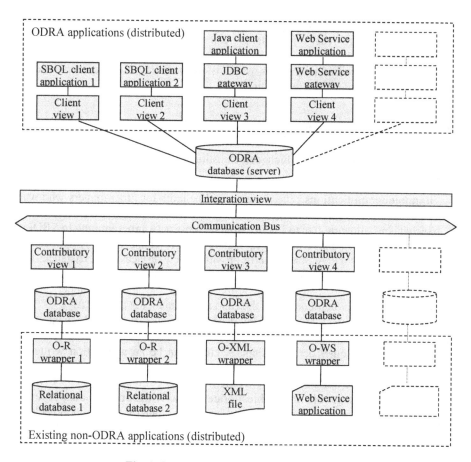

Fig. 1. ODRA Virtual Repository Architecture

mechanisms can often be expressed in one line of code. The repository has a highly decentralized architecture. In order to get access to the integration view, clients do not send queries to any centralized location in the network. Instead, every client possesses its own copy of the global view, which is automatically downloaded from the integration server after successful authentication to the repository. A query executed on the integration view is to be optimized using such techniques as rewriting, pipelining, global indexing and global caching.

Local sites are fully autonomous, which means it is not necessary to change them in order to make their content visible to the global user of the repository. Their content is visible to global clients through a set of contributory views which must conform to the integration view (be a subset of it). Non-ODRA data sources are available to global clients through a set of wrappers, which map data stored in them to the canonical object model assumed for ODRA. We are developing wrappers for several popular databases, languages and middleware technologies. Despite of their diversity, they can all be made available to global users of the repository. The global user may not only query local data sources, but also update their content using SBQL.

Instead of exposing raw data, the repository designer may decide to expose only procedures. Calls to such procedures can be executed synchronously and asynchronously. Together with SBQL's support for semi-structured data, this feature enables document-oriented interaction, which is characteristic of current technologies supporting Service Oriented Architecture (SOA).

4 Deployment Scenarios

ODRA is a flexible system which can substitute many current technologies. In particular, it can be used to build:

- **Standalone applications.** There are several benefits for programmers who want to create applications running under ODRA. Since SBQL delivers a set of mechanisms allowing one to use declarative constructs, programming in SBQL is more convenient than in languages usually used to create business applications (such as Java). ODRA transparently provides such services as persistence; thus applications do not have to use additional DBMS-s for small and medium data sets. Other databases can be made visible through a system of wrappers, which transparently map their content to the canonical data model of ODRA.

- **Database systems.** ODRA can be used to create a traditional client-server database system. In this case, one installation of ODRA plays the role of a database server, other act as clients. A client can be an application written in SBQL and can run in ODRA, as well as a legacy application connected by a standard database middleware (e.g. JDBC). The system may be used as a DBMS designed to manage relational, object-oriented and XML-oriented data. In every case all database operations (querying, updating, transforming, type checking, etc.) can be accomplished using SBQL. Technologies, such as SQL, OQL, XQuery [15], XSLT, XML Schema, can be fully substituted by the SBQL capabilities.

- **Object request brokers.** ODRA-based middleware is able to provide functionality known from distributed objects technologies (e.g. CORBA). In ODRA, the global integrator provides a global view on the whole distributed system, and the communication protocol transports requests and responses between distributed objects. Local applications can be written in any programming language, as the system of wrappers maps local data to ODRA's canonical model. There are several advantages of ODRA comparing to traditional ORB technologies. Firstly, the middleware is defined using a query language, which speeds up middleware development and facilitates its maintenance. Secondly, in CORBA it is assumed that resources are only horizontally partitioned, not replicated and not redundant. ODRA supports horizontal and vertical fragmentation, can resolve replications and can make choice from redundant or replicated data.

- **Application servers.** A particular installation of ODRA can be chosen to store application logic. By doing so, the developers can exert increased control over the application logic through centralization. The application server can also take several existing enterprise systems, map them into ODRA's canonical model and expose them through a Web-based user interface. It is possible to specify exactly how such an application server behaves, so the developer can implement such mechanisms as clustering, or load balancing by him/her own using declarative constructs or already implemented components.

- **Integration servers.** Integration servers can facilitate information movement between two or more resources, and can account for differences in application semantics and platforms. Apart from that, integration servers provide such mechanisms as: message transformation, content based routing, rules processing, message warehousing, directory services, repository services, etc. The most advanced incarnation of this technology (called Enterprise Service Bus, ESB) is a highly decentralized architecture combining concepts of Message Oriented Middleware (MOM), XML, Web Services and workflow technologies. Again, this technology is one of the forms that our updatable view-based middleware can take.

- **Grid Computing infrastructure.** Grid Computing is a technology presented by its advocates as integration of many computers into one big virtual computer, which combines all the resources that particular computers possess. People involved in grid research usually think of resources in terms of hardware (computation, storage, communications, etc.), not data. It is a result of their belief that "in a grid, the member machines are configured to execute programs rather than just move data" [3]. However, all business applications are data-intensive and in our opinion distribution of computation depends almost always on data location. Moreover, responsibility, reliability, security and complexity of business applications imply that distribution of data must be a planned phase of a disciplined design process. ODRA supports this point of view and provides mechanisms enabling grid technology for businesses.

5 Conclusion

We have presented an overview of the ODRA system, which is used as our research platform aiming future database application development tools. ODRA comprises a very-high-level object-oriented query/programming language SBQL, its runtime environment integrated with a DBMS, and a novel infrastructure designed to integration of distributed applications. The core of the prototype is already operational and is used experimentally for various test cases. The project is still under way and focuses on adding new functionalities, improving existing ones, implementing new interoperability modules (wrappers), implementing specialized network protocols and elaborating various optimization techniques.

References

[1] Albano, A., Bergamini, R., Ghelli, G., Orsini, R.: An Object Data Model with Roles. In: Proc. VLDB Conf., pp. 39–51 (1993)
[2] Atkinson, M., Morrison, R.: Orthogonally Persistent Object Systems. The VLDB Journal 4(3), 319–401 (1995)

[3] Berstis, V.: Fundamentals of Grid Computing. IBM Redbooks paper. IBM Corp. (2002), http://www.redbooks.ibm.com/redpapers/pdfs/redp3613.pdf

[4] Cattell, R.G.G., Barry, D.K. (eds.): The Object Data Standard: ODMG 3. Morgan Kaufmann, San Francisco (2000)

[5] Cook, W.R., Rosenberger, C.: Native Queries for Persistent Objects A Design White Paper (2006), http://www.db4o.com/about/ productinformation/ whitepapers/ Native% 20 Queries%20Whitepaper.pdf,

[6] Hibernate -Relational Persistence for Java and.NET (2006), http://www.hibernate.org/

[7] Jodlowski, A., Habela, P., Plodzien, J., Subieta, K.: Objects and Roles in the Stack-Based Approach. In: Hameurlain, A., Cicchetti, R., Traunmüller, R. (eds.) DEXA 2002. LNCS, vol. 2453, Springer, Heidelberg (2002)

[8] Kozankiewicz, H.: Updateable Object Views. PhD Thesis (2005), http://www.ipipan. waw.pl/˜subieta/Finished PhD-s-Hanna Kozankiewicz

[9] Kozankiewicz, H., Stencel, K., Subieta, K.: Integration of Heterogeneous Resources through Updatable Views. In: Workshop on Emerging Technologies for Next Generation GRID (ETNGRID-2004), June 2004, IEEE Computer Society Press, Los Alamitos (2004)

[10] Kozankiewicz, H., Leszczylowski, J., Subieta, K.: Updateable XML Views. In: Kalinichenko, L.A., Manthey, R., Thalheim, B., Wloka, U. (eds.) ADBIS 2003. LNCS, vol. 2798, pp. 385–399. Springer, Heidelberg (2003)

[11] Plodzien, J., Kraken, A.: Object Query Optimization in the Stack-Based Approach. In: Eder, J., Rozman, I., Welzer, T. (eds.) ADBIS 1999. LNCS, vol. 1691, pp. 3003–3316. Springer, Heidelberg (1999)

[12] Stencel, K.: Semi-strong Type Checking in Database Programming Languages (in Polish), p. 207. PJIIT - Publishing House (2006) ISBN 83-89244-50-0

[13] Subieta, K.: Theory and Construction of Object-Oriented Query Languages (in Polish), p. 522. PJIIT - Publishing House (2004) ISBN 83-89244-28-4

[14] Subieta, K.: Stack-Based Approach (SBA) and Stack-Based Query Language (SBQL) (2006), http://www.sbql.pl

[15] World Wide Web Consortium (W3): XML Query specifications, http://www.w3. org/ XML/Query/

An Object-Oriented Based Algebra for Ontologies and Their Instances

Stéphane Jean, Yamine Ait-Ameur, and Guy Pierra

Laboratoire d'Informatique Scientifique et Industrielle
LISI - ENSMA and University of Poitiers
BP 40109, 86961 Futuroscope Cedex, France
{jean,yamine,pierra}@ensma.fr

Abstract. Nowadays, ontologies are used in a lot of diverse research fields. They provide with the capability to describe a huge set of information contents. Therefore, several approaches for storing ontologies and their instances in databases have been proposed. We call Ontology Based Database (OBDB) a database providing such a capability. Several OBDB have been developed using different ontology models and different representation schemas to store the data. This paper proposes a data model and an algebra of operators for OBDB which can be used whatever are the used ontology model and representation schema. By extending the work done for object oriented databases (OODB), we highlight the differences between OODB and OBDB both in terms of data model and query languages.

Keywords: Ontology, Database, Query Algebra, OWL, PLIB, RDF-S.

1 Introduction

Nowadays, ontologies are used in a lot of diverse research fields including natural language processing, information retrieval, electronic commerce, Semantic Web, software component specification, information systems integration and so on. In these diverse domains, they provide with the capability to describe a huge set of information contents. Therefore, the need to manage ontologies as well as the data they describe in a database emerged as a crucial requirement.

We call Ontology Based Database (OBDB) a database that stores data together with the ontologies defining the semantics of these data. During the last decade, several OBDB have been proposed. They support different ontology models such as PLIB [1], RDFS [2] or OWL [3] for describing ontologies and they use different logical schemas for persistancy: unique table of triples [4], vertical representation [5] or table-like structure [6,7] for representing the huge sets of data described by the ontologies.

In parallel to this work, ontology query languages like SPARQL [8] for RDF, RQL [9] for RDF-Schema, or OntoQL [10] for PLIB and a subset of OWL have been defined. Because of the lack of a common data model for OBDBs, dealing with the heterogeneity of OBDB data models in order to implement these languages on top of OBDBs, is a major concern of current research activities. In this paper, we propose a data model for OBDBs which can be used whatever are the used ontology models and the representation schema.

Y. Ioannidis, B. Novikov, and B. Rachev (Eds.): ADBIS 2007, LNCS 4690, pp. 141–156, 2007.
© Springer-Verlag Berlin Heidelberg 2007

Our work started by trying to answer to the following question: since ontologies use extensively object oriented concepts, why don't we use Object Oriented Database (OODB) models as the kernel of such a model? The answer to this question is that the OODB model is not usable without a necessary tuning effort. This answer led us to (1) highlight the existing differences between OODBs and OBDBs either from the conceptual or from the structural points of view and (2) propose another algebra of operators (to provide OBDBs with a formal semantics) extending the algebra defined for OODBs and (3) clarify the differences between OODBs and OBDBs query languages.

The objective of this paper is two-fold. On the one hand, we study and show the differences between OODB and OBDB models. For building our comparison and proposal, we have chosen *ENCORE* [11] as the OODB data model and its corresponding algebra of operators. On the other hand, we propose a generic algebra of operators defining a generic formal semantics for OBDB and show how these operators are used to describe queries of specific OBDB languages. Three languages based on different OBDB models illustrate this work: OntoQL [10], RQL [9] and SPARQL [8].

Compared to the OntoQL definition presented in [10], where we have presented the concrete syntax of OntoQL and its use on the OntoDB OBDB model [6], this paper presents a generic algebra of operators for OBDB models plus extensions and contributions like:

- support of the multi-instanciation paradigm;
- discussions of the differences between OODB and OBDB models and the corresponding query languages;
- the capability of the proposed model and algebra to overcome the heterogeneity of OBDBs data models;
- presentation of the query algebra at the different querying levels enabled by an OBDB: data, ontology and both data and ontology.

This paper is structured as follows. Next section presents a formal representation of the OBDB data model proposed in this paper. The differences between this data model and the OODB data model are highlighted as well. Section 3 presents the *ENCORE* algebra for OODBs and shows its insufficiencies to manage OBDBs. Then, an algebra based on the OBDB data model is presented as an extension of the *ENCORE* algebra. Section 4 discusses related work. Finally, section 5 concludes and introduces future work.

2   Data Models

2.1   The *ENCORE* Formal Data Model for OODBs

Formally, an OODB in the *ENCORE* data model is defined as a 8-tuple $<$ ADT, O, P, SuperTypes, TypeOf, PropDomain, PropRange, Value $>$, where:

- ADT is a set of available abstract data types. It provides with atomic types (Int, String, Boolean), a global super type Object and user-defined abstract data types;

- O is the set of objects available in the database or that can be constructed by a query. All objects have an unique identifier;
- P is the set of properties used to describe the state of each object;
- SuperTypes : $ADT \rightarrow 2^{ADT1}$ is a partial function. It associates a set of super types to a type. This function defines a lattice of types. Its semantics is inheritance and it ensures substitutability;
- TypeOf : $O \rightarrow ADT$ associates to each object the lower (strongest) type in the hierarchy it belongs to;
- PropDomain : $P \rightarrow ADT$ defines the domain of each property;
- PropRange : $P \rightarrow ADT$ defines the range of each property;
- Val : $O \times P \rightarrow O$ gives the value of a property of an object. This property must be defined for the datatype of the object.

This data model supports collections of objects by providing the parameterized ADT named Set$[T]$. Set$[T]$ denotes a collection type of objects of type T. $\{o_1, \ldots, o_n\}$ denotes an object of this type where the o_i's are objects of type T. Another parameterized ADT, called Tuple, is provided to create relationships between objects. A Tuple type is constructed by providing a set of attribute names (A_i) and attribute types (T_i). Tuple$[< (A_1, T_1), \ldots, (A_n, T_n) >]$ denotes a type tuple constructed using the A_i's attribute's name and T_i's attribute types. $< A_1 : o_1, \ldots, A_n : o_n >$ denotes an object of this type where the o_i's are objects of the corresponding type T_i. The Tuple type is equipped with the Get_A_i_value functions to retrieve the value of a Tuple object o for the attribute A_i. The application of this function may be abbreviated using the dot-notation (o.A_i). The Tuple type construct is fundamental for building new data types. In particular, it is useful for describing new data types that are not available in the core database schema and that may be built by expressions of the algebra.

2.2   Definition of the OBDB Data Model

The OBDB data model is based on the definition of two main parts : ontology and content. Instances are stored in the content part while ontologies are stored in the ontology part. The description of these two parts use extensively object-oriented database features. Let us describe these two parts and their relationships.

Ontology. The ontology part, can be formally defined like an OODB by a 7-Tuple as $< E, OC, A, SuperEntities, TypeOf, AttDomain, AttRange, Val >$. Here, abstract data types (ADT) are replaced by entities (E), properties (P) by attributes (A) and objects (O) by concepts of ontologies (OC). To define the built-in entities and attributes of this part, we have considered the constructors shared by the standard ontology models PLIB [1], RDF-Schema [2] and OWL [3]. Thus, in addition to atomic types, the global super type ObjectE, and the parameterized types Tuple and Set, E provides the predefined entities C and P. Instances of C and P are respectively the classes and properties of the ontologies. Each

---

[1] We use the symbol 2^C to denote the power set of C.

class and each property has an identifier defined in the context of a names-pace. This is represented by the attribute `namespace : C ∪ P → String`. Entity C also defines the attribute `SuperClasses : C → SET[C]` and entity P defines the attributes `PropDomain : P → C` and `PropRange : P → C`. The description of these attributes is similar to the definition given for an OODB. Moreover, a global super class `ObjectC` is predefined and the parameterized types `Tuple` and `Set` are also available for C. Thus, an ontology is similar to an OODB schema. How-ever, an ontology gives a precise definition of concepts which means that many more attributes (name, comment, version ...) are available to describe classes and properties of ontologies. These predefined entities and attributes constitute the kernel of the ontology models we have considered. User-defined entities (re-striction, objectProperty ...) and attributes (isSymetric, unionOf, remark ...) may be added to this kernel in order to take into account specific features of a given ontology model. This capability is illustrated in the following example.

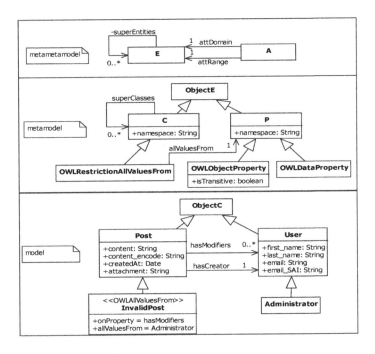

Fig. 1. An illustration of our data model based on the MOF architecture

An Example of the Ontology Part of an OBDB. Using the UML notation, figure 1 presents the data model defined for handling our ontologies. Let us comment this figure in a top-down manner. The upper part presents the subset of the data model defined to describe the used ontology model. In the MOF terminology [12], this part is the *metametamodel*.

The middle part presents the used ontology model. This part corresponds to the level M_2 of the MOF namely the *metamodel*. Each element of this *metamodel* is an instance of an element of the *metametamodel*. In this part, we have added to the predefined entities C, P, and ObjectE, some specific constructors of the OWL ontology model: OWLRestrictionAllValuesFrom, OWLObjectProperty and OWL-DataProperty. Due to space limitations, we did not represent the whole OWL ontology model. However, this can be handled using the OWL metamodel proposed by [13].

Finally, the lower part presents a toy example of an ontology inspired from the SIOC Ontology (http://sioc-project.org/). This part constitutes the level M_1 of the MOF. Each concept of the ontology is an instance of the ontology model defined at level M_2. The SIOC ontology describes the domain of online communities by defining concepts such as Community, Usergroup or Forum. In this example, we have represented the classes User and Post and refined them by the classes Administrator and InvalidPost. The class InvalidPost is defined as an OWLRestrictionAllValuesFrom on the property hasModifiers whose values must be instances of the class Administrator. Thus, an invalid post is a post which has been modified by an administrator only. Notice that the UML notation doesn't allow us to show the whole description of the classes and of the properties of this ontology (labels, comments, documents ...).

Content. The content part allows to store instances of ontology classes. Different representation schemas have been proposed and used by OBDBs (see [14] for a survey). The simplest and more general one uses an unique table of triples [4] where the data is stored in triples of the form (subject, predicate, object). Another representation schema is the vertical structure [5] where a binary table is used for each property and a unary table for each ontology class. Recently, horizontal approaches have been proposed [6,7]. They associate, to each class, a table which columns represent those properties that are valued for at least one instance of the class. Our formalization is based on this latter approach.

In the proposed data model for OBDBs, the content part is a 5-tuple < EXTENT, I, TypeOf, SchemaProp, Val > where:

- EXTENT is a set of extensional definitions of ontology classes;
- I is the set of instances of the OBDB. Each instance has an identity;
- TypeOf : I → 2^{EXTENT} associates to each instance the extensional definitions of the classes it belongs to;
- SchemaProp : EXTENT → 2^P gives the properties used to describe the instances of an extent (the set of properties valued for its instances);
- Val : I × P → I gives the value of a property occurring in a given instance. This property must be used in one of the extensional definitions of the class the instance belongs to. Since Val is a function, an instance can only have a unique value (which can be a collection) for a given property. Thus, if the same property is defined on different classes the instance belongs to, this property must have the same value in each extent associated to these classes.

Relationship Between Each Part. The relationship between ontology and its instances (content) is defined by the partial function Nomination : C → EXTENT. It associates a definition by intension with a definition by extension of a class.

Classes without extensional definition are said *abstract*. The set of properties used in an extensional definition of a class must be a subset of the properties defined in the intensional definition of a class ($\texttt{propDomain}^{-1}(\texttt{c}) \supseteq \texttt{SchemaProp}(\texttt{nomination(c)})$).

An Example of the Content Part of an OBDB. Figure 2 illustrates the OBDB data model on the content part. The horizontal representation of the extents of the four classes of our toy ontology are presented on figure 2(A). As shown on this example, some of the properties of an ontology class may not be used in class extent. For example, the property $\texttt{content\_encoded}$, described in the ontology as "the encoded content of the post, contained in CDATA areas", is not used in the extent of the class $\texttt{Post}$. This example also demonstrates that properties used in the extent of a class may not be used in the extent of one of its subclasses. This is the case for the properties $\texttt{first\_name}$ and $\texttt{last\_name}$ which are used in the extent of $\texttt{User}$ but not in the extent of $\texttt{Administrator}$.

On figure 2(B), the two main representations proposed for the content of an OBDB, i.e. the vertical and triple schemas are showed. Because of space limitations, the vertical representation corresponding to the extent of the class $\texttt{User}$ is solely represented.

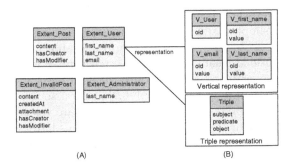

(A)   (B)

Fig. 2. Example of the content of an OBDB

A Generic Approach. Although, we have been using this approach on specific OBDBs with specific query languages (like OntoQL, RQL or SPARQL in this paper), this approach is generic and can be applied to other OBDBs thanks to the following capabilities.

1. The *metametamodel* offers the capability to add other attributes encoding specific information of a given ontology model.
2. The content or extent is associated to a class whatever is the logical model used to represent it (vertical, table-like, etc). A view can be associated to represent the extent of a class in order to hide the specific logical model.
3. From an implementation point of view, the operators of the algebra, we are discussing in this paper, define a generic Application Programming Interface allowing to access any logical model for contents provided that an implementation of these operators is supplied.

From these three features, we notice that our approach is flexible enough to handle different ontology models either from a conceptual modeling point of view (1,2) or from an implementation point of view (2, 3).

2.3   Differences Between OODB and OBDB Data Models

This section describes the differences between the OODB and OBDB data model. Indeed, one can ask why another database model. Below, we describe a list of main identified differences from a structural and logical point of views.

Ontology part. From a functional point of view, the OBDB data model differs from the OODB model in the sense that the former stores not only the logical database model but it also stores the ontology which provides (1) data definition and descriptions independently of any context of use, while the latter stores the logical model of a database for a given application and (2) a formal theory which allows to check some level of consistency and to perform some level of automatic reasoning over the ontology-defined concepts.

Incomplete information. An extent of an ontology class is similar to a typed table associated to a user-defined type in the relational-object data model or to an extent of an object class in the ODL object-oriented data model. However, there is a crucial difference between ontologies and conceptual models of classical databases. Indeed, while a conceptual model *prescribes* which attributes *shall* be represented for each instance of each particular class, an ontology just *describes* which properties *may* be associated with instances of a class. Thus, the extent of an ontology class gathers only the set of properties valued for its instances.

Subsumption relationships. As we have pointed out in the previous section, applicable properties are distinguished from used properties in the OBDB data model. If the applicable properties are inherited through the subsumption relationship as in the object-oriented data model, this is not the case for used properties. Ontology classes may be linked by a subsumption relationship without implicit inheritance of valued properties (partial inheritance).

Universal identification of classes and properties. The identifier of classes and properties are defined in the context of a namespace allowing to universally refer to this concept from any other environment, independently of the particular system where this concept was defined.

3   Query Algebras

On top of the proposed data model, an algebra which can be the basis of an ontology query language whatever is the used ontology model is built. From the listing of the previous differences, it appears that the algebras defined for OODBs are not fully adequate for OBDBs although, as stated previously, the OBDB data model uses extensively OODB features. As a consequence, we have chosen to tune, specialize, extend and reuse the operators of the *ENCORE* algebra in order to get benefits of their properties (e.g., closure, completeness and equivalence rules). Next subsection reviews the main operators of this algebra and subsection 3.2 presents the *OntoAlgebra* algebra we propose.

3.1   Main Operators of the *ENCORE* Query Algebra

Each operator of the *ENCORE* algebra takes a collection of objects whose type is an ADT T and returns a collection of objects whose type is an ADT Q. Thus, the signature of such an operator is $\mathtt{Set}[\mathtt{T}] \to \mathtt{Set}[\mathtt{Q}]$. Following our formalisation, we use the signature $\mathtt{ADT} \times 2^0 \to \mathtt{ADT} \times 2^0$ in order to record the datatype and its set of objects. Let us briefly describe the main operators of the *ENCORE* algebra.

Image. The *Image* operator returns the collection of objects resulting from applying a function to a collection of objects. Its signature is $\mathtt{ADT} \times 2^0 \times \mathtt{Function} \to \mathtt{ADT} \times 2^0$. Function contains all the properties in P and all properties that can be defined by composing properties of P (path expressions). It is easy to extend the domain of $\mathtt{PropDomain}, \mathtt{PropRange}$ and $\mathtt{Val}$ from P to Function. So, this operator is defined by:

$$\mathtt{Image}(\mathtt{T}, \{o_1, \ldots, o_n\}, \mathtt{f}) = (\mathtt{PropRange}(\mathtt{f}), \{\mathtt{Val}(o_1, \mathtt{f}), \ldots, \mathtt{Val}(o_n, \mathtt{f})\}) \; .$$

Project. The *Project* operator extends *Image* allowing the application of more than one function to an object. The result type is a Tuple whose attribute names are taken as parameter. Its signature is $\mathtt{ADT} \times 2^0 \times 2^{\mathtt{String} \times \mathtt{Function}} \to \mathtt{ADT} \times 2^0$ and it is defined by:

$$\mathtt{Project}(\mathtt{T}, O_t, \{(A_1, f_1), \ldots (A_n, f_n)\}) =$$
$$(\mathtt{Tuple}[< (A_1, \mathtt{PropRange}(f_1)), \ldots, (A_n, \mathtt{PropRange}(f_n)) >],$$
$$\{< A_1 : \mathtt{Val}(o, f_1), \ldots, A_n : \mathtt{Val}(o, f_n) > | o \in O_t\}) \; .$$

Select. The *Select* operator creates a collection of objects satisfying a selection predicate. Its signature is $\mathtt{ADT} \times 2^0 \times \mathtt{Predicate} \to \mathtt{ADT} \times 2^0$ and it is defined by:

$$\mathtt{Select}(\mathtt{T}, O_t, \mathtt{pred}) = (\mathtt{T}, \{o | o \in O_t \wedge \mathtt{pred}(o)\}) \; .$$

OJoin. The *OJoin* operator creates relationships between objects of two input collections. Its signature is $\mathtt{ADT} \times 2^0 \times \mathtt{ADT} \times 2^0 \times \mathtt{String} \times \mathtt{String} \times \mathtt{Predicate} \to \mathtt{ADT} \times 2^0$ and it is defined by:

$$\mathtt{OJoin}(\mathtt{T}, O_t, \mathtt{R}, O_r, A_1, A_2, \mathtt{pred}) =$$
$$(\mathtt{Tuple}[< (A_1, \mathtt{T}), (A_2, \mathtt{R}) >], \{< A_1 : t, A_2 : r > | t \in O_t \wedge r \in O_r \wedge \mathtt{pred}(t, r)\}) \; .$$

The definition of this operator is modified when it takes a type Tuple as parameter. Indeed, it becomes necessary to flatten the resulting nested tuples in order to preserve composition. In this case, the flattening operation allows the preservation of the associativity of this operator.

In addition to these main operators, the *ENCORE* algebra includes set operations (*Union*, *Difference*, and *Intersection*) and collection operations (*Flatten*, *Nest* and *UnNest*). All these operators define an algebra allowing to query OODBs. Next, we show how the definitions of these operators could be reused and extended for querying an OBDB.

3.2 *OntoAlgebra*: Adaptation of Encore Query Algebra to the OBDB Data Model

Signatures of the operators defined on the OBDB data model are $(E \cup C) \times 2^{OC \cup I} \to (E \cup C) \times 2^{OC \cup I}$. The main operators of this algebra are *OntoImage*, *OntoProject*, *OntoSelect* and *OntoOJoin*. Next subsections present the semantics of these operators for each part of the OBDB data model.

Ontology Part. Signatures of the operators defined on the ontology part of the OBDB data model are restricted to $E \times 2^{OC} \to E \times 2^{OC}$. Since the data model of this part is similar to the OODB data model, the semantics of the operators of the *ENCORE* algebra is well adapted to the *OntoAlgebra* operators on this part. To illustrate the *OntoAlgebra* operators, we show how several queries are decomposed into calls to the operators of this algebra. These queries are expressed in OntoQL (a), RQL (b) and SPARQL (c). To simplify, the namespaces used in these examples are not explicitly specified.

Example 1. Retrieve the superclasses of the class named `Administrator`.

```
a. SELECT c.#superClasses FROM #class c WHERE c.#name = 'Administrator'
b. SELECT superClassOf($C) FROM $C WHERE $C = 'Administrator'
c. SELECT ?csup WHERE { ex:Administrator rdfs:subClassOf ?csup }
```

$ext^* : E \to 2^{OC}$ denotes the function returning the instances of an entity. Using this notation and the lambda notation, this query is expressed by applying the following *OntoAlgebra* operators:

$$\mathtt{ClassAdministrator} = \mathtt{OntoSelect}(C, ext^*(C), \lambda c \bullet c.name =' Administrator')$$
$$\mathtt{Result} = \mathtt{OntoImage}(\mathtt{ClassAdministrator}, \lambda c \bullet c.superClasses)$$
$$\mathtt{ResultSPARQL} = \mathtt{OntoFlatten}(\mathtt{Result})$$

The *OntoSelect* operator is applied to find the class named `Administrator`. Thus, the type of `ClassAdministrator` is SET[C]. Then, the *OntoImage* operator applies the attribute `superClasses` to this class. The type of `Result` is SET[SET[C]]. Contrariwise to OntoQL and RQL, SPARQL doesn't support collections. Thus, we need to flatten the result using the *OntoFlatten* operator defined by:

$$\mathtt{OntoFlatten}(\mathtt{Set}[T], O_{setT}) = (T, \{r | \exists t \in O_{setT} \wedge r \in t\}) \ .$$

As a consequence, the type of `SPARQLResult` is SET[C].

Example 2. List the properties with their domain.

```
a. SELECT p, c FROM #propery as p, #class as c WHERE p.#scope = c.#oid
b. SELECT @P, $C FROM @P, $C WHERE domain(@P)=$C
c. SELECT ?p, ?c WHERE { ?p rdfs:domain ?c }
```

This query is expressed by applying the *OntoOJoin* operator:

$$\texttt{Result} := \texttt{OntoOJoin}(P, \texttt{ext}(P), C, \texttt{ext}(C), p, c, \lambda p \lambda c \bullet p.\texttt{propDomain} = c)$$

The type of `Result` is $\texttt{SET}[\texttt{Tuple} < (p : P), (c : C) >]$.

Content Part. Signatures of the operators defined on the content part of the OBDB data model are restricted to $C \times 2^I \rightarrow C \times 2^I$. The data model of the content part presents some particularities which impose to redefine the *ENCORE* operators on this part.

OntoImage. Contrariwise to the OODB data model, one of the properties occurring in the function parameter may not be valued in the extensional definition of the class. Thus, we can not use the `Val` function to define the semantics of this operator as it is done in the definition of the *Image* operator. It becomes necessary to extend its domain to the properties defined on the intensional definition of a class but not valued on its extensional definition. This novelty of our algebra requires the introduction of the `UNKNOWN` value. We call `OntoVal` the extension of `Val`. It is defined by:

$$\texttt{OntoVal}(i, p) = \texttt{Val}(i, p), \text{ if } \exists e \in \texttt{TypeOf}(i) \text{ such that } p \in \texttt{SchemaProp}(e)$$
$$\text{else, } \texttt{UNKNOWN} .$$

`UNKNOWN` is a special instance of `ObjectC` like `NULL` is a special value for SQL. Whereas `NULL` may have many different interpretations like value unknown, value inapplicable or value withheld, the only interpretation of `UNKNOWN` is value unknown, i.e., there is a value, but we don't know what it is. To preserve composition, `OntoVal` applied to a property whose value is `UNKNOWN` returns `UNKNOWN`. So, *OntoImage* is defined by:

$$\texttt{OntoImage}(T, \{i_1, \ldots, i_n\}, f) =$$
$$(\texttt{PropRange}(f), \{\texttt{OntoVal}(i_1, f), \ldots, \texttt{OntoVal}(i_n, f)\}) .$$

Example 3. List the first names of users.

```
a. SELECT u.first_name FROM User u
b. SELECT fn FROM User{u}.first_name{fn}
c. SELECT ?fn WHERE { ?u rdf:type ex:User .
                      OPTIONAL { ?u ex:first_name ?fn } }
```

This query is expressed by applying the *OntoImage* operator:

$$\texttt{Result} := \texttt{OntoImage}(\texttt{User}, \texttt{ext}^*(\texttt{User}), \lambda u \bullet u.\texttt{first\_name})$$

The type of `Result` is `SET[String]`. Since the property `first_name` is not valued for the class `Administrator`, this expression returns the value `UNKNOWN` for each administrator. In the SPARQL vocabulary, the variable is said *unbound*. This is not the case for the RQL query because this language doesn't allow to express optional patterns. As a result, this query doesn't return a value for administrators.

OntoProject. *Project* is also extended to *OntoProject* using the `OntoVal` operator previously defined :

$$\mathtt{OntoProject}(\mathtt{T}, \mathtt{I_t}, \{(\mathtt{A_1}, \mathtt{f_1}), \ldots (\mathtt{A_n}, \mathtt{f_n})\}) =$$
$$(\mathtt{Tuple}[< (\mathtt{A_1}, \mathtt{PropRange}(\mathtt{f_1})), \ldots, (\mathtt{A_n}, \mathtt{PropRange}(\mathtt{f_n})) >],$$
$$\{< \mathtt{A_1} : \mathtt{OntoVal}(\mathtt{i}, \mathtt{f_1}), \ldots, \mathtt{A_n} : \mathtt{OntoVal}(\mathtt{i}, \mathtt{f_n}) > | \mathtt{i} \in \mathtt{I_t}\}) \ .$$

OntoSelect. The semantics of *OntoSelect* is similar to the one of *Select*:

$$\mathtt{OntoSelect}(\mathtt{T}, \mathtt{I_t}, \mathtt{pred}) = (\mathtt{T}, \{\mathtt{i} | \mathtt{i} \in \mathtt{I_t} \wedge \mathtt{pred}(\mathtt{i})\}) \ .$$

If the predicate taken as parameter of *OntoSelect* contains function applications, then `OntoVal` must be used. So, operations involving `UNKNOWN`, that may appear in a predicate, must be extended to handle this value. Because `UNKNOWN` is often interpreted by `NULL`, the same semantics as `NULL` is given to `UNKNOWN`. Thus, arithmetic operators like $\times$ or $+$ applied to `UNKNOWN` return `UNKNOWN`, and comparing `UNKNOWN` to any instance using a comparison operator like $=$ or $>$ returns `UNKNOWN`.

Example 4. List the posts created by an user whose email end with '@ensma.fr'.

```
a. SELECT p FROM Post p WHERE p.hasCreator.email LIKE '%@ensma.fr'
b. SELECT p FROM Post{p}.hasCreator.email{e} WHERE e LIKE '%@ensma.fr'
c. SELECT ?p WHERE { ?p rdf:type ex:Post . ?p ex:has_creator ?c .
                     ?c ex:email ?e . FILTER (?e LIKE '%@ensma.fr') }
```

This query is expressed by applying the *OntoSelect* operator:

$$\mathtt{Result} := \mathtt{OntoSelect}(\mathtt{Post}, \mathtt{ext}^*(\mathtt{Post}),$$
$$\lambda \mathtt{p}_\bullet \mathtt{p.hasCreator.email} \ \mathtt{LIKE} \ '\%@ensma.fr')$$

The type of `Result` is `SET[Post]`. For each post created by an administrator, the value `UNKNOWN` is returned for the property `email`. As a consequence, only post created by users who are not administrators may be returned as result.

OntoOJoin. The semantics of *OntoOJoin* is similar to the one of *OJoin*:

$$\mathtt{OntoOJoin}(\mathtt{T}, \mathtt{I_t}, \mathtt{R}, \mathtt{I_r}, \mathtt{A_1}, \mathtt{A_2}, \mathtt{pred}) =$$
$$(\mathtt{Tuple}[< (\mathtt{A_1}, \mathtt{T}), (\mathtt{A_2}, \mathtt{R}) >], \{< \mathtt{A_1} : \mathtt{t}, \mathtt{A_2} : \mathtt{r} > | \mathtt{t} \in \mathtt{I_t} \wedge \mathtt{r} \in \mathtt{I_r} \wedge \mathtt{pred}(\mathtt{t}, \mathtt{r})\}) \ .$$

The predicate taken in parameter of *OntoOJoin* is treated as for *OntoSelect*.

Operator *. In the *ENCORE* algebra, a class C refers to instances of C and instances of all subclasses of C. The *ENCORE* algebra doesn't supply a built-in operator to write a non polymorphic query. Thus, we define the explicit polymorphic operator, named *, to distinguish between local queries on instances of a class C and instances of all the classes denoted C^* subsumed by C. We denote $ext : C \rightarrow 2^I$ the function which returns the instances of a class and we overload the function ext^* for the signature $C \rightarrow 2^I$ to denote the deep extent of a class. If c is a class and $c_1, \ldots c_n$ are the direct sub-classes of c, ext and ext^* are derived recursively[2] in the following way on the OBDB data model:

$$ext(c) = \text{TypeOf}^{-1}(\text{Nomination}(c)) .$$
$$ext^*(c) = ext(c) \cup ext^*(c_1) \cup \ldots \cup ext^*(c_n) .$$

The ext and ext^* make it possible to define the * operator as $* : C \rightarrow C \times 2^I$ where $^*(T) = (T, ext^*(T))$.

Support of the multi-instanciation paradigm. In the *ENCORE* algebra, the *Image* operator can only be applied with a property defined on the class taken in parameter. Because of the multi-instanciation paradigm, an instance of a class can provide a value for a property not defined on this class but on an another class the instance belongs to. As a consequence, this paradigm raises the need to extend the *OntoImage* operator. We denote *OntoImage'* the definition of *OntoImage* when this operator is applied to a property not defined on the class taken in parameter. When *OntoImage'* is applied to a class C_1 and a property p not defined on C_1 but defined on the class C_2, this operator is defined by:

$$\text{OntoImage}'(C_1, I_{C_1}, p) =$$
$$\text{OntoImage}(\text{OntoOJoin}(C_1, I_{C_1}, C_2, ext^*(C_2), \lambda i_{c_1} \lambda i_{c_2 \bullet} i_{c_1} = i_{c_2}), i_{c_2}.p) .$$

The other operators of *OntoAlgebra* are extended in the same way to handle the multi-instanciation paradigm.

Example 5. List the file size of the posts.

```
a. Not supported
b. SELECT f FROM Post{p}.file_size{f}
c. SELECT ?p WHERE { ?p rdf:type ex:Post . ?p ex:file_size ?c }
```

Let's suppose that the property file_size is defined on a class ExtResource. This query is expressed by applying the *OntoImage'* operator:

$$\text{Result} := \text{OntoImage}'(\text{Post}, ext^*(\text{Post}), \text{file\_size})$$
$$:= \text{OntoImage}(\text{OntoOJoin}(\text{Post}, ext^*(\text{Post}),$$
$$\text{ExtResource}, ext^*(\text{ExtResource}), p, e, \lambda p \, \lambda e_{\bullet} p = e), e.\text{file\_size})$$

---

[2] To simplify notation, we extend all functions f by $f(\emptyset) = \emptyset$.

Ontology and Content Parts. *OntoAlgebra* provides the capability to query simultaneously ontology and content parts.

Example 6. For each ontology class, whose name contains the word Post, *list the properties applicable (defined and inherited) on this class and the values of the instances of this class for these properties.*

```
a. SELECT p.#name, i.p
     FROM #class as C, C as i, unnest(C.#properties) as p
     WHERE C.#name like '%Post%'
b. SELECT @P, V FROM {i;$C}@P{V} WHERE $C like '%Post%'
c. applicable properties can not be expressed
```

This query is expressed by applying the following *OntoAlgebra* operators:

$$\text{ClassPost} := \text{OntoSelect}(C, \text{ext}(C), \lambda c_\bullet c.\text{name like } '\%Post\%')$$

$$\text{ClassInst} := \text{OntoOJoin}(\text{ClassPost}, *(\text{ObjectC}), C, i, \lambda C \, \lambda i_\bullet i \in \text{ext}^*(C))$$

$$\text{ClassPropInst} := \text{OntoProject}(\text{ClassInst}, \lambda ci_\bullet$$
$$< (C, ci.C), (i, ci.i), (p, ci.C.\text{properties}) >$$

$$\text{UClassPropInst} := \text{OntoUnNest}(\text{ClassPropInst}, p)$$

$$\text{Result} := \text{OntoProject}(\text{UClassPropInst}, \lambda cip_\bullet$$
$$< (n, cip.p.\text{name}), (v, cip.i.(cip.p)) >$$

The first selection finds the classes whose names contain the word Post. The result is ClassPost of type Set[C]. The *OntoOJoin* operator is then used to join the classes of ClassPost and all the instances, i.e the polymorphic instances of the root class (*(ObjectC)). The result of this operation is ClassInst of type SET[Tuple $< (c : C), (i : \text{ObjectC}) >$]. The function properties is then applied to the classes contained in ClassInst using the *OntoProject* operator. The function properties returns the applicable properties of a class as a set. As a consequence, the result of this step is ClassProp of type SET[Tuple $< (c : C), (i : \text{ObjectC}), (p, \text{SET}[P]) >$]. The next step consists in unnesting the set of properties contained in each of the tuple of ClassProp. This is achieved using the OntoUnNest operator defined by:

$$\text{OntoUnNest}(\text{Tuple}[< (A_1, T_1), \ldots, (A_i, \text{SET}[T_i]), \ldots, (A_n, T_n) >], I_t, A_i) =$$
$$(\text{Tuple}[< (A_1, T_1), \ldots, (A_i, T_i), \ldots, (A_n, T_n) >],$$
$$\{< A_1 : s.A_1, \ldots, A_i : t, \ldots, A_n : s.A_n > | s \in I_t \wedge t \in s.A_i\}) \; .$$

The result of this operation is UClassProp of type SET[Tuple $< (c : C), (i : \text{ObjectC}), (p, P) >$]. Finally the result is obtained by applying the *OntoProject* operator to retrieve, for each tuple, the name of the property referenced by the attribute name p and to apply this property to the instances referenced by the attribute name i. The final result is of type SET[Tuple $< (\text{name} : \text{String}), (\text{value} : \text{ObjectC}) >$].

3.3 Differences Between OODB and OBDB Languages

In this section, we describe a list of the identified main differences between the query languages issued from the *ENCORE* and *OntoAlgebra* algebras.

Two levels language. An OBDB query language offers the capability to query ontology, data and both ontology and data. Each of these querying levels corresponds to a specific need. Querying ontology may be useful to discover concepts of an ontology. Querying data from the concepts of an ontology allows to query the data independently of the structure of the data (semantic querying). Querying both ontology and data is useful to extract a subset of an ontology together with its instances (from the ontology to the content part of an OBDB) or to discover how a given instance of an ontology class may be described by some other classes (from the content to the ontology part of an OBDB). In a number of OODB implementations, *metadata* are recorded in the system catalog. Using the object query language provided, one can query these *metadata*. However, object-oriented algebras define how to query the data of an OODB only. As a consequence, it is difficult to combine querying both *metadata* and data.

Unknown value. OBDB query languages may return a special value for properties defined on the intensional definition of a class but not used in its extensional definition (see section 3.2). Contrary to the NULL value, introduces in classical algebra, there is only one interpretation of this value: a value exists, but we don't know what it is. In *OntoAlgebra*, we have chosen to give the same semantics to this value as the one of the NULL value in order to remain compatible with classical database languages. As shown in [15], this is not the case of the SPARQL semantics which has introduced some mismatches with the processing of the NULL value in classical databases.

Path expression. OBDB query languages extend the capability of path expressions introduced by OODB query languages. Indeed, a path expression in an OBDB query language can be composed with a property not defined on the previous element of the path. This capability is introduced to handle the multi-instanciation paradigm. Moreover, a path expression can be composed with properties determined at runtime (generalized path expression). This capability is introduced to allow querying both ontologies and data.

Parametric language. OBDB query language may use environment variables such as the used natural language or the namespace of the ontology queried to restrict the search space in the OBDB and to allow users to define queries in different natural languages.

4 Related Work

To our knowledge, the SOQA Ontology Meta Model [16] is the only other proposition of an independent data model of a given ontology model. It incorporates constructors not supported by some ontology languages (e.g., methods or relationships) but it can not be extended. Our approach is dual, we have decided to incorporate only the shared constructors but to allow the extension of this core model thanks to the *metametamodel* level. This approach is much more flexible since it allows to represent all the constructors of a given ontology model. This capability is not offered by the SOQA Ontology Meta Model. For example, restrictions of OWL or documents of PLIB are not included in this model. As a consequence managing ontologies which use these constructors with SOQA Ontology Meta Model based tools is not possible without loss of data.

Concerning the query algebra, formal semantics defined for ontology query languages [15,17] or more generally for an ontology model [18,19] can be regarded as related work. Close to our work is the relational algebra for SPARQL presented in [15]. It presents a correspondence between SPARQL and relational algebra queries. Based on this analysis, author points out several differences between SPARQL and SQL semantics. With this work, we share the idea of defining ontology query languages starting from the well known algebra of classical database languages. However, we do not address the same kind of data. While the operators defined in its algebra regard RDF as triple data without schema or ontology information, our algebra proposes operators to exploit the ontology level (e.g. computation of the transitive closure of the subsumption relationship ...). Thus, while its algebra has the expressive power of the relational algebra, our algebra has the expressive power of object-oriented algebra to query the ontology, the data and both the ontology and the data of an OBDB.

5 Conclusion and Future Work

In this paper, we have formally defined a data model for OBDBs independent of the used ontology model and representation schema. Using this data model, we have discussed and shown the differences existing between classical databases and OBDBs. These formalization and comparison are a sound basis for engineers willing to implement ontology databases using classical databases.

As a second step, we have proposed a formal algebra of operators for querying OBDBs. We have built this algebra by extending the *ENCORE* algebra proposed for OODB. As a consequence, our algebra clarifies the differences between object-oriented query languages (e.g., SQL2003, OQL ...) and ontology query languages (e.g., RQL, OntoQL ...) in terms of semantics and expressive power.

For the future, we plan to use the proposed algebra to study optimization of OBDBs. By reusing the ENCORE algebra, we hope to benefit from most of the equivalence rules defined in this algebra. The main challenge is to find new equivalence rules deriving from the specific features of the OBDB data model.

References

1. ISO13584-42: Industrial automation systems and integration. Parts Library Part 42. Description methodology: Methodology for structuring parts families. Technical report, International Standards Organization (1998)
2. Brickley, D., Guha, R.V.: RDF Vocabulary Description Language 1.0: RDF Schema. World Wide Web Consortium (2004)

3. Dean, M., Schreiber, G.: OWL Web Ontology Language Reference. World Wide Web Consortium (2004)

4. Harris, S., Gibbins, N.: 3store: Efficient bulk rdf storage. In: Proceedings of the First International Workshop on Practical and Scalable Semantic Systems (PPP'03) (2003)

5. Broekstra, J., Kampman, A., van Harmelen, F.: Sesame: A generic architecture for storing and querying rdf and rdf schema. In: Horrocks, I., Hendler, J. (eds.) ISWC 2002. LNCS, vol. 2342, pp. 54–68. Springer, Heidelberg (2002)

6. Dehainsala, H., Pierra, G., Bellatreche, L.: Ontodb: An ontology-based database for data intensive applications. In: Proceedings of the 12th International Conference on Database Systems for Advanced Applications (DASFAA'07), pp. 497–508 (2007)

7. Park, M.J., Lee, J.H., Lee, C.H., Lin, J., Serres, O., Chung, C.W.: An efficient and scalable management of ontology. In: DASFAA 2007, LNCS, vol. 4443, Springer, Heidelberg (2007)

8. Prud'hommeaux, E., Seaborne, A.: SPARQL Query Language for RDF. World Wide Web Consortium (2006)

9. Karvounarakis, G., Alexaki, S., Christophides, V., Plexousakis, D., Scholl, M.: Rql: a declarative query language for rdf. In: Proceedings of the Eleventh International World Wide Web Conference, pp. 592–603 (2002)

10. Jean, S., Aït-Ameur, Y., Pierra, G.: Querying ontology based database using ontoql (an ontology query language). In: Proceedings of On the Move to Meaningful Internet Systems 2006: CoopIS, DOA, GADA, and ODBASE, OTM Confederated International Conferences (ODBASE'06), pp. 704–721 (2006)

11. Shaw, G.M., Zdonik, S.B.: A query algebra for object-oriented databases. In: Proceedings of the Sixth International Conference on Data Engineering (ICDE'90), pp. 154–162. IEEE Computer Society Press, Los Alamitos (1990)

12. Object Management Group: Meta Object Facility (MOF), formal/02-04-03 (2002)

13. Object Management Group: Ontology Definition Metamodel (ODM) Final Adopted Specification ptc/06-10-11 (2006)

14. Theoharis, Y., Christophides, V., Karvounarakis, G.: Benchmarking database representations of rdf/s stores. In: Gil, Y., Motta, E., Benjamins, V.R., Musen, M.A. (eds.) ISWC 2005. LNCS, vol. 3729, pp. 685–701. Springer, Heidelberg (2005)

15. Cyganiak, R.: A relational algebra for sparql. Technical Report HPL-2005-170, HP-Labs (2005)

16. Ziegler, P., Sturm, C., Dittrich, K.R.: Unified querying of ontology languages with the sirup ontology query api. In: Proceedings of Business, Technologie und Web (BTW'05), pp. 325–344 (2005)

17. Pérez, J., Arenas, M., Gutierrez, C.: Semantics and complexity of sparql. In: Cruz, I., Decker, S., Allemang, D., Preist, C., Schwabe, D., Mika, P., Uschold, M., Aroyo, L. (eds.) ISWC 2006. LNCS, vol. 4273, Springer, Heidelberg (2006)

18. Frasincar, F., Houben, G.J., Vdovjak, R., Barna, P.: Ral: An algebra for querying rdf. World Wide Web 7, 83–109 (2004)

19. Gutierrez, C., Hurtado, C., Mendelzon, A.O.: Foundations of semantic web databases. In: Proceedings of the twenty-third ACM SIGMOD-SIGACT-SIGART symposium on Principles of database systems (PODS '04), pp. 95–106. ACM Press, New York (2004)

The MM-Tree: A Memory-Based Metric Tree Without Overlap Between Nodes

Ives R.V. Pola, Caetano Traina Jr., and Agma J.M. Traina

Computer Science Department - ICMC
University of Sao Paulo at Sao Carlos – Brazil
{ives,caetano,agma}@icmc.usp.br

Abstract. Advanced database systems offer similarity queries on complex data. Searching by similarity on complex data is accelerated through the use of metric access methods (MAM). These access methods organize data in order to reduce the number of comparison between elements when answering queries. MAM can be categorized in two types: disk-based and memory-based. The disk-based structures limit the partitioning of space forcing nodes to have multiple elements according to disk page sizes. However, memory-based trees allows more flexibility, producing trees faster to build and to perform queries. Although recent developments target disk-based methods on tree structures, several applications benefits from a faster way to build indexes on main memory. This paper presents a memory-based metric tree, the MM-tree, which successively partitions the space into non-overlapping regions. We present experiments comparing MM-tree with existing high performance MAM, including the disk-based Slim-tree. The experiments reveal that MM-tree requires up to one fifth of the number of distance calculations to be constructed when compared with Slim-tree, performs range queries requiring 64% less distance calculations and KNN queries requiring 74% less distance calculations.

1 Introduction

Existing DBMS were developed to handle data in numeric and short text domains, not being able to handle complex data efficiently, because its internal structures require the total ordering property on the data domains. in this way, the comparison operators available in traditional Database Management Systems (DBMS) (i.e., the relational $<, \leq, >$ or $\geq$ and identity $=$ or $\neq$ operators) are not adequate to compare complex data such as images.

As the complexity of the data stored in modern DBMS grows, new access methods (AM), tailored to deal with the properties of the new data types, need to be designed. The majority of complex data domains do not possesses the total ordering property, precluding the use of the traditional AM to index data from these domains. Fortunately, these data domains allow the definition of similarity relations among pairs of elements. Therefore, a new class of AM was developed, the Metric Access Methods (MAM), which is well-suited to answer similarity queries over complex data domains.

Y. Ioannidis, B. Novikov, and B. Rachev (Eds.): ADBIS 2007, LNCS 4690, pp. 157–171, 2007.
© Springer-Verlag Berlin Heidelberg 2007

A similarity query returns elements $\{s_i \in S, S \subset \mathbb{S}\}$ that meet a given similarity criterion, expressed through a reference element $s_q \in \mathbb{S}$. For example, for image databases one may ask for images that are similar to a given one, according to a specific criterion. There are two main types of similarity queries: the range and the k-nearest neighbor queries.

- **Range query - Rq:** given a maximum query distance r_q, the query $Rq(s_q, r_q)$ retrieves every element $s_i \in S$ such that $d(s_i, s_q) \leq r_q$. An example is: "Select the images that are similar to the image P by up to five similarity units", and it is represented as $Rq(P, 5)$;
- **k-Nearest Neighbor query - $kNNq$:** given a quantity $k \geq 1$, the KNN query $kNNq(s_q, k)$ retrieves the k elements in S that are the nearest from the query center s_q. An example is: "Select the 3 images most similar to the image P", which is represented as $kNNq(P, 3)$.

Besides those types of queries, the point query, which probes if an element is stored in the dataset, is another useful query over sets of complex objects. Several access methods were proposed to speed up the similarity queries. Some spatial access methods (SAM) were adapted to index multi dimensional data, but they become useless when the number of dimensions increases because of the "dimensionality curse". Metric access methods (MAM) were designed to index complex data, including high-dimensional and non-dimensional datasets, indexing the features extracted from data into a metric space. A MAM is associated with a metric that measures the similarity between pairs of elements as the distance between them. The metric is then used to build the tree, causing it to be referenced also as "distance-based trees".

The similarity relationships are usually calculated by a distance function. The data domain and the distance function defines a metric domain, or metric space. To guarantee repeatability when ranking the elements that answer a similarity query, the metric must hold some properties. Formally, a metric space is a pair $< \mathbb{S}, d() >$, where $\mathbb{S}$ is the domain of data elements and $d : \mathbb{S} \times \mathbb{S} \to \mathbb{R}^+$ is metric that given $s_1, s_2, s_3 \in \mathbb{S}$, it holds the following properties: Symmetry ($d(s_1, s_2) = d(s_2, s_1)$); Non-negativity ($0 < d(s_1, s_2) < \infty$ if $s_1 \neq s_2$ and $d(s_1, s_1) = 0$) and Triangular inequality ($d(s_1, s_2) \leq d(s_1, s_3) + d(s_3, s_2)$).

Complex elements usually require extracting predefined features from them that are used in place of the elements to define the distance function. For example, images are preprocessed by specific feature extraction algorithms to retrieve their color and/or texture histograms, polygonal contours of the pictured objects, etc., which are used to define the corresponding distance functions.

Several applications require the DBMS to quickly build indexes on data that fits on main memory. With the increase of the main memory capacity and its lowering costs, it is becoming useful to index data in main-memory, mainly when the index needs to be often rebuilt. In these situations, the extra requirements of a disk-based MAM design force the nodes to have many elements, excluding good approaches to the partitioning of the space. Moreover, every disk-based MAM worries about reducing the disk accesses, because the disk access itself is much slower then accessing the main memory, affecting the performance of queries.

Once they are not limited on how to partition the space, memory-based MAM are faster to built, do not require disk accesses and answer similarity queries faster. Existing memory-based MAM are built upon the concept that the tree must be balanced, in order to guarantee that it will not degenerate into a bad structure. However, the processing required to assure the balancing is always overwelmingly high, turning the construction of a memory-based tree too high for practical purposes. Moreover, these MAM only warranty the balancing when the tree is created at once, without new insertions so they are static structures.

This paper presents the MM-tree (**M**etric-**M**emory tree), an easy to build memory-based MAM whose partitioned regions do no overlap. Moreover, a low cost semi-balancing insertion is presented in order to control the balancing of the tree.

The remainder of this paper is structured as follow. Section 2 presents related works. Section 3 presents the new MM-tree structure and the algorithm to build the trees. Section 4 presents the searching algorithms for the MM-tree to perform both range and k-nn queries. Section 5 presents the experiments and results achieved in our tests. Finally the section 6 concludes this work.

2   Background

One of the underpinnings of a database system is the indexing structures, and the design of efficient access methods has long been pursued in the database area.

Spatial access methods can index muti-dimensional data, considering each object as a point in a multidimensional domain. However, they still suffer from the dimensionality curse. Several methods were proposed in the literature dealing with high-dimensional data, but they are still limited to the space where the data are embedded [1] [2] [3] [4]. An excelent survey on multidimensional methods can be found in the work of Gaede [5].

Unfortunately, many complex data do not have a defined dimensionality, but a distance function can be used to compare pairs of such elements. Thus, they are considered occupying a metric domain. The pioneering work of Burkhard and Keller [6] introduced approaches to index data in metric spaces using discrete distance functions. A survey on MAM can be found in [7].

The GH-tree of Uhlmann [8] recursively partitions the space into two regions, by electing two elements and associating the others to the side of closest elected element, creating an generalized hyperplane separating each subset. The ball decomposition partitioning scheme [8] partitions the data set based on distances from one distinguished element. This element is called a vantage point in the VP-tree [9]. The VP-tree construction process is based on finding the median element of a sorted sample list of the elements, thus recursively creating the tree. Some disk-based MAMs have been proposed, as the MVP-tree [10], where if there are two pivots per node, the first pivot partitions the data set in t regions and the second partitions each new region into other t regions, giving a fanout of t^2 per node. The Geometric Near Access Tree (GNAT) of Brin [11] can be viewed as a refinement of the second technique presented by Burkhard and Keller [6].

The memory-based MAM presented above are considered static, where no further insertions, deletions or updates are handled, without compromising the tree structure. Overcoming this inconvenience, dynamic MAM structures were proposed in the literature, as disk-based MAM [12][13]. Disk-based MAM force the tree structure to hold many elements per node, in order to decrease disk accesses when accessing nodes. These MAM employ the bottom up strategy to construct the tree, leading to a dynamic tree. However, they present overlap on nodes in the same level affecting even point queries. This means that a point query on the tree will access more than one node per level on the tree scan, increasing the number of disk accesses and distance calculations.

The M-tree [13] is the first balanced dynamic metric tree proposed, followed by the Slim-tree [12], which was proposed with different strategies for splitting nodes and choosing-subtree plus the Slim-down algorithm that optimizes (reduces) the overlap between nodes, post-processing the structure. The OMNI concept proposed in [14] optimizes the pruning power on searches by using strategically positioned elements, the foci set. The OMNI concept applied on Slim-tree led to the DF-tree proposed in [15]. These methods store some distances between elements, so the triangular inequality property held by the distance function can be used to prune nodes, in order to reduce subtree accesses.

This paper introduces a new memory–based MAM, the MM-tree, which partitions the space forming disjoint regions. The MM-tree proposed here don't impose a rigid height-balancing, being computacionally easy to build. Nevertheless, it has a balancing control, that prevents it to degenerate. Moreover, it allows both range and k-NN queries, and the point query can be answered without node overlapping.

3   The MM-Tree

The general idea of the structure of MM-tree is to select two elements $s_1, s_2 \in S$ as pivots to partition the space into four disjoint regions. Figure 1 presents an example considering eight elements, where elements 'a' and 'b' are pivots.

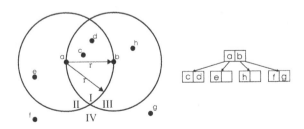

Fig. 1. An example of a MM-tree indexing 8 elements

Table 1. Regions of space where element s_i can be assigned

$d(s_i, s_1) \; \theta \; r$	$d(s_i, s_2) \; \theta \; r$	Region
$<$	$<$	I
$<$	$\geq$	II
$\geq$	$<$	III
$\geq$	$\geq$	IV

Each element s_i is assigned to one partition, so there is no node overlap for point queries. Partitioning of regions is based on distances from elements to the pivots and the distance between pivots r as indicated in Table 1. The same procedure is recursively applied to each sub-region, creating the tree. Note that only two distance calculations per tree level are required to determine the appropriate region.

Every node of an MM-tree has the same structure, which is defined as follows:

$$node[s_1, s_2, d(s_1, s_2), Ptr_1, Ptr_2, Ptr_3, Ptr_4]$$

where s_1 and s_2 are the node pivots, $d(s_1, s_2)$ is the distance between the pivots and Ptr_1 to Ptr_4 are the pointers to the four subtrees storing the elements in regions 1 to 4, respectively.

3.1 Building the MM-Tree

The MM-tree is designed as a main-memory dynamic tree, so new elements can be inserted in the MM-tree at any time. To insert a new element s_n, the algorithm traverses the tree performing a point query, searching for an appropriated node to hold the element. At each node, s_n is compared to pivots (s_1, s_2) and the correct region is determined according to Table 1. When the element s_n is not indexed in the tree, that is, if a leaf node reached does not hold s_n, then it is inserted in that leaf node. Note that there is no overlap between regions at any level of the tree, so the choice of the next subtree in each step of the point query algorithm is always directly achieved.

Figure 1 graphically exemplifies the structure of the first level of a MM-tree indexing the eight elements {a,b,c,d,e,f,g,h}, taking (a,b) as pivots. Note that the $distance(a, b) = r$ defines the ball radius of each pivot selected, forming the disjoint regions I, II, III and IV. The MM-tree is built in a top down scheme, where each node contains up to two elements, partitioning the subspace according the same rule.

Starting from the root node, if the visited node has one element, s_n is stored and the algorithm terminates. If not, the distances from s_n to the pivots s_1 and s_2 are calculated and the proper region is determined comparing these distances with the distance between the pivots $d(s_1, s_2)$, called r for short. Such comparisons are specified in table 1. This process is recursively applied to all levels of the tree.

The Algorithm 1 implements the MM-tree insertion method. Note that only two distance calculations are needed per node to decide which of the four sub-trees the element will belong to, giving the MM-tree a fanout of 4 with only 2 distance calculations per node, in contrast to the disk-based MAM where each subtree decision requires one distance calculation per element in node, so it would demand 4 distance calculations for the same fanout.

Algorithm 1. Insert (s_n, root N)

Let N be the root, s_1 and s_2 the pivots in N and r the $distance(s_1, s_2)$.
while N is full **do**
    $d_1 = distance(s_1, s_n)$
    $d_2 = distance(s_2, s_n)$
    if $d_1 < r$ **then**
        if $d_2 < r$ **then**
            N = region I
        else
            N = region II
    else
        if $d_2 < r$ **then**
            N = region III
        else
            N = region IV
Insert s_n on node N

Balancing Control. The performance of the MM-tree can be affected if many elements are concentrated on the same node at a given level. This section presents a technique to improve the basic insertion algorithm, the semi-balancing algorithm, which controls the unbalancing on leaf nodes.

To give an intuition of the proposed technique, a graphical example is shown in figure 2. In Figure 2(a), the elements a to h were inserted using a and b as pivots, leading to an unbalanced tree. Note that a two-level sub-tree holds up to 10 elements, so the creation of the third level of the tree in Figure 2(a) should be avoided. While finding another pair of pivots among up to nine elements is not expensive, our proposed technique aims to choose a new pair of pivots which the new subtree can acomodate the existing and the newly inserted object in just two levels. Figure 2(b) show the same elements a to h elements indexed now by pivots (e,d).

The semi-balancing algorithm (Algorithm 2) aims at replacing the pivots in the parent node when the leaf node that is target of insertion cannot hold the new element. This replacement may rearrange elements in such way that the region that was full before can hold the new element (lines 1 to 6). If none of the pivots combination frees space for the new element, the subtree gains a level (lines 7 and 8).

This procedure is applied only on leaf nodes in order to avoid the need of rebuilding subtrees when changing pivots in the index level. For statistical reasons it is only performed when the 2-level subtree has at most 8 elements stored, because the probability of getting a combination of pivots that rearrange 10 or 9 elements in a 2-level is low.

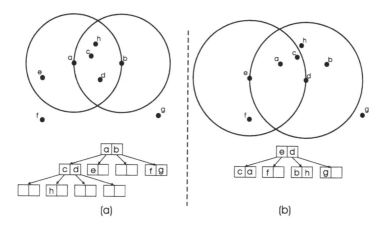

Fig. 2. An example of the semi-balancing algorithm

Algorithm 2. Semi-balancing (N, s_n)

INPUT: pointer to subtree node N, new element s_n.
OUTPUT: pointer to a new subtree.

1: Let O be the list of the children elements of N adding s_n and the pivots of N.
2: Let NN be a new node.
3: **for** every $s_i \in O$ **do**
4:    **for** every $s_j \in O$, $s_j \neq s_i$ **do**
5:       NN.clear()
6:       NN.addPivots(s_i, s_j)
7:       Let El be the list of elements in $(O - s_i - s_j)$
8:       NN.insertElements(El)
9:       **if** elements El were distributed in the same level under N **then**
10:          Return new root NN.
11: N.insertElement(s_n).
12: Return node N.

---

4   Similarity Queries on MM-Tree

Besides answering point queries, the MM-tree must support the two most com-mom similarity queries: range and nearest neighbors. The basic idea here is to define an active radius where the answer must be found, and to visit each region that overlaps with the active radius. The active radius of range queries is the query radius, and the active radius of k-NN queries is a dynamic value that starts at the maximum distance possible, and is reduced as the algorithm proceeds and finds successively nearer elements.

Algorithm 3 corresponds to the range query in a MM-tree. It receives a query radius r_q and the center query element s_q. At each node it verifies if there are interceptions of the region centered at s_q with radius r_q with the four regions

Table 2. Conditions to occur intersection between a node and a query region

Region I	$(d(s_q, s_2) < r_q + r)$	$\wedge$	$(d(s_q, s_1) < r_q + r)$
Region II	$(d(s_q, s_2) + r_q \geq r)$	$\wedge$	$(d(s_q, s_1) < r_q + r)$
Region III	$(d(s_q, s_2) < r_q + r)$	$\wedge$	$(d(s_q, s_1) + r_q \geq r)$
Region IV	$(d(s_q, s_2) + r_q \geq r)$	$\wedge$	$(d(s_q, s_1) + r_q \geq r)$

defined in the node (lines 6 and 7), and it visits the covered regions (line 8) recursively, based on the distance from s_q to the pivots (s_1, s_2). At each node visited, if the distance from s_q to any pivots is less than r_q then that pivot is added to the result (lines 4 and 5). At each node an intersection occurs if any of the conditions presented in Table 2 holds.

Algorithm 3. Range Query $(s_q, r_q, root)$

INPUT: query center s_q, query radius r_q, pointer to subtree $root$.
OUTPUT: answer list $List$.

1: **if** root == NULL **then**
2:     return
3: Calculate the distances $d_1 = d(s_1, s_q)$ and $d_2 = d(s_2, s_q)$ in current node $root$
4: **if** $d_1 \leq r_q$ **then** $List.add(s_1)$
5: **if** $d_2 \leq r_q$ **then** $List.add(s_2)$
6: **for** every region R_j in (I, II, III, IV) of root **do**
7:     **if** query radius intersects region R_j **then**
8:         RangeQuery(s_q, r_q, R_j)

Algorithm 4 corresponds to the k-nearest neighbor query algorithm over a MM-tree. It keeps the answer list of the k elements nearest to s_q, ordered according to their distance from s_q. When the algorithm starts, the list is empty and the active radius r_q is set to ∞. The active radius is reduced every time the list is updated, becoming the distance from s_q to the kth farthest element in the list. The answer list is created with the first k elements found navigating the MM-tree. Thereafter, the list is updated whenever an element closer to s_q is found. A region covering occurs if any of the conditions presented in Table 2 holds, now r_q being the dynamic active radius. Lines 14 to 16 test the intersection conditions and visit the appropriated nodes. Note that if s_q lies on region P, that region has to be visited and no condition test needs to be performed.

Guided Technique. The active radius over the k-nearest neighbor algorithm reduces as elements closer to s_q are found during the process. As this radius becomes smaller, more subtrees will be pruned, reducing the distance calculations. Therefore, using a different search order when visiting the covered regions can reduce the active radius more quickly and prune subtrees more effectively.

Algorithm 4. Nearest Query (s_q, k, $root$)

INPUT: query center s_q, number of nearest neighbors K, pointer to subtree $root$.
OUTPUT: answer list $List$.

```
 1: if root == NULL then
 2:     return
 3: Calculate the distances d₁ = d(s₁, s_q) and d₂ = d(s₂, s_q) in current node root
 4: if List.size() < k then
 5:     r_q = ∞
 6: else
 7:     r_q = List[k].distance
 8: for i = 1..2 do
 9:     if d_i < r_q then
10:         Add s_i to List, keeping it sorted
11: Let P be the region where s_q belong.
12: NearestQuery(s_q, r_q, P)
13: if node has two pivots then
14:     for every region R_j in (I, II, III, IV) of root, R_j ≠ P do
15:         if query intercepts R_j then
16:             NearestQuery(s_q, k, R_j)
```

The guided nearest neighbor algorithm chooses a better sequence order when visiting nodes in the MM-tree, in order to reduce the query radius quickly. The algorithm first visits the region where the query element s_q lies, then visits the next regions in a sequence according to the distance of s_q to other regions. The sequence is then formed by first visiting the nearest regions to the query element.

Although such sequence depends on data distribution and insertion order, we assume that the datasets have clusters and are inserted randomly (the most commom situation regarding real world datasets). The sequence presented in Table 3 leads to a better prunning ability experimentally confirmed on datasets tested.

Table 3. Determining the next region to visit in the guided nearest neighbor query

region s_q lies	condition C	visit order	
		C is true	C is false
I	$d_1 \leq d_2$	I→II→(III,IV)	I→III→(II,IV)
II	$d_2 - d \leq d - d_1$	II→I→IV→III	II→IV→IV→II
III	$d_1 - d \leq d - d_2$	III→I→IV→II	III→IV→I→II
IV	$d_1 \leq d_2$	IV→II→I→III	IV→III→I→II

5   Experiments

This section shows experimental results comparing the MM-tree with the Slim-tree and VP-tree. Although the Slim-tree belongs to the disk-based MAM category, we developed a version to run only in main memory (although preserving the use of fixed size blocks of memory like disk records). Therefore, in our experiments the time spent to query and to construct the Slim-tree were consistent when comparing it with other memory-based MAM. The memory-based Slim-tree was achieved by keeping nodes in main memory and the disk pointers become memory pointers.

All compared MAM were implemented in C++ language at the arboretum framework [1] and the tests were done in an AMD Athlon XP 3200+ with 1Gb of RAM. Three datasets were used in our experiments, being two real and one synthetic. They are explained in the Table 5. The euclidean distance function was used for the Points, Cities and Color Histograms datasets.

Table 4. Datasets used on tests

Dataset name	Description
Points	10 thousand points randomly created in a 6 dimension space.
Cities	5507 latitude and longitude coordinates from cities in Brazil.
Color histograms	10 thousand color histograms in a 32 dimension space extracted from an image database.

The Slim-tree was built using the min-occupation and MST (minimum spanning tree) policies, considered the best configuration by its authors. The page size of Slim-tree is different for each dataset, in order to keep 50 elements per node in average. The binary VP-tree was built using the sampling algorithm at the rate of 10% in order to select the vantage points. This section compares both the construction process and the similarity queries performance.

Construction

The speed of building a main-memory tree is one of the most important aspects in a database system. An experiment measuring the number of distance calculations and the time in mileseconds needed to build every MAM over all datasets were done, and the results are showed in Table 5. From the table, we can see that VP-tree takes the longest time to be built among the others, performing also the largest number of distance calculations over all datasets. This behavior occurs because during the construction process, the best vantage point is chosen by finding the best spread of the median element in each sample of elements (a quadratic order algorithm on the number of elements involved in the operations), sorted by the distances to each vantage point candidate, causing too many comparisons between elements. From the table can be noted that MM-tree has the lower time and the lower number of distance calculations to be built also, overcomming the disk-based representant, the Slim-tree, even indexed in memory.

A scalability experiment measures the behavior of the tree when indexing datasets of different size. This is important to identify patterns on the performance

---

[1] http://gbdi.icmc.usp.br/arboretum

Table 5. Construction statistics of tested MAM over the datasets

MAM	Points		Cities		Color Histograms	
	Dist	Time (ms)	Dist	Time (ms)	Dist	Time (ms)
MM-tree	161143	190	89783	126	167705	737
Slim-tree	633374	297	451830	156	665453	1234
VP-tree	2381532	1625	1203897	640	2346300	6188

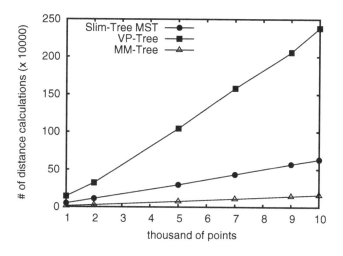

Fig. 3. Scalability of distance calculations on referred MAM using Points dataset

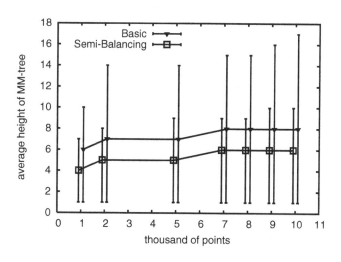

Fig. 4. MM-tree average heights using Points dataset

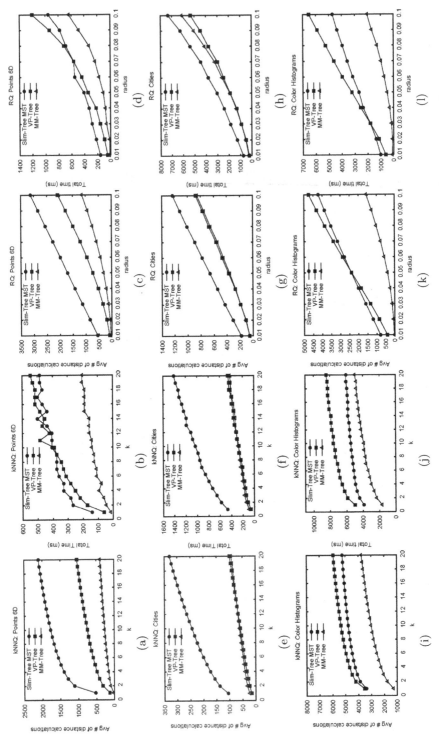

Fig. 5. Performance of range queries and nearest neighbor queries over Points, City and Color Histograms datasets

of queries on different sizes of datasets. The scalability test results are showed in Figure 3, comparing the number of distances calculations to build each tree using fractions of the Points dataset. Note that the number of distance calculations performed by VP-tree increases following a superlinear behavior when the number of indexed points raises. The MM-tree scales well as the indexed elements number increases, becoming the fastest MAM to build among others.

The heights of metric trees directly affect the query performance when searching on non-overlapping regions. The MM-tree has the semi-balancing algorithm that controls its height. Figure 4 shows the behavior of the MM-tree over different insertion policies, with and without the semi-balancing technique enabled, measuring how the average height of MM-tree varies. The average height was measured from experiments using the Points dataset. From this figure it can be seen that the semi-balancing algorithm leads to shorten trees.

Similarity Queries

The experiments on range and nearest-neighbor queries were made using all the datasets, measuring the number of distance calculations and time. The plots measuring the number of distance calculations represents the average of the values obtained from a set of 500 queries, while the plots measuring time represents the total time spent to answer all the 500 queries. The query elements are indexed in the structures.

Figure 5 shows the results obtained from the tests. The radius of range queries is the percentile of similar elements obtained by the query, i. e., a range query of radius 0.1 retrieves approximately 10% of the elements from the dataset. Considering the all datasets it can be seen that MM-tree outperforms the other MAM performing lesser distance calculations and being much faster.

6 Conclusion

There are emerging applications that require the DBMS to provide fast ways to build indexes on data that fits in main memory. The increase of the main memory capacity and its lowering cost allow to index data in main-memory leading to light and faster metric access methods. A disk-based MAM design forces the nodes to have many elements, excluding good approaches to the partitioning of the space. Memory-based MAM are faster to built, do not perform disk access to perform their operations and can provide faster similarity queries, once they are not limited on how to partition the space.

This paper presented a new memory-based metric access method, which partitions the space into non-overlapping regions. The MM-tree has a balancing control that produces shorter trees and allows both K-NN and range queries.

Experiments shown that the MM-tree has better performance than the VP-tree and the Slim-tree to index complex data. The construction process of the MM-tree is much faster than the other MAM, requiring up to one fifth of the number of distance calculations to be constructed when compared to Slim-tree and up to one fifteenth times less when compared to VP-tree.

When compared to the Slim-tree, the MM-tree achieved the reduction on the number of distance calculations up to 74% for k-NN queries leading to an improvement in time up to 43%. For range queries, MM-tree achieved the reduction on the number of distance calculations up to 64% leading to improvement in time up to 32%. When compared to VP-tree, MM-tree needs 58% less distance calculations for nearest neighbors queries, leading to improvement in time up to 57%, and needs 67% less distance calculations for range queries, leading to improvement in time up to 69%.

Acknowledgements

This research has been partially suported by CAPES, CNPq and by FAPESP.

References

1. Berchtold, S., Keim, D.A., Kriegel, H.P.: The x-tree: An index structure for high-dimensional data. In: VLDB, Bombay, India, pp. 28–39. Morgan Kaufmann, San Francisco (1996)
2. Chakrabarti, K., Mehrotra, S.: The hybrid tree: An index structure for high dimensional feature spaces. In: IEEE (ICDE), Sydney, Australia, pp. 440–447. IEEE Computer Society Press, Los Alamitos (1999)
3. Katayama, N., Satoh, S.: The sr-tree: An index structure for high-dimensional nearest neighbor queries. In: Peckham, J. (ed.) ACM SIGMOD, Tucson, Arizona, USA, pp. 369–380. ACM Press, New York (1997)
4. Lin, K.I.D., Jagadish, H.V., Christos, F.: The tv-tree: An index structure for high-dimensional data. VLDB Journal 3(4), 517–542 (1994)
5. Gaede, V., Gunther, O.: Multidimensional access methods. ACM Computing Surveys 30(2), 170–231 (1998)
6. Burkhard, W.A., Keller, R.M.: Some approaches to best-match file searching. Communications of the ACM (CACM) 16(4), 230–236 (1973)
7. Chávez, E., Navarro, G., Baeza-Yates, R.A., Marroquín, J.L.: Searching in metric spaces. ACM Computing Surveys 33(3), 273–321 (2001)
8. Uhlmann, J.K.: Satisfying general proximity/similarity queries with metric trees. Information Processing Letters 40(4), 175–179 (1991)
9. Yianilos, P.N.: Data structures and algorithms for nearest neighbor search in general metric spaces. In: ACM/SIGACT-SIAM (SODA), Austin, TX, pp. 311–321 (1993)
10. Bozkaya, T., Özsoyoglu, Z.M.: Distance-based indexing for high-dimensional metric spaces. In: ACM SIGMOD, Tucson, AZ, pp. 357–368. ACM Press, New York (1997)
11. Brin, S.: Near neighbor search in large metric spaces. In: Dayal, U., Gray, P.M.D., Nishio, S. (eds.) VLDB, Zurich, Switzerland, pp. 574–584. Morgan Kaufmann, San Francisco (1995)
12. Traina Jr., C., Traina, A.J.M., Faloutsos, C., Seeger, B.: Fast indexing and visualization of metric datasets using slim-trees. IEEE (TKDE) 14(2), 244–260 (2002)
13. Ciaccia, P., Patella, M., Rabitti, F., Zezula, P.: Indexing metric spaces with m-tree. In: Atti del Quinto Convegno Nazionale SEBD, Verona, Italy, pp. 67–86 (1997)

14. Santos Filho, R.F., Traina, A.J.M., Traina Jr., C., Faloutsos, C.: Similarity search without tears: The omni family of all-purpose access methods. In: IEEE (ICDE), Heidelberg, Germany, pp. 623–630. IEEE Computer Society Press, Los Alamitos (2001)
15. Traina Jr., C., Traina, A.J.M., Santos Filho, R.F., Faloutsos, C.: How to improve the pruning ability of dynamic metric access methods. In: CIKM, McLean, VA, USA, pp. 219–226. ACM Press, New York (2002)

Improving the Performance of M-Tree Family by Nearest-Neighbor Graphs

Tomáš Skopal and David Hoksza

Charles University in Prague, FMP, Department of Software Engineering
Malostranské nám. 25, 118 00 Prague, Czech Republic
{tomas.skopal,david.hoksza}@mff.cuni.cz

Abstract. The M-tree and its variants have been proved to provide an efficient similarity search in database environments. In order to further improve their performance, in this paper we propose an extension of the M-tree family, which makes use of nearest-neighbor (NN) graphs. Each tree node maintains its own NN-graph, a structure that stores for each node entry a reference (and distance) to its nearest neighbor, considering just entries of the node. The NN-graph can be used to improve filtering of non-relevant subtrees when searching (or inserting new data). The filtering is based on using "sacrifices" – selected entries in the node serving as pivots to all entries being their reverse nearest neighbors (RNNs). We propose several heuristics for sacrifice selection; modified insertion; range and kNN query algorithms. The experiments have shown the M-tree (and variants) enhanced by NN-graphs can perform significantly faster, while keeping the construction cheap.

1 Introduction

In multimedia retrieval, we usually need to retrieve objects based on similarity to a query object. In order to retrieve the relevant objects in an *efficient* (quick) way, the similarity search applications often use *metric access methods* (MAMs) [15], where the similarity measure is modeled by a metric distance δ. The principle of all MAMs is to employ the triangle inequality satisfied by any metric, in order to partition the dataset $\mathbb{S}$ among classes (organized in a *metric index*). When a query is to be processed, the metric index is used to quickly filter the non-relevant objects of the dataset, so that the number of distance computations needed to answer a query is minimized.[1] In the last two decades, there were many MAMs developed, e.g., gh-tree, GNAT, (m)vp-tree, SAT, (L)AESA, D-index, M-tree, to name a few (we refer to monograph [15]). Among them, the M-tree (and variants) has been proved as the most universal and suitable for practical database applications.

In this paper we propose an extension to the family of M-tree variants (implemented for M-tree and PM-tree), which is based on maintaining an additional structure in each tree node – the nearest-neighbor (NN) graph. We show that utilization of NN-graphs can speed up the search significantly.

---

[1] The metric is often supposed expensive, so the number of distance computations is regarded as the most expensive component of the overall runtime costs.

Y. Ioannidis, B. Novikov, and B. Rachev (Eds.): ADBIS 2007, LNCS 4690, pp. 172–188, 2007.

2   M-Tree

The *M-tree* [7] is a dynamic (easily updatable) index structure that provides good performance in secondary memory (i.e. in database environments). The M-tree index is a hierarchical structure, where some of the data objects are selected as centers (references or local *pivots*) of ball-shaped regions, and the remaining objects are partitioned among the regions in order to build up a balanced and compact hierarchy of data regions, see Figure 1a. Each region (subtree) is indexed recursively in a B-tree-like (bottom-up) way of construction.

The inner nodes of M-tree store *routing entries*

$$rout_l(O_i) = [O_i, r_{O_i}, \delta(O_i, Par(O_i)), ptr(T(O_i))]$$

where $O_i \in \mathbb{S}$ is a data object representing the center of the respective ball region, r_{O_i} is a *covering radius* of the ball, $\delta(O_i, Par(O_i))$ is so-called *to-parent* distance (the distance from O_i to the object of the parent routing entry), and finally $ptr(T(O_i))$ is a pointer to the entry's subtree. The data is stored in the leaves of M-tree. Each leaf contains *ground entries*

$$grnd(O_i) = [O_i, \delta(O_i, Par(O_i))]$$

where $O_i \in \mathbb{S}$ is the indexed data object itself, and $\delta(O_i, Par(O_i))$ is, again, the to-parent distance. See an example of routing and ground entries in Figure 1a.

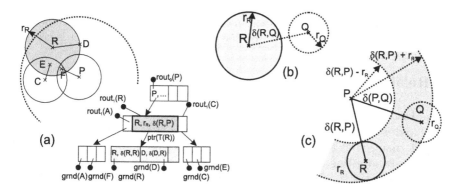

Fig. 1. (a) Example of an M-tree (b) Basic filtering (c) Parent filtering

2.1   Query Processing

The range and k nearest neighbors (kNN) queries are implemented by traversing the tree, starting from the root[2]. Those nodes are accessed, the parent region (R, r_R) of which (described by the routing entry) is overlapped by the query ball (Q, r_Q). In case of a kNN query (we search for k closest objects to Q), the radius r_Q is not known in advance, so we have to additionally employ a heuristic to dynamically decrease the radius during the search (initially set to ∞).

---

[2] We just outline the main principles, the algorithms are described in Section 4, including the proposed extensions. The original M-tree algorithms can be found in [7].

Basic Filtering. The check for region-and-query overlap requires an explicit distance computation $\delta(R, Q)$, see Figure 1b. In particular, if $\delta(R, Q) \leq r_Q + r_R$, the data ball R overlaps the query ball, thus the child node has to be accessed. If not, the respective subtree is filtered from further processing.

Parent Filtering. As each node in the tree contains the distances from the routing/ground entries to the center of its parent node, some of the non-relevant M-tree branches can be filtered out without the need of a distance computation, thus avoiding the "more expensive" basic overlap check (see Figure 1c). In particular, if $|\delta(P, Q) - \delta(P, R)| > r_Q + r_R$, the data ball R cannot overlap the query ball, thus the child node has not to be re-checked by basic filtering. Note $\delta(P, Q)$ was computed in the previous (unsuccessful) parent's basic filtering.

2.2   M-Tree Construction

Starting at the root, a new object O_i is recursively inserted into the best subtree $T(O_j)$, which is defined as the one for which the covering radius r_{O_j} must increase the least in order to cover the new object. In case of ties, the subtree whose center is closest to O_i is selected. The insertion algorithm proceeds recursively until a leaf is reached and O_i is inserted into that leaf. A node's overflow is managed in a similar way as in the B-tree – two objects from the overflowed node are selected as new centers, the node is split, and the two new centers (forming two routing entries) are promoted to the parent node. If the parent overflows, the procedure is repeated. If the root overflows, it is split and a new root is created.

3   Related Work

In the last decade, there have been several modifications/successors of M-tree introduced, some improving the performance of M-tree, some others improving the performance but restricting the general metric case into vector spaces, and, finally, some adjusting M-tree for an extended querying model.

In the first case, the *Slim-tree* [14] introduced two features. First, a new policy of node splitting using the minimum spanning tree was proposed to reduce internal CPU costs during insertion (but not the number of distance computations). The second feature was the slim-down algorithm, a post-processing technique for redistribution of ground entries in order to reduce the volume of bottom-level data regions. Another paper [12] extended the slim-down algorithm to the general case, where also the routing entries are redistributed. The authors of [12] also introduced the multi-way insertion of a new object, where an optimal leaf node for the object is found. A major improvement in search efficiency has been achieved by the proposal of *PM-tree* [13,10] (see Section 5.1), where the M-tree was combined with pivot-based techniques (like LAESA).

The second case is represented by the M^+-*tree* [16], where the nodes are further partitioned by a hyper-plane (where a key dimension is used to isometrically partition the space). Because of hyper-plane partitioning, the usage of M^+-tree is limited just to Euclidean vector spaces. The BM^+-tree [17] is a generalization of M^+-tree where the hyper-plane can be rotated, i.e. it has not to be parallel to the coordinate axes (the restriction to Euclidean spaces remains).

The last category is represented by the M^2-tree [5], where the M-tree is generalized to be used with an arbitrary aggregation of multiple metrics. Another structure is the M^3-tree [3], a similar generalization of M-tree as the M^2-tree, but restricted just to linear combinations of partial metrics, which allows to effectively compute the lower/upper bounds when using query-weighted distance functions. The last M-tree modification of this kind is the QIC-M-tree [6], where a user-defined query distance (even non-metric) is supported, provided a lower-bounding metric to the query distance (needed for indexing) is given by user.

In this paper we propose an extension belonging to the first category (i.e. improving query performance of the general case), but which could be utilized in all the M-tree modifications (M-tree family) overviewed in this section.

4   M*-Tree

We propose the M*-tree, an extension of M-tree having each node additionally equipped by *nearest-neighbor graph* (NN-graph). The motivation for employing NN-graphs is related to the advantages of methods using *global pivots* (objects which all the objects in dataset are referenced to). Usually, the global pivots are used to filter out some non-relevant objects from the dataset (e.g. in LAESA [9]). In general, the global pivots are good if they are far from each other and provide different "viewpoints" with respect to the dataset [2]. The M-tree, on the other hand, is an example of a method using *local pivots*, where the local pivots are the parent routing entries of nodes. As a hybrid structure, the PM-tree is combining the "local-pivoting strategies" of M-tree with the "global-pivoting strategies" of LAESA.

We can also state a "closeness criterion" for good pivots, as follows. A pivot is good in case it is close to a data object or, even better, close to the query object. In both cases, a close pivot provides tight distance approximation to the data/query object, so that filtering of data object by use of precomputed pivot-to-query or pivot-to-data distances is more effective. Unfortunately, this criterion cannot be effectively utilized for global pivots, because there is only a limited number of global pivots, while there are many objects referenced to them (so once we move a global pivot to become close to an object, many other objects become handicapped). Nevertheless, the closeness criterion can be utilized in local-pivoting strategies, because only a limited number of dataset objects are referenced to a local pivot and, conversely, a dataset object is referenced to a small number of pivots. Hence, a replacement of local pivot would impact only a fraction of objects within the entire dataset.

4.1   Nearest-Neighbor Graphs

In M-tree node, there exists just one local pivot, the parent (which moreover plays the role of center of the region). The parent filtering described in Section 2.1

is, actually, pivot-based filtering where the distance to the pivot is precomputed so we can avoid computation of the query-to-object distance (see Figure 1c). However, from the closeness criterion point of view, the parent is not guaranteed to be close to a particular object in the node, it is rather a compromise which is "close to all of them".

Following the closeness criterion, in this paper we propose a technique where all the objects in an M-tree node mutually play the roles of local pivots. In order to reduce space and computation costs, each object in a node is explicitly referenced just to its nearest neighbor. This fits the closeness criterion; we obtain a close pivot for each of the node's objects. In this way, for each node N we get a list of triplets $\langle O_i, NN(O_i, N), \delta(O_i, NN(O_i, N)) \rangle$, which we call the *nearest-neighbor graph* (NN-graph) of node N, see Figure 2a.

The result is *M*-tree*, an extension of M-tree, where the nodes are additionally equipped by NN-graphs, i.e. a routing entry is defined as

$$rout_l(O_i) = [O_i, r_{O_i}, \delta(O_i, Par(O_i)), \langle NN(O_i), \delta(O_i, NN(O_i)) \rangle, ptr(T(O_i))]$$

while the ground entry is defined as

$$grnd(O_i) = [O_i, \delta(O_i, Par(O_i)), \langle NN(O_i), \delta(O_i, NN(O_i)) \rangle]$$

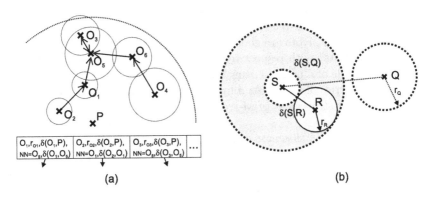

(a)   (b)

Fig. 2. (a) NN-graph in M*-tree node (b) Filtering using sacrifice pivot

4.2   Query Processing

In addition to M-tree's basic and parent filtering, in M*-tree we can utilize the NN-graph when filtering non-relevant routing/ground entries from the search.

NN-Graph Filtering. The filtering using NN-graph is similar to the parent filtering, however, instead of the parent, we use an object S from the node. First, we have to select such object S; then its distance to the query object Q is explicitly computed. We call the object S a *sacrifice pivot* (or simply *sacrifice*), since to "rescue" other objects from basic filtering, this one must be "sacrificed" (i.e. the distance to the query has to be computed).

Lemma 1. *(NN-graph filtering)*

Let entry(R) be a routing or ground entry in node N to be checked against a range query (Q, r_Q). Let entry(S) be a sacrifice entry in node N, which is R's nearest neighbor, or R is nearest neighbor to S. Then if

$$|\delta(S, Q) - \delta(S, R)| > r_R + r_Q$$

the entry(R) does not overlap the query and we can safely exclude entry(R) from further processing (for a ground entry $r_R = 0$).

Proof: Follows immediately from the triangle inequality. (a) Since $\delta(S, R) + r_R$ is the upper-bound distance of any object in (R, r_R) to S, we can extend or reduce the radius of rout(S) to $\delta(S, R) + r_R$ and then perform the standard overlap check between (Q, r_Q) and $(S, \delta(S, R) + r_R)$, i.e. $\delta(S, Q) - \delta(S, R) > r_R + r_Q$. (b) Since $\delta(S, R) - r_R$ is the lower-bound distance of any object in (R, r_R) to S, we can check whether the query (Q, r_Q) lies entirely inside the hole $(S, \delta(S, R) - r_R)$, i.e. $\delta(S, R) - \delta(S, Q) > r_R + r_Q$. ■

When using a sacrifice S, all objects which are *reverse nearest neighbors* (RNNs) to S can be passed to the NN-graph filtering (in addition to the single nearest neighbor of S). The reverse nearest neighbors are objects in N having S as their nearest neighbor. Note that for different sacrifices S_i within the same node N, their sets of RNNs are disjoint. The operator RNN can be defined as $RNN :$ $\mathbb{S} \times Nodes \mapsto 2^{\mathbb{S} \times \mathbb{R}^+}$, where for a given sacrifice S and a node N, the operator RNN returns a set of pairs $\langle O_i, d_i \rangle$ of reverse nearest neighbors.

The objects returned by RNN operator can be retrieved from the NN-graph without a need of distance computation (i.e. "for free"). See the NN-graph filtering pseudocode in Listing 1.

Listing 1. *(NN-graph filtering)*

```
set FilterByNNGraph(Node N, sacrifice ⟨S, δ(S, Q)⟩, RQuery (Q, rQ)) {
    set notFiltered = ∅
    for each entry(Oj) in RNN(S, N) do
        if |δ(S, Q) − δ(S, Oj)| > rQ + rOj then
            filtered[entry(Oj)] = true;
        else
            add entry(Oj) to notFiltered
    return notFiltered
}
```

Range Query. When implementing a query processing, the tree structure is traversed such that non-overlapping nodes (their parent regions do not overlap the query ball) are excluded from further processing (filtered out). In addition to the basic and parent filtering, in M*-tree we can use the NN-graph filtering step (inserted after the step of parent filtering and before the basic filtering), while we hope some distance computations needed by basic filtering after an unsuccessful parent filtering will be saved.

When processing a node N, there arises a question how to choose the ordering of N's entries passed to NN-graph filtering as sacrifices. We will discuss various orderings in Section 4.2, however, now suppose we already have a heuristic function $\mathcal{H}(N)$ which returns an ordering on entries in N. In this order the potential sacrifices are initially inserted into a queue SQ.

In each step of a node processing, the first object S_i from SQ (a sacrifice candidate) is fetched. Then, the sacrifice candidate is checked whether it can be filtered by parent filtering. If not, the sacrifice candidate becomes a regular sacrifice, so distance from Q to S_i is computed, while all S_i's RNNs (objects in $RNN(S_i, N)$) are tried to be filtered out by NN-graph filtering. Note the non-filtered RNNs (remaining also in SQ) surely become sacrifices, because RNN sets are disjoint, i.e., an object is passed at most once to the NN-graph filtering. Hence, we can immediately move the non-filtered RNNs to the beginning of SQ – this prevents some "NN-filterable" objects in SQ to become sacrifices. See the range query algorithm in Listing 2.

Listing 2. *(range query algorithm)*

```
QueryResult RangeQuery(Node N, RQuery (Q, rQ), ordering heuristic H) {
    let P be the parent routing object of N
                /* if N is root then δ(Oᵢ,P)=δ(P,Q)=0 */
    let filtered be an array of boolean flags, size of filtered is |N|
    set filtered[entry(Oᵢ)]=false, ∀entry(Oᵢ) ∈ N
    let SQ be a queue filled with all entries of N, ordered by H(N)

    if N is not a leaf then {
        /* parent filtering */
        for each rout(Oᵢ) in N do
            if |δ(P,Q) − δ(Oᵢ,P)| > rQ + rOᵢ then
                filtered[rout(Oᵢ)] = true;
        /* NN-graph filtering */
        while SQ not empty
            fetch rout(Sᵢ) from the beginning of SQ
            if not filtered[rout(Sᵢ)] then
                compute δ(Sᵢ,Q)
                NF = FilterByNNGraph(N, ⟨Sᵢ,δ(Sᵢ,Q)⟩, (Q,rQ))
                move all entries in SQ ∩ NF to the beginning of SQ
                /* basic filtering */
                if δ(Sᵢ,Q) ≤ rQ + rSᵢ then
                    RangeQuery(ptr(T(Sᵢ)), (Q,rQ), H)
    } else {
        /* parent filtering */
        for each grnd(Oᵢ) in N do
            if |δ(P,Q) − δ(Oᵢ,P)| > rQ then
                filtered[grnd(Oᵢ)] = true;
        /* NN-graph filtering */
        while SQ not empty
            fetch grnd(Sᵢ) from the beginning of SQ
            if not filtered[grnd(Sᵢ)] then
                compute δ(Sᵢ,Q)
                NF = FilterByNNGraph(N, ⟨Sᵢ,δ(Sᵢ,Q)⟩, (Q,rQ))
                move all entries in SQ ∩ NF to the beginning of SQ
                /* basic filtering */
                if δ(Sᵢ,Q) ≤ rQ then
                    add Sᵢ to the query result
    }
}
```

kNN Query. The kNN algorithm is a bit more difficult, since the query radius r_Q is not known at the beginning of kNN search. In [7] the authors applied a modification of the well-known heuristic (based on priority queue) to the M-tree. Due to lack of space we omit the listing of kNN pseudocode, however, its form can be easily derived from the original M-tree's kNN algorithm and the M*-tree's range query implementation presented above (for the code see [11]).

Choosing the Sacrifices. The order in which individual entries are treated as sacrifices is crucial for the algorithms' efficiency. Virtually, all entries of a node can serve as sacrifices, however, when a good sacrifice is chosen at the beginning, many of others can be filtered out, so the node processing requires less sacrifices (distance computations, actually). We propose several heuristic functions $\mathcal{H}$ which order all the entries in a node, thus setting a priority in which individual entries should serve as sacrifices when processing a query.

– **hMaxRNNCount**
  An ordering based on the number of RNNs belonging to an entry e_i,
  i.e. $|RNN(N, e_i)|$.
  Hypothesis: The more RNNs, the greater probability the sacrifice would filter more entries.
– **hMinRNNDistance**
  An ordering based on the entry's closest NN or RNN,
  i.e. $min\{\delta(e_i, O_i)\}, \forall O_i \in RNN(N, e_i) \cup NN(N, e_i)$.
  Hypothesis: An entry close to (R)NN stands for a close pivot, so there is a greater probability of effective filtering (following the closeness criterion).
– **hMinToParentDistance**
  An ordering based on the entry's to-parent distance.
  Hypothesis: The lower to-parent distance, the greater probability that the entry is close to the query; for such an entry the basic filtering is unavoidable, so we can use it as a sacrifice sooner.

4.3   M*-Tree Construction

The M*-tree construction consists of inserting new data objects into leaves, and of splitting overflowed nodes (see Listing 3). When splitting, the NN-graphs of the new nodes must be created from scratch. However, this does not imply additional computation costs, since we use M-tree's `MinMaxRad` splitting policy (originally denoted `mM_UB_RAD`) which computes all the pairwise distances among entries of the node being split (the `MinMaxRad` has been proved to perform the best [7]). Thus, these distances are reused when building the NN-graphs.

The search for the target leaf can use NN-graph filtering as well (see Listing 4). In the first phase, a candidate node is tried to be found, which spatially covers the inserted object. If more such candidates are found, the one with shortest distance to the inserted object is chosen. In case no candidate was found in the first phase, in the second phase the closest routing entry is determined. The search for candidate nodes recursively continues until a leaf is reached.

Listing 3. *(dynamic object insertion)*

```
Insert(Object O_i) {
    let N be the root node
    node = FindLeaf(N,O_i)
    store ground entry grnd(O_i) in the the target leaf node
    update radii in the insertion path
    while node is overflowed then
        /* split the node, create NN-graphs, produce two routing entries, and return the parent node */
        node = Split(node)
        insert (update) the two new routing entries into node
}
```

When inserting a new ground entry into a leaf (or a routing entry into an inner node after splitting), the existing NN-graph has to be updated. Although searching for the nearest neighbor of the new entry could utilize the NN-graph filtering (in a similar way as in kNN query), a check whether the new entry became the new nearest neighbor to some of the old entries would lead to computation of all the distances needed for a NN-graph update. Thus, we consider simple update of the NN-graph, where all the distances between the new entry and the old entries are computed.

Listing 4. *(navigating to the leaf for insertion)*

```
Node FindLeaf(Node N, Object New) {
    if N is a leaf node then
        return N
    let P be the parent routing object of N        /* if N is root then δ(O_i, P)=δ(P, New)=0 */
    let filtered be an array of boolean flags, size of filtered is |N|
    set filtered[entry(O_i)]=false, ∀entry(O_i) ∈ N
    let usedSacrifices = ∅ be an empty set
    let SQ be a queue filled with all entries of N, ordered by H(N)
        set candidateEntry = null
        minDist = ∞
    while SQ not empty {
        fetch rout(S_i) from the beginning of SQ
        /* parent filtering */
        if |δ(P, New) − δ(S_i, P)| > minDist + r_{S_i} then
            filtered[rout(S_i)] = true;
        if not filtered[rout(S_i)] then {
            compute δ(S_i, New)
            insert ⟨S_i, δ(S_i, New)⟩ into usedSacrifices
            /* NN-graph filtering */
            NF = ∅
            for each S_j in usedSacrifices do
                NF = NF ∪ FilterByNNGraph(N, ⟨S_j, δ(S_j, New)⟩, (New,minDist))
            move all entries in SQ ∩ NF to the beginning of SQ
            if δ(S_i, New) ≤ r_{S_i} and δ(S_i, New) ≤ minDist then
                candidateEntry = S_i
                minDist = δ(S_i, New)
        }
    }
    if candidateEntry = null then {
        do the same as in the previous while cycle, the difference is just in the condition when updating
        a candidate entry, which is relaxed to:
            if δ(S_i, New) ≤ minDist then
                candidateEntry = S_i
                minDist = δ(S_i, New)
    }
    return FindLeaf(ptr(T(candidateEntry)), New)
}
```

4.4   Analysis of Computation Costs

The worst case time complexities of single object insertion into M*-tree as well as of querying are the same as in the M-tree, i.e. $O(log\ n)$ in case of insertion and $O(n)$ in case of querying, where n is the dataset size. Hence, we are rather interested in typical reduction/increase of absolute computation costs[3] exhibited by M*-tree, with respect to the original M-tree.

M*-Tree Construction Costs. When inserting a new object into M*-tree, the navigation to the target leaf makes use of NN-graph filtering, so we achieve faster navigation. However, for insertion into the leaf itself the update of leaf's NN-graph is needed, which takes m distance computations for M*-tree instead of no computation for M-tree (where m is the maximum number of entries in a leaf). On the other side, the expensive splitting of a node does not require any additional distance computation, since all pairwise distances have to be computed to partition the node, regardless of using M-tree or M*-tree.

M*-Tree Querying Costs. The range search is always more efficient in M*-tree than in M-tree, because only such entries are chosen as sacrifices which cannot be filtered by the parent, so for them distance computation is unavoidable. On the other side, due to NN-graph filtering some of the entries can be filtered before they become a sacrifice, thus distance computations are reduced in this case.

5   PM*-Tree

To show the benefits of NN-graph filtering also on other members of the M-tree family, we have implemented *PM*-tree*, a NN-graph-enhanced extension of the PM-tree (see the next section for a brief overview). The only structural extension over the PM-tree are the NN-graphs in nodes, similarly like M*-tree extends the M-tree.

5.1   PM-Tree

The idea of PM-tree [10,13] is to combine the hierarchy of M-tree with a set of p global pivots. In a PM-tree's routing entry the original M-tree-inherited ball region is further cut off by a set of rings (centered in the global pivots), so the region volume becomes smaller (see Figure 3a). Similarly, the PM-tree ground entries are extended by distances to the pivots (which are also interpreted as rings due to approximations). Each ring stored in a routing/ground entry represents a distance range (bounding the underlying data) with respect to a particular pivot. The combination of all the p entry's ranges produces a p-dimensional minimum bounding rectangle (MBR), hence, the global pivots actually map the

---

[3] The NN-graph filtering is used just to reduce the computation costs, the I/O costs are the same as in the case of M-tree.

metric regions/data into a "pivot space" of dimensionality p (see Figure 3b). The number of pivots can be defined separately for routing and ground entries – we typically choose less pivots for ground entries to reduce storage costs.

Prior to the standard M-tree "ball filtering" (either basic or parent filtering), a query ball mapped into a hyper-cube in the pivot space is checked for an overlap with routing/ground entry's MBRs – if they do not overlap, the entry is filtered out without a distance computation (otherwise needed in M-tree's basic filtering). Actually, the overlap check can be also understood as L_∞ filtering (i.e. if the L_∞ distance from a region to the query object Q is greater than r_Q, the region is not overlapped by query).

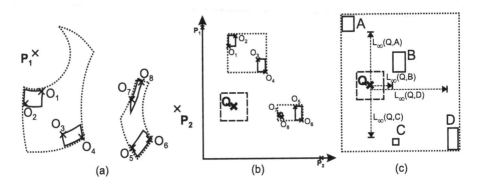

Fig. 3. Hierarchy of PM-tree regions (using global pivots P_1, P_2): (a) metric space view (b) pivot space view. (c) L_∞ distance from Q to MBRs of routing/ground entries.

In PM*-tree we suppose the following ordering of steps when trying to filter a non-relevant node:

$$\text{parent filtering} \rightarrow L_\infty \text{ filtering} \rightarrow \textbf{NN-filtering} \rightarrow \text{basic filtering}$$

5.2 Choosing the Sacrifices

The idea of L_∞ filtering can be also used to propose a PM*-tree-specific heuristics for choosing the sacrifice ordering (in addition to the ones proposed for M*-tree).

– **hMinLmaxDistance**
  An ordering based on the minimum L_∞ distance from Q to the entry's MBR (see Figure 3c).
  Hypothesis: The smaller L_∞ distance, the greater probability that also the δ distance is small, so that the entry has to be filtered by basic filtering (requiring a distance computation).
– **hMaxLmaxDistance**
  An ordering based on the maximum L_∞ distance from Q to the entry's MBR.
  Hypothesis: The greater L_∞ distance, the greater probability that also the δ distance is great, so the entry's RNNs are far from the query and could be filtered by NN-graph.

6   Experimental Results

To verify the impact of NN-graph filtering, we have performed extensive experimentation with M*-tree and PM*-tree on three datasets. The sets of query objects were selected as 500 random objects from each dataset, while the query sets were excluded from indexing. We have monitored the computation costs (number of distance computations) required to index the datasets, as well as costs needed to answer range and kNN queries. Each query test unit consisted of 500 query objects and the results were averaged. The computation costs for querying on PM(*)-tree do not include the external distance computations needed to map the query into the pivot space (this overhead is equal to the number of pivots used, and cannot be affected by filtering techniques).

The datasets were indexed for varying dimensionality, capacity of entries per inner node, and the number of pivots (in case of PM(*)-tree). Unless otherwise stated, the PM-tree and PM*-tree indexes were built using 64 pivots (64 in inner nodes and 32 in leaf nodes). Moreover, although the M-tree and PM-tree indexes are designed for database environments, in this paper we are interested just in computation costs (because the NN-graph filtering cannot affect I/O costs). Therefore, rather than fixing a disk page size used for storage of a node, we specify the inner node capacity (maximum of entries stored within an inner node – set to 50 in most experiments).

6.1   Corel Dataset

The first set of experiments was performed on the *Corel dataset* [8], consisting of 65,615 feature vectors of images (we have used the color moments). As indexing metric the L_1 distance was used. Unless otherwise stated, we have indexed the first 8 out of 32 dimensions. In Table 1 see the statistics obtained when indexing the Corel dataset – the numbers of distance computations needed to index all the vectors and the index file sizes. The computation costs and index sizes of M*-tree/PM*-tree are represented as percentual growth with respect to the M-tree/PM-tree values.

Table 1. Corel indexing statistics

index type	construction costs	index file size
M-tree	3,708,968	4.7MB
M*-tree	+22%	+25.5%
PM-tree(64,32)	18,522,252	8.8MB
PM*-tree(64,32)	+25.6%	+0%

In Figure 4a see the M-tree and M*-tree querying performance with respect to increasing capacity of nodes. We can see the M-tree performance improves up to the capacity of 30 entries, while M*-tree steadily improves up to the capacity of 100 entries – here the M*-tree is by up to 45% more efficient than M-tree. The most effective heuristic is the hMaxRNNCount.

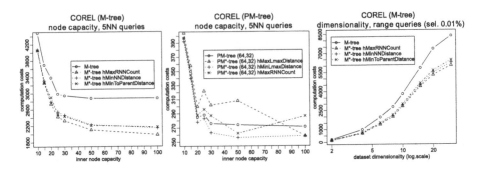

Fig. 4. 5NN queries depending on varying node capacity (a) M*-tree (b) PM*-tree. (c) Range queries depending on increasing dataset dimensionality.

The impact of increasing PM*-tree node capacities is presented in Figure 4b. The PM*-tree with `hMinLmaxDistance` heuristic performs the best, while the other heuristics are even outperformed by the PM-tree. The gain in efficiency is not so significant as in case of M*-tree (5%).

In Figure 4c see a comparison of M-tree and M*-tree when querying datasets of increasing dimensionality. Again, the `hMaxRNNCount` heuristic performed the best, the efficiency gain of M*-tree was significant – about 40% on average.

6.2   Polygons Dataset

The second set of experiments was carried out on the *Polygon* dataset, a synthetic dataset consisting of 1,000,000 randomly generated 2D polygons, each having 5-10 vertices. As the indexing metric the Hausdorff set distance was employed (where L_2 distance was used as partial distance on vertices). In Table 2 see the statistics obtained for the Polygons indexes.

Table 2. Polygons indexing statistics

index type	construction costs	index file size
M-tree	70,534,350	148.4MB
M*-tree	+12,1%	+5%
PM-tree(64,32)	291,128,463	202.7MB
PM*-tree(64,32)	+17%	+0%

In Figure 5a the M*-tree 1NN performance with respect to the increasing dataset size is presented. To provide a better comparison, the computation costs are represented in proportion of distance computations needed to perform full sequential search on the dataset of given size. The efficiency gain of M*-tree is about 30% on average.

The costs for kNN queries are shown in Figure 5b, we can observe the efficiency gain ranges from 30% in case of 1NN query to 23% in case of 100NN query. As

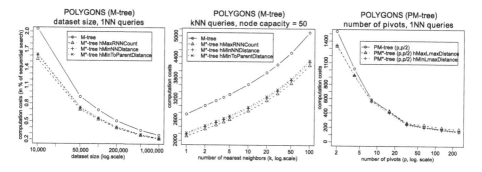

Fig. 5. (a) 1NN queries depending on increasing dataset size. (b) kNN queries (c) 1NN queries depending on increasing number of pivots used in PM-tree.

usual, the M*-tree equipped with the `hMaxRNNCount` heuristic performed the best.

The performance of 1NN queries on PM*-tree is presented in Figure 5c, considering increasing number of pivots used. The PM*-tree performance improvement with respect to PM-tree is quite low, ranging from 15% (2 and 4 pivots, heuristic `hMaxLmaxDistance`) to 6% ($\geq$ 8 pivots, heuristic `hMinLmaxDistance`).

6.3   GenBank Dataset

The last dataset in experiments was created by sampling 250,000 strings of protein sequences (of lengths 50-100) from the *GenBank* file `rel147` [1]. The edit distance was used to index the GenBank dataset. In Table 3 see the indexing statistics obtained for the GenBank dataset.

Table 3. GenBank indexing statistics

index type	construction costs	index file size
M-tree	17,726,084	54.5M
M*-tree	+38,9%	+18,2%
PM-tree(64,32)	77,316,482	66.8MB
PM*-tree(64,32)	+20.6%	+0%

In Figure 6 see the results for kNN queries on M*-tree and PM*-tree, respectively. Note that the GenBank dataset is generally hard to index, the best achieved results for 1NN by M*-tree and PM*-tree are 136,619 (111,086 respectively) distance computations, i.e. an equivalent of about half of the sequential search. Nevertheless, in these hard conditions the M*-tree and PM*-tree have outperformed the M-tree and PM-tree quite significantly, by up to 20%. As in the previous experiments, the `hMaxRNNCount` heuristic on M*-tree and `hMinLmaxDistance` heuristic on PM*-tree performed the best.

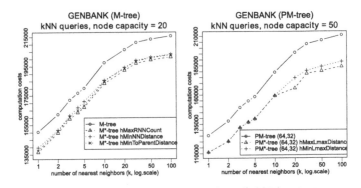

Fig. 6. kNN queries (a) M-tree (b) PM-tree

6.4 Summary

The construction costs and index file sizes of all M*-tree and PM*-tree indexes exhibited an increase, ranging from 5% to 25%. For PM*-tree the increase in index file size was negligible in all cases.

The results on the small-sized Corel dataset have shown the M*-tree can significantly outperform the M-tree. However, the PM*-tree performed only slightly better than PM-tree. We suppose the superior performance of PM-tree simply gives only a little room for improvements. Note the dataset of size $\approx 65,000$ can be searched by the PM-tree for less than 300 distance computation – this corresponds, for example, to six paths of length 3 in the PM-tree where 15 entries per node must be filtered by basic filtering.

We suppose the extent of M*-tree/PM*-tree performance gain is related to the *intrinsic dimensionality* [4] of the respective dataset. The high-dimensional GenBank dataset is hard to index, so any "help" by additional filtering techniques (like the NN-graph filtering) would result in better pruning of index subtrees. On the other side, when considering the low-dimensional Corel and Polygons datasets, the PM-tree alone is extremely successful (up to 10x faster than M-tree), so we cannot achieve a significant gain in performance.

7 Conclusions

We have proposed an extension of M-tree family, based on utilization of nearest-neighbor graphs in tree nodes. We have shown the NN-graphs can be successfully implemented into index structures designed for a kind of local-pivot filtering. The improvement is based on using so-called "sacrifices" – selected entries in the tree node which serve as local pivots to all entries being reverse nearest neighbors (RNNs) to a sacrifice. Since the distances from a sacrifice to its RNNs are pre-computed in the NN-graph, we could prune several subtrees for just one distance computation. We have proposed several heuristics on choosing the sacrifices, and the modified insertion, range and kNN query algorithms.

7.1 Future Work

The properties of NN-graph filtering open possibilities for other applications. Generally, the metric access methods based on compact partitioning using local pivots could be extended by NN-graphs. In the future we would like to integrate the NN-graph filtering into other metric access methods, as a supplement to their own filtering techniques.

Acknowledgments

This research has been supported in part by grants GAČR 201/05/P036 provided by the Czech Science Foundation, and "Information Society program" grant number 1ET100300419.

References

1. Benson, D.A., Karsch-Mizrachi, I., Lipman, D.J., Ostell, J., Rapp, B.A., Wheeler, D.L.: Genbank. Nucleic Acids Res 28(1), 15–18 (2000)
2. Bustos, B., Navarro, G., Chávez, E.: Pivot selection techniques for proximity searching in metric spaces. Pattern Recognition Letters 24(14), 2357–2366 (2003)
3. Bustos, B., Skopal, T.: Dynamic Similarity Search in Multi-Metric Spaces. In: Proceedings of ACM Multimedia, MIR workshop, pp. 137–146. ACM Press, New York (2006)
4. Chávez, E., Navarro, G.: A Probabilistic Spell for the Curse of Dimensionality. In: Buchsbaum, A.L., Snoeyink, J. (eds.) ALENEX 2001. LNCS, vol. 2153, pp. 147–160. Springer, Heidelberg (2001)
5. Ciaccia, P., Patella, M.: The M^2-tree: Processing Complex Multi-Feature Queries with Just One Index. In: DELOS Workshop: Information Seeking, Searching and Querying in Digital Libraries, Zurich, Switzerland (June 2000)
6. Ciaccia, P., Patella, M.: Searching in metric spaces with user-defined and approximate distances. ACM Database Systems 27(4), 398–437 (2002)
7. Ciaccia, P., Patella, M., Zezula, P.: M-tree: An Efficient Access Method for Similarity Search in Metric Spaces. In: VLDB'97, pp. 426–435 (1997)
8. Hettich, S., Bay, S.: The UCI KDD archive (1999), http://kdd.ics.uci.edu
9. Mic ó, M.L., Oncina, J., Vidal, E.: An algorithm for finding nearest neighbour in constant average time with a linear space complexity. In: Int. Cnf. on Pattern Recog. (1992)
10. Skopal, T.: Pivoting M-tree: A Metric Access Method for Efficient Similarity Search. In: Proceedings of the 4th annual workshop DATESO, Desná, Czech Republic. CEUR, vol. 98, pp. 21–31 (2004), http://www.ceur-ws.org/Vol-98 ISBN 80-248-0457-3, ISSN 1613-0073
11. Skopal, T., Hoksza, D.: Electronic supplement for this paper (2007), http://siret.ms.mff.cuni.cz/skopal/pub/suppADBIS07.pdf
12. Skopal, T., Pokorný, J., Krátký, M., Snášel, V.: Revisiting M-tree Building Principles. In: Kalinichenko, L.A., Manthey, R., Thalheim, B., Wloka, U. (eds.) ADBIS 2003. LNCS, vol. 2798, pp. 148–162. Springer, Heidelberg (2003)
13. Skopal, T., Pokorný, J., Snášel, V.: Nearest Neighbours Search using the PM-tree. In: Zhou, L.-z., Ooi, B.-C., Meng, X. (eds.) DASFAA 2005. LNCS, vol. 3453, pp. 803–815. Springer, Heidelberg (2005)

14. Traina Jr., C., Traina, A., Seeger, B., Faloutsos, C.: Slim-Trees: High performance metric trees minimizing overlap between nodes. In: Zaniolo, C., Grust, T., Scholl, M.H., Lockemann, P.C. (eds.) EDBT 2000. LNCS, vol. 1777, Springer, Heidelberg (2000)
15. Zezula, P., Amato, G., Dohnal, V., Batko, M.: Similarity Search: The Metric Space Approach (Advances in Database Systems). Springer, New York (2005)
16. Zhou, X., Wang, G., Xu, J.Y., Yu, G.: M^+-tree: A New Dynamical Multidimensional Index for Metric Spaces. In: Proceedings of the Fourteenth Australasian Database Conference - ADC'03, Adelaide, Australia (2003)
17. Zhou, X., Wang, G., Zhou, X., Yu, G.: BM+-Tree: A Hyperplane-Based Index Method for High-Dimensional Metric Spaces. In: Zhou, L.-z., Ooi, B.-C., Meng, X. (eds.) DASFAA 2005. LNCS, vol. 3453, pp. 398–409. Springer, Heidelberg (2005)

Indexing Mobile Objects on the Plane Revisited

S. Sioutas, K. Tsakalidis, K. Tsihlas, C. Makris, and Y. Manolopoulos

Department of Informatics, Ionian University, Corfu, Greece
`sioutas@ionio.gr`
Computer Engineering and Informatics Department, University of Patras, Greece
`{tsakalid,tsihlas,makri}@ceid.upatras.gr`
Department of Informatics, Aristotle University of Thessaloniki, Greece
`manolopo@csd.auth.gr`

Abstract. We present a set of time-efficient approaches to index objects moving on the plane to efficiently answer range queries about their future positions. Our algorithms are based on previously described solutions as well as on the employment of efficient data structures. Finally, an experimental evaluation is included that shows the performance, scalability and efficiency of our methods.

Keywords: Spatio-Temporal Databases, Indexing.

1 Introduction

This paper focuses on the problem of indexing mobile objects in two dimensions and efficiently answering range queries over the objects locations in the future. This problem is motivated by a set of real-life applications such as intelligent transportation systems, cellular communications, and meteorology monitoring. There are two basic approaches used when trying to handle this problem; those that deal with discrete and those that deal with continuous movements.

In a discrete environment the problem of dealing with a set of moving objects can be considered to be equivalent to a sequence of database snapshots of the object positions/extents taken at time instants $t_1 < t_2 < \ldots$, with each time instant denoting the moment where a change took place. From this point of view, the indexing problems in such environments can be dealt with by suitably extending indexing techniques from the area of temporal [30] or/and spatial databases [11]; in [21] it is elegantly exposed how these indexing techniques can be generalized to handle efficiently queries in a discrete spatiotemporal environment. When considering continuous movements there exists a plethora of efficient data structures [2,14,17,22,23,28,29,33].

The common thrust behind these indexing structures lies in the idea of abstracting each object's position as a continuous function $f(t)$ of time and updating the database whenever the function parameters change; accordingly an object is modeled as a pair consisted of its extent at a reference time (design parameter) and of its motion vector. One categorization of the aforementioned structures is according to the family of the underlying access method used. In

Y. Ioannidis, B. Novikov, and B. Rachev (Eds.): ADBIS 2007, LNCS 4690, pp. 189–204, 2007.

particular, there are approaches based either on R-trees or on Quad-trees as explained in [25,26,27]. On the other hand, these structures can be also partitioned into (a) those that are based on geometric duality and represent the stored objects in the dual space [2,17,23] and (b) those that leave the original representation intact by indexing data in their native n-d space [4,22,28,29,33]. The *geometric duality transformation* is a tool heavily used in the Computational Geometry literature, which maps hyper-planes in R^n to points and vice-versa. In this paper we present and experimentally evaluate techniques using the duality transform that are based on previous approaches [17,22] to efficiently index the future locations of moving points on the plane.

In Section 2 we give a formal description of the problem. In Sections 3 and 4 we present our new solutions that outperform the solution presented in [17,22] since they use more efficient indexing schemes. In particular, Section 4 presents two alternative solutions. The first one is very easily implemented and has many practical merits. The second one has only theoretical interest since it uses clever but very complicated data structures, the implementation of which is very difficult and constitutes an open future problem. Section 5 presents an extended experimental evaluation and Section 6 concludes the paper.

2   Definitions and Problem Description

We consider a database that records the position of moving objects in two dimensions on a finite terrain. We assume that objects move with a constant velocity vector starting from a specific location at a specific time instant. Thus, we can calculate the future object position, provided that its motion characteristics remain the same. Velocities are bounded by $[u_{min}, u_{max}]$. Objects update their motion information, when their speed or direction changes. The system is dynamic, i.e. objects may be deleted or new objects may be inserted.

Let $P_z(t_0) = [x_0, y_0]$ be the initial position at time t_0 of object z. If object z starts moving at time $t > t_0$, its position will be $P_z(t) = [x(t), y(t)] = [x_0 + u_x(t - t_0), y_0 + u_y(t - t_0)]$, where $U = (u_x, u_y)$ is its velocity vector. For example, in Figure 1 the lines depict the objects trajectories on the (t, y) plane.

We would like to answer queries of the form: "Report the objects located inside the rectangle $[x_{1_q}, x_{2_q}] \times [y_{1_q}, y_{2_q}]$ at the time instants between t_{1_q} and t_{2_q} (where $t_{now} \leq t_{1_q} \leq t_{2_q}$), given the current motion information of all objects."

3   Indexing Mobile Objects in Two Dimensions

3.1   Indexing Mobile Objects in One Dimension

The Duality Transform. The duality transform, in general, maps a hyperplane h from R^n to a point in R^n and vice-versa. In this subsection we briefly describe how we can address the problem at hand in a more intuitive way, by using the duality transform on the 1-d case.

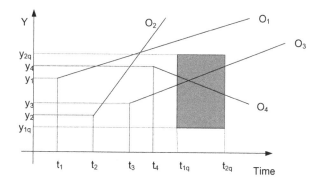

Fig. 1. Trajectories and query in (t, y) plane

Hough-X Transform. One duality transform for mapping the line with equation $y(t) = ut + a$ to a point in R^2 is by using the dual plane, where one axis represents the slope u of an objects trajectory (i.e. velocity), whereas the other axis represents its intercept a. Thus we get the dual point (u, a) (this is the so called *Hough-X transform* [17,22]). Accordingly, the 1-d query $[(y_{1_q}, y_{2_q}), (t_{1_q}, t_{2_q})]$ becomes a polygon in the dual space. By using a linear constraint query [12], the query in the dual Hough-X plane is expressed as follows (see Figure 2):

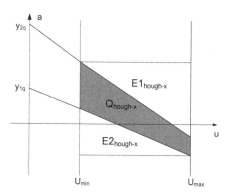

Fig. 2. Query in the Hough-X dual plane

Thus, the initial query $[(t_{1_q}, t_{2_q}), (y_{1_q}, y_{2_q})]$ in the (t, y) plane is transformed to the following rectangular query $[(u_{min}, u_{max}), (y_{1_q} - t_{1_q}u_{max}, y_{2_q} - t_{2_q}u_{min})]$ in the (u, a) plane.

Hough-Y Transform. By rewriting the equation $y = ut + a$ as $t = \frac{1}{u}y - \frac{a}{u}$, we can arrive to a different dual representation (the so called *Hough-Y transform* in [17,22]). The point in the dual plane has coordinates (b, n), where $b = -\frac{a}{u}$ and $n = \frac{1}{u}$. Coordinate b is the point where the line intersects the line $y = 0$ in the primal space. By using this transform horizontal lines cannot be represented.

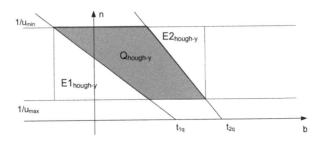

Fig. 3. Query on the Hough-Y dual plane

Similarly, the Hough-X transform cannot represent vertical lines. Nevertheless, since in our setting lines have a minimum and maximum slope (velocity is bounded by $[u_{min}, u_{max}]$), both transforms are valid.

Similarly, the initial query $[(t_{1_q}, t_{2_q}), (y_{1_q}, y_{2_q})]$ in the (t, y) plane (see Figure 3) can be transformed to the following rectangular query in the (b, n) plane:
$[(t_{1_q} - \frac{y_{2_q}}{u_{min}}, t_{2_q} - \frac{y_{1_q}}{u_{max}}), (\frac{1}{u_{max}}, \frac{1}{u_{min}})]$.

3.2 The Proposed Algorithm for Indexing Mobile Objects in Two Dimensions

In [17,22], motions with small velocities in the Hough-Y approach are mapped into dual points (b, n) having large n coordinates ($n = 1/u$). Thus, since few objects can have small velocities, by storing the Hough-Y dual points in an index structure such as an R*-tree, MBR's with large extents are introduced, and the index performance is severely affected. On the other hand, by using a Hough-X for the small velocities' partition, this effect is eliminated, since the Hough-X dual transform maps an object's motion to the (u, a) dual point. The query area in Hough-X plane is enlarged by the area E, which is easily computed as $E_{Hough-X} = (E1_{hough-X} + E2_{hough-X})$. By $Q_{Hough-X}$ we denote the actual area of the simplex query. Similarly, on the dual Hough-Y plane, $Q_{Hough-Y}$ denotes the actual area of the query, and $E_{Hough-Y}$ denotes the enlargement. According to these observations the solution in [17,22] proposes the choice of that transformation which minimizes the following criterion: $c = \frac{E_{Hough-X}}{Q_{Hough-X}} + \frac{E_{Hough-Y}}{Q_{Hough-Y}}$.

The procedure for building the index follows:

1. Decompose the 2-d motion into two 1-d motions on the (t, x) and (t, y) planes.
2. For each projection, build the corresponding index structure.

   Partition the objects according to their velocity:

   - Objects with small velocity are stored using the Hough-X dual transform, while the rest are stored using the Hough-Y dual transform.
   - Motion information about the other projection is also included.

The outline of the algorithm for answering the exact 2-d query follows:

1. Decompose the query into two 1-d queries, for the (t, x) and (t, y) projection.
2. For each projection get the dual - simplex query.
3. For each projection calculate the criterion c and choose the one (say p) that minimizes it.
4. Search in projection p the Hough-X or Hough-Y partition.
5. Perform a refinement or filtering step "on the fly", by using the whole motion information. Thus, the result set contains only the objects that satisfy the query.

In [17,22], $Q_{Hough-X}$ is computed by querying a 2-d partition tree, whereas $Q_{Hough-Y}$ is computed by querying a B$^+$-tree that indexes the b parameters of Figure 3. Our construction instead is based: (a) on the use of the Lazy B-tree [15] instead of the B$^+$-tree when handling queries with the Hough-Y transform and (b) on the employment of a new index that outperforms partition trees in handling polygon queries with the Hough-X transform. In the next section we present the main characteristics of our proposed structures.

4   The Access Methods

4.1   Handling Polygon Queries When Using the Hough-Y Transform

As described in [17,22], polygon queries when using the Hough-Y transform can be approximated by a constant number of 1-d range queries that can be handled by a classical B-tree [9]. Our construction is based on the use of a B-tree variant, which is called *Lazy B-tree* and has better dynamic performance as well as optimal I/O complexities for both searching and update operations [15]. An orthogonal effort towards developing another yet B-tree variant under the same name has been proposed in [20]. The Lazy B-tree of [15] is a simple but non-trivial externalization of the techniques introduced in [24]. In simple words, it is a typical case of a two-level access method as depicted in Figure 4.

The Lazy B-tree operates on the external memory model of computation. The first level consists of an ordinary B-tree, while the second one consists of buckets of size $O(\log^2 n)$, where n is approximately equal to the number of elements stored in the access method. Each bucket consists of two list layers, L and L_i respectively, where $1 \leq i \leq O(\log n)$, each of which has $O(\log n)$ size. The rebalancing operations are guided by the *global rebalancing lemma* given in [24] (see also [10,19]). In this scheme, each bucket is assigned a *criticality* indicating how close this bucket is to be fused or split. Every $O(\log_B n)$ updates we choose the bucket with the largest criticality and make a rebalancing operation (fusion or split). The update of the Lazy B-tree is performed incrementally (i.e., in a step-by-step manner) during the next $O(\log_B n)$ update operations and until the next rebalancing operation. The global rebalancing lemma ensures that the size of the buckets will never be larger than $O(\log^2 n)$.

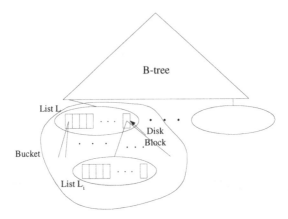

Fig. 4. The Lazy B-tree

Let n be approximately equal to the number of elements stored in the access method, and B be the size of blocks in the external memory. Then:

Theorem 1. *The Lazy B-Tree supports the search operation in $O(\log_B n)$ worst-case block transfers and update operations in $O(1)$ worst-case block transfers, provided that the update position is given.*

4.2   Handling Polygon Queries When Using the Hough-X Transform

Our construction is based on an interesting geometric observation that the polygon queries are a special case of the general simplex query and hence can be handled more efficiently without resorting to partition trees.

Let us examine the polygon (4-sided) indexability of Hough-X transformation. Our crucial observation is that the query polygon has the nice property of being divided into orthogonal objects, i.e. orthogonal triangles or rectangles, since the lines $X = U_{min}$ and $X = U_{max}$ are parallel.

We depict schematically the three basic cases that justify the validity of our observation.

Case I. Figure 5 depicts the first case where the polygon query has been transformed to four range queries employing the orthogonal triangles $(P_1 P_2 P_5)$, $(P_2 P_7 P_8)$, $(P_4 P_5 P_6)$, $(P_3 P_4 P_7)$ and one range query for querying the rectangle $(P_5 P_6 P_7 P_8)$.

Case II. The second case is depicted in the Figure 6. In this case the polygon query has been transformed to two range queries employing the orthogonal triangles $(P_1 P_4 P_5)$ and $(P_2 P_3 P_6)$ and one range query for querying the rectangle $(P_2 P_5 P_4 P_6)$.

Case III. The third case is depicted in the Figure 7. In this case the polygon query has been transformed to two range queries employing the orthogonal

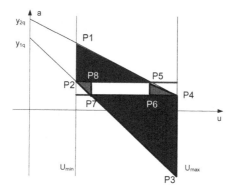

Fig. 5. Orthogonal triangulations: Case I

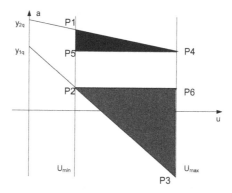

Fig. 6. Orthogonal triangulations: Case II

triangles $(P_1P_4P_5)$ and $(P_2P_3P_6)$ and one range query for querying the rectangle $(P_2P_1P_5P_6)$.

The problem of handling orthogonal range search queries has been handled in [3], where an optimal solution was presented to handle general (4-sided) range queries in $O((N/B)(\log(N/B))\log\log_B N)$ disk blocks and could answer queries in $O(\log_B N + T/B)$ I/O's ; the structure also supports updates in $O((\log_B N)(\log(N/B)))/\log\log_B N)$ I/O's.

Let us now consider the problem of devising an access method for handling orthogonal triangle range queries; in this problem we have to determine all the points from a set S of n points on the plane lying inside an orthogonal triangle. Recall that a triangle is orthogonal if two of its edges are axis-parallel. A basic ingredient of our construction will be a structure for handling half-plane range queries, i.e. queries that ask for the reporting all the points in a set S of n points in the plane that lie on a given side of a query line L.

A main memory solution presented in [6] and achieves optimal $O(\log n + A)$ query time and linear space using the notion of duality. The above main memory

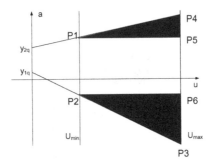

Fig. 7. Orthogonal triangulations: Case III

construction was extended to external memory in [1], where an access method was presented that was the first optimal one for answering 2-d halfpspace range queries in the worst case, based on the geometric technique called *filtering search* [7]. It uses $O(n)$ blocks of space and answers a query using $O(\log_B n + A)$ I/Os, where A is the answer size. We will use these methods to satisfy orthogonal triangle range queries on points.

Let us now return to our initial problem, i.e the devise of a structure suitable for handling orthogonal triangle range queries. Recall, a triangle is orthogonal if two of its edges are axis-parallel. Let T be an orthogonal triangle defined by the point (x_q, y_q) and the line L_q that is not axis-parallel (see Figure 8). A retrieval query for this problem can be supported efficiently by the following 3-layered access method.

To set up the access method, we first sort the n points according to their x-coordinates and then store the ordered sequence in a leaf-oriented balanced binary search tree of depth $O(\log n)$. This structure answers the query: "determine the points having x-coordinates in the range $[x_1, x_2]$ by traversing the two paths to the leaves corresponding to x_1, x_2". The points stored as leaves at the subtrees of the nodes which lie between the two paths are exactly these points in the range $[x_1, x_2]$. For each subtree, the points stored at its leaves are organized further to a second level structure according to their y-coordinates in the same way. For each subtree of the second level structure, the points stored at its leaves

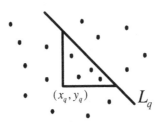

Fig. 8. The query triangle

are organized further to a third level structure as in [1,6] for half-plane range queries. Thus, each orthogonal triangle range query is performed through the following steps:

1. In the tree storing the pointset S according to x-coordinates, traverse the path to x_q. All the points having x-coordinate in the range $[x_q, \infty)$ are stored at the subtrees on the nodes that are right sons of a node of the search path and do not belong to the path. There are at most $O(\log n)$ such disjoint subtrees.
2. For every such subtree traverse the path to y_q. By a similar argument as in the previous step, at most $O(\log n)$ disjoint subtrees are located, storing points that have y-coordinate in the range $[y_q, \infty)$.
3. For each subtree in Step 2, apply the half-plane range query of [1,6] to retrieve the points that lie on the side of line L_q towards the triangle.

The correctness of the above algorithm follows from the structure used. In each of the first two steps we have to visit $O(\log n)$ subtrees. If in step 3 we apply the main memory solution of [6], then the query time becomes $O(\log^3 n + A)$, whereas the required space is $O(n \log^2 n)$. Otherwise, if we apply the external memory solution of [1], then our method above requires $O(\log^2 n \log_B n + A)$ I/O's and $O(n \log^2 n)$ disk blocks. Although the space becomes superlinear the $O(\log^2 n \log_B n + A)$ worst-case I/O complexity of our method is better than the $O(\sqrt{n/B} + A/B))$ worst-case I/O complexity of a partition tree.

5 Experimental Evaluation

The structure presented in [1] is very complicated and thus it is not easily implemented neither efficient in practice. For this reason, the solution presented in Subsection 4.2 is interesting only from a theoretical point of view. On the other hand, as implied by the following experiments, the solution presented in Subsection 4.1 is very efficient in practice.

This section compares the query/update performance of our solution with those ones that use B$^+$-trees and TPR*-tree [33], respectively. For all experiments, the disk size is set to 1 Kbyte, the key length is 8 bytes, whereas the pointer length is 4 bytes. This means that the maximum number of entries ($< x >$ or $< y >$, respectively) in both Lazy B-trees and B$^+$-trees is $1024/(8+4)=85$. In the same way, the maximum number of entries (2-d rectangles or $< x1, y1, x2, y2 >$ tuples) in TPR*-tree is $1024/(4*8+4)=27$. We use a small page size so that the number of nodes in an index simulates realistic situations. Similar methodology was used in [4]. We deploy spatio-temporal data that contain insertions at a single timestamp 0. In particular, objects' MBRs (Maximum Bounded Rectangles) are taken from the real spatial dataset LA (128971 MBRs) [Tiger], where each axis of the space is normalized to [0,10000]. For the TPR*-tree, each object is associated with a VBR (Velocity Bounded Rectangle) such that (a) the object does not change spatial extents during its movement, (b) the velocity value distribution is skewed (Zipf) towards 0

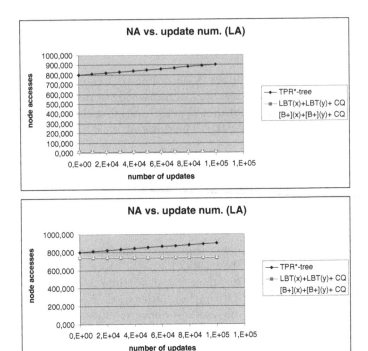

Fig. 9. $q_V len = 5$, $q_T len = 50$ $q_R len = 100$ (top), $q_R len = 2500$ (bottom)

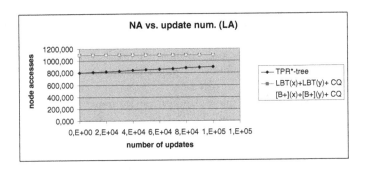

Fig. 10. $q_R len = 3000$, $q_V len = 5$, $q_T len = 50$

in range [0,50], and (c) the velocity can be either positive or negative with equal probability. For each dataset, all indexes have similar sizes. Specifically, for LA, each tree has 4 levels and around 6700 leaves. Each query q has three parameters: $q_R len$, $q_V len$, and $q_T len$, such that (a) its MBR q_R is a square, with length $q_R len$, uniformly generated in the data space, (b) its VBR is $q_V = -q_V len/2, q_V len/2, -q_V len/2, q_V len/2$, and (c) its query interval is $q_T = [0, q_T len]$. The query cost is measured as the average number of node accesses in executing a workload of 200 queries with the same parameters. Implementations were carried out in C++ including particular libraries from SECONDARY LEDA v4.1. The main performance metric is measured in number of I/Os.

Query Cost Comparison. We measure the performance of our technique earlier described (two Lazy B-trees, one for each projection, plus the query processing between the two answers), the traditional technique (two B$^+$-trees, one for each projection, plus the query processing between the two answers) and that one of TPR*-tree, using the same query workload, after every 10000 updates. The following figures show the query cost (for datasets generated from LA as described above) as a function of the number of updates, using workloads with different parameters. In figures concerning query costs our solution is almost the same efficient as the solution using B$^+$-trees ((B+)(x), (B+)(y) plus CQ). This fact is an immediate result of the same time complexity of searching procedures in both structures B$^+$-tree and Lazy B-trees, respectively. In particular, we have to index the appropriate b parameters in each projection and then to combine the two answers by detecting and filtering all the pair permutations. Obviously, the required number of block transfers depends on the answer's size and is exactly the same in both solutions for all conducted experiment.

Figure 9 depicts the efficiency of our solution toward that one of TPR*-tree. The performance of our solution degrades as the length of the query rectangle grows from 100 to 2500. It is almost equally efficient to the solution of B$^+$-trees.

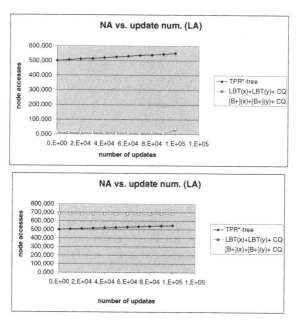

Fig. 11. $q_V len = 10$, $q_T len = 50$, $q_R len = 400$ (top), $q_R len = 2500$ (bottom)

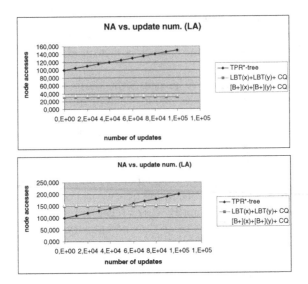

Fig. 12. $q_V len = 5$, $q_T len = 1$, $q_R len = 400$ (top), $q_R len = 1000$ (bottom)

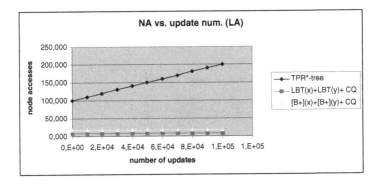

Fig. 13. $q_R len = 400$, $q_V len = 5$, $q_T len = 100$

In Figure 10 the TPR*-tree outperforms the other two solutions since the length of the query rectangle became too large (3000).

Figure 11 depicts the efficiency of our solution towards that one of TPR*-tree in case the velocity vector grows up. The performance of our solution degrades as the length of the query rectangle grows from 400 to 2500. It is almost the same efficient with the solution of B$^+$-trees.

Figure 12 depicts the efficiency of our solution toward that one of TPR*-tree in case the length of time interval extremely degrades to value 1. The performance of our solution outperforms the TPR*-tree after 50.000 updates have been occurred. It is almost the same efficient as the solution of B$^+$-trees is.

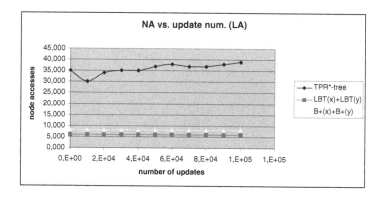

Fig. 14. Update Cost Comparison

Figure 13 depicts the efficiency of our solution toward that one of TPR*-tree in case the length of time interval enlarges to value 100. Apparently, the length of the query rectangle remains in sensibly realistic levels. It is almost the same efficient with the solution of B$^+$-trees.

Update Cost Comparison. Figure 14 compares the average cost (amortized over each insertion and deletion) as a function of the number of updates. The Lazy B-trees for the x- and y-projections (LBT(x) and LBT(y) respectively) have nearly optimal update performance and consistently outperform the TPR*-tree by a wide margin. They also outperform the update performance of B$^+$-trees by a logarithmic factor but this is not depicted clearly in Figure 14 due to small datasets.

For this reason we performed another experiment with gigantic synthetic data sets of size $n_0 \in [10^6, 10^{12}]$. In particular, we initially have 10^6 mobile objects and during the experiment we continuously insert new till their number becomes 10^{12}. For each object we considered a synthetic linear function where the velocity value distribution is skewed (zipf) towards 30 in the range [30,50]. The velocity can be either positive or negative with equal probability. For simplicity, all objects are stored using the Hough-Y dual transform. This assumption is also realistic, since in practice the number of mobile objects, which are moving with very small velocities, is negligible.

Due to gigantic synthetic dataset we increased the page size from 1024 to 4096 bytes. Since the length of each key is 8 bytes and the length of each pointer is 4 bytes the block size now becomes 341. We have not measured the performance of the initialization bulk-loading procedure. In particular, we have measured the performance of update only operations.

Figure 15 establishes the overall efficiency of our solution. It is also expected that the block transfers for the update operations will remain constant even for gigantic data sets. This fact is an immediate result of the time complexity of update procedures in the Lazy B-tree.

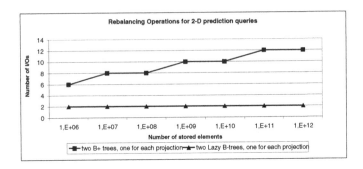

Fig. 15. Rebalancing Operations for the particular problem of 2-D Prediction Queries

6 Conclusions

We presented access methods for indexing mobile objects that move on the plane to efficiently answer range queries about their location in the future. The performance evaluation illustrates the applicability of our first solution since the second solution has only theoretical interest. Our future plan is to simplify the second complicated solution to be more implementable and as a consequence more applicable in practice.

References

1. Agarwal, P.K., Arge, L., Erickson, J., Franciosa, P.G., Vitter, J.S.: Efficient Searching with Linear Constraints. Journal of Computer and System Sciences 61(2), 194–216 (2000)
2. Agarwal, P.K., Arge, L., Erickson, J.: Indexing Moving Points. In: Proceedings 19th ACM Symposium on Principles of Database Systems (PODS), Dallas, TX, pp. 175–186 (2000)
3. Arge, L., Samoladas, V., Vitter, J.S.: On Two-Dimensional Indexability and Optimal Range Search Indexing. In: Proceedings 18th ACM Symposium on Principles of Database Systems (PODS), Philadelphia, PA, pp. 346–357 (1999)
4. Beckmann, N., Kriegel, H.P., Schneider, R., Seeger, B.: The R*-tree: an Efficient and Robust Access Method for Points and Rectangles. In: Proceedings ACM International Conference on Management of Data (SIGMOD), Atlantic City, NJ, pp. 322–331 (1990)
5. Chazelle, B.: Optimal Algorithms for Computing Depths and Layers, Brown University, Technical Report CS-83-13 (1983)
6. Chazelle, B., Guibas, L., Lee, D.L.: The Power of Geometric Duality. In: Proceedings 24th IEEE Annual Symposium on Foundations of Computer Science (FOCS), Tucson, AZ, pp. 217–225 (1983)
7. Chazelle, B.: Filtering Search: a New Approach to Query Answering. SIAM Journal on Computing 15(3), 703–724 (1986)
8. Chazelle, B., Cole, R., Preparata, F.P., Yap, C.K.: New Upper Bounds for Neighbor Searching. Information and Control 68(1-3), 105–124 (1986)

9. Comer, D.: The Ubiquitous B-Tree. ACM Computing Surveys 11(2), 121–137 (1979)
10. Dietz, P., Raman, R.: A Constant Update Time Finger Search Tree. Information Processing Letters 52(3), 147–154 (1994)
11. Gaede, V., Gunther, O.: Multidimensional Access Methods. ACM Computing Surveys 30(2), 170–231 (1998)
12. Goldstein, J., Ramakrishnan, R., Shaft, U., Yu, J.B.: Processing Queries by Linear Constraints. In: Proceedings 16th ACM Symposium on Principles of Database Systems (PODS), Tucson, AZ, pp. 257–267 (1997)
13. Guttman, A.: R-trees: a Dynamic Index Structure for Spatial Searching. In: Proceedings ACM International Conference on Management of Data (SIGMOD), Boston, MA, pp. 47–57 (1984)
14. Jensen Christian, S., Lin, D., Ooi, B.C.: Query and Update Efficient B+-Tree Based Indexing of Moving Objects. In: VLDB 2004, pp. 768–779 (2004)
15. Kaporis, A., Makris, C., Sioutas, S., Tsakalidis, A., Tsichlas, K., Zaroliagis, K.: ISB-Tree: a New Indexing Scheme with Efficient Expected Behaviour. In: Proceedings International Symposium on Algorithms and Computation (ISAAC), Sanya, Hainan, China (2005)
16. Kollios, G., Gunopulos, D., Tsotras, V.: Nearest Neighbor Queries in a Mobile Environment. In: Proceedings 1st Workshop on Spatio-Temporal Database Management (STDBM), Edinburgh, Scotland, pp. 119–134 (1999)
17. Kollios, G., Gunopulos, D., Tsotras, V.: On Indexing Mobile Objects. In: Proceedings 18th ACM Symposium on Principles of Database Systems (PODS), Philadelphia, PA, pp. 261–272 (1999)
18. Kollios, G., Tsotras, V.J., Gunopulos, D., Delis, A., Hadjieleftheriou, M.: Indexing Animated Objects Using Spatiotemporal Access Methods. IEEE Transactions on Knowledge and Data Engineering 13(5), 758–777 (2001)
19. Levcopoulos, S., Overmars, M.H.: Balanced Search Tree with O(1) Worst-case Update Time. Acta Informatica 26(3), 269–277 (1988)
20. Manolopoulos, Y.: B-trees with Lazy Parent split. Information Sciences 79(1-2), 73–88 (1994)
21. Manolopoulos, Y., Theodoridis, Y., Tsotras, V.: Advanced Database Indexing. Kluwer Academic Publishers, Dordrecht (2000)
22. Papadopoulos, D., Kollios, G., Gunopulos, D., Tsotras, V.J.: Indexing Mobile Objects on the Plane. In: Hameurlain, A., Cicchetti, R., Traunmüller, R. (eds.) DEXA 2002. LNCS, vol. 2453, pp. 693–697. Springer, Heidelberg (2002)
23. Patel, J., Chen, Y., Chakka, V.: STRIPES: an Efficient Index for Predicted Trajectories. In: Proceedings ACM International Conference on Management of Data (SIGMOD), Paris, France, pp. 637–646 (2004)
24. Raman, R.: Eliminating Amortization: on Data Structures with Guaranteed Response Time", Ph.D. Thesis, Technical Report TR-439, Department of Computer Science, University of Rochester, NY (1992)
25. Raptopoulou, K., Vassilakopoulos, M., Manolopoulos, Y.: Towards Quadtree-based Moving Objects Databases. In: Benczúr, A.A., Demetrovics, J., Gottlob, G. (eds.) ADBIS 2004. LNCS, vol. 3255, pp. 230–245. Springer, Heidelberg (2004)
26. Raptopoulou, K., Vassilakopoulos, M., Manolopoulos, Y.: Efficient Processing of Past-future Spatiotemporal Queries. In: Proceedings 21st ACM Symposium on Applied Computing (SAC), Minitrack on Advances in Spatial and Image-based Information Systems (ASIIS), Dijon, France, pp. 68–72 (2006)
27. Raptopoulou, K., Vassilakopoulos, M., Manolopoulos, Y.: On Past-time Indexing of Moving Objects. Journal of Systems and Software 79(8), 1079–1091 (2006)

28. Saltenis, S., Jensen, C., Leutenegger, S., Lopez, M.A.: Indexing the Positions of Continuously Moving Objects. In: Proceedings ACM International Conference on Management of Data (SIGMOD), Dallas, TX, pp. 331–342 (2000)
29. Saltenis, S., Jensen, C.S.: Indexing of Moving Objects for Location-Based Services. In: Proceedings 18th IEEE International Conference on Data Engineering (ICDE), San Jose, CA, pp. 463–472 (2002)
30. Salzberg, B., Tsotras, V.J.: A Comparison of Access Methods for Time-Evolving Data. ACM Computing Surveys 31(2), 158–221 (1999)
31. Samet, H.: The Design and Analysis of Spatial Data Structures. Addison Wesley, Reading (1990)
32. Sellis, T., Roussopoulos, N., Faloutsos, C.: The R^+-tree: a Dynamic Index for Multi- Dimensional Objects. In: Proceedings 13th International Conference on Very Large Data Bases (VLDB), Brighton, England, pp. 507–518 (1987)
33. Tao, Y., Papadias, D., Sun, J.: The TPR*-Tree: an Optimized Spatio-Temporal Access Method for Predictive Queries. In: Proceedings 29th. International Conference on Very Large Data Bases (VLDB), Berlin, Germany, pp. 790–801 (2003)

A Clustering Framework for Unbalanced Partitioning and Outlier Filtering on High Dimensional Datasets

Turgay Tugay Bilgin[1] and A. Yilmaz Camurcu[2]

[1] Department of Computer Engineering, Maltepe University
Maltepe, Istanbul, Turkey
`ttbilgin@maltepe.edu.tr`
[2] Department of Electronics and Computer Education, Marmara University
Kadikoy, Istanbul, Turkey
`camurcu@marmara.edu.tr`

Abstract. In this study, we propose a better relationship based clustering framework for dealing with unbalanced clustering and outlier filtering on high dimensional datasets. Original relationship based clustering framework is based on a weighted graph partitioning system named METIS. However, it has two major drawbacks: no outlier filtering and forcing clusters to be balanced. Our proposed framework uses Graclus, an unbalanced kernel k-means based partitioning system. We have two major improvements over the original framework: First, we introduce a new space. It consists of tiny unbalanced partitions created using Graclus, hence we call it micro-partition space. We use a filtering approach to drop out singletons or micro-partitions that have fewer members than a threshold value. Second, we agglomerate the filtered micro-partition space and apply Graclus again for clustering. The visualization of the results has been carried out by CLUSION. Our experiments have shown that our proposed framework produces promising results on high dimensional datasets.

Keywords: Data Mining, Dimensionality, Clustering, Outlier filtering.

1 Introduction

One of the important problems of Data mining (DM) community is mining high dimensional datasets. As dimensionality increases, the performance of the clustering algorithms sharply decreases. Traditional clustering algorithms are susceptible to the well-known problem of the curse of dimensionality [1], which refers to the degradation in algorithm performance as the number of features increases. As a result, these methods will often fail to identify coherent clusters when applied to high dimensional data.

In this paper we introduce a better framework for unbalanced partitioning and visualization based on Strehl and Ghosh's relationship based clustering framework [2]. Their framework has two fundamental components named CLUSION and OPOSSUM. CLUSION is a similarity matrix based visualization technique and

Y. Ioannidis, B. Novikov, and B. Rachev (Eds.): ADBIS 2007, LNCS 4690, pp. 205–216, 2007.

OPOSSUM is a balanced partitioning system which uses a graph partitioning tool called METIS [3]. OPOSSUM produces either sample or value balanced clusters. Neither OPOSSUM/CLUSION nor METIS has the capability of outlier detection, outlier filtering or unbalanced partitioning. Unlike Strehl and Ghosh's original framework, our proposed framework can discover the clusters of variable size and it can filter or detect outliers on high dimensional datasets.

Our framework employs Graclus [4], a kernel based k-means clustering [5] system, instead of METIS. Weighted form of kernel based k-means approach is mathematically equivalent to general weighted graph clustering approach of which METIS uses. However, kernel k-means approach is extremely fast and gives high-quality partitions. Unlike METIS, it does not force clusters to be nearly equal size. We also introduce an intermediate space called micro-partition space as an input space for the Graclus. Details of our approach have been explained in the following sections.

2 Related Work

Cluster analysis and visualization are two of the difficulties of high dimensional clustering. Cluster analysis [6,7] divides data into meaningful or useful groups (clusters). Clustering has been widely studied in several disciplines [6,8]. Some classic approaches include partitional methods such as k-means, k-medoids, hierarchical agglomerative clustering, unsupervised Bayes or EM based techniques. Above all, graph based partitioning methods frequently used on very high dimensional datasets recently. Currently, the most popular tools for graph partitioning are CHACO [9] and METIS [3,10].

Kernel K-means based Graclus system provide a mathematical connection to relate the two seemingly different approaches of kernel k-means and graph clustering. In multilevel graph partitioning algorithms, the input graph is repeatedly coarsened level by level until only a small number of vertices remain. An initial clustering is performed on this graph, and then the clustering is refined as the graph is uncoarsened level by level. However, well-known methods such as METIS[3] and CHACO [9] force the clusters to be of nearly equal size and are all based on optimizing the Kerninghan-Lin objective[11]. In contrast, Graclus removes the restriction of equal size and uses the weighted kernel k-means algorithm during the refinement phase to directly optimize a wide class of graph clustering objectives.

There are many visualization techniques for high dimensional datasets. Keim and Kriegel [12] grouped visual data exploration techniques for multivariate, multidimensional data into six classes, namely, geometric projection, icon based, pixel-oriented, hierarchical, graph-based, and hybrid. Although these techniques perform well on visualizing multidimensional datasets, they become almost useless on the datasets that have dimensions above some hundreds.

To overcome the drawbacks of dimensionality, matrix based visualization techniques [13] can be used on very high dimensional datasets. In matrix based visualization techniques, similarity in each cell is represented using a shade to indicate the similarity value: greater similarity is represented with dark shading, while lesser similarity with light shading. The dark and light cells may initially be scattered over the matrix. To bring out the potential clusterings, the rows and columns need to be reorganized so that the similar objects are put on adjacent positions. If 'real' clusters exist in the data, they should appear as symmetrical dark squares along the diagonal [13]. CLUSION, a matrix based visualization technique, is used in both Strehl and Ghosh's and our framework as explained in the following sections.

3   Relationship-Based Clustering Approach

Strehl A. and Ghosh J. proposed a different approach in [2] very high dimensional data mining. In their framework the focus was on the similarity space rather than the feature space. Most standard algorithms spend little attention on the similarity space. Generally, similarity computations are directly integrated into the clustering algorithms which proceed straight from feature space to the output space. The key difference between relationship-based clustering and regular clustering is the focus on the similarity space S instead of working directly in the feature domain F. Once similarity space is computed, a modified clustering algorithm, which can operate on the similarity space, is used to partition the similarity space. The resulting space is reordered in such a way that the points within the same cluster are put on adjacent positions. The final step is the visualization of the similarity matrix and visually inspecting the clusters.

3.1   Relationship-Based Clustering Framework

A brief overview of the general relationship-based framework which is shown in figure 1 χ is a collection of n data source. Extracting features from pure data source yields X feature space. In most cases, some data preprocessing applied to the data source to obtain feature space. Similarities are computed, using e.g. euclidean, cosine, jaccard which is denoted by Ψ yields the $n \times n$ similarity matrix S. Once the similarity matrix is computed, further clustering algorithms run on similarity space. Clustering algorithm Φ yields cluster labels λ.

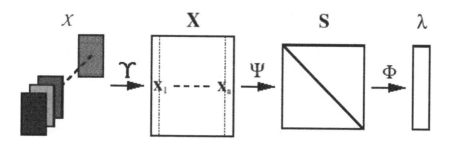

Fig. 1. Relationship-based clustering framework [2]

3.2   OPOSSUM System

Relationship-based clustering framework employs METIS for clustering. METIS is a graph partitioning tool, so it can operate on similarity space. Strehl & Ghosh call

METIS based balanced partitioning visualization system as OPOSSUM (Optimal Partitioning of Sparse Similarities Using METIS) [2]. OPOSSUM differs from other graph-based clustering techniques by sample or value balanced clusters and visualization driven heuristics for finding an appropriate k.

3.3 High Dimensional Data Visualization with CLUSION

CLUSION (CLUSter visualizatION toolkit) is a matrix based visualization tool which is used for visualizing the results. CLUSION employs the human vision system to explore the relationships in the data, guide the clustering process, and verify the quality of the results.

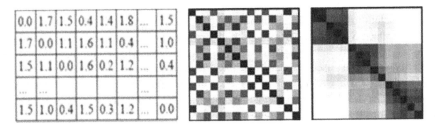

Fig. 2. a) Similarity matrix b) Matrix visualization c) CLUSION output of similarity matrix [14]

CLUSION looks at the output of the clustering routine (λ index), reorders the data points so that points with the same cluster label are contiguous, and then visualizes the resulting permuted similarity matrix, S'. More formally, the original $n{\times}n$ similarity matrix S is permuted with a $n{\times}n$ permutation matrix P. Mathematical background of the whole process can be found on [2]. Clusters appear as symmetrical dark squares along the main diagonal on the CLUSION graphs. Visualization process is shown in figure 2.

3.4 Problems with Strehl and Ghosh's Framework

The computing performance and the quality of the clusters produced by OPOSSUM/CLUSION framework is quiet impressive. However it is not perfect. It has two major drawbacks:

a) *Produces balanced clusters only:* OPOSSUM/CLUSION framework forces clusters to be of equal size. In some type of datasests balanced clusters could be important, because it avoids trivial clusterings such as k-1 singletons and 1 big clusters. But in most real world datasets forcing clusters to be balanced can result in incorrect knowledge discovery environment.

b) *No outlier filtering :* Partitioning based clustering algorithms generally suffer from outliers. As it is denoted in previous sections, OPOSSUM system produces either value or sample balanced clusters. On this kind of systems, outliers can reduce the quality and the validity of the clusters depending on the resolution and distribution of the dataset.

4 Our Proposed Graclus-Based Framework

The architecture of our framework is shown in figure 3. It consists of the following three major improvements:

a) An intermediate space is introduced. We call it 'micro-partition space' and it is denoted by P in figure 3. Similarity space has been transformed into the micro-partition space by means of Graclus. This process is represented by ρ in the figure. Graclus creates unbalanced micro-partitions consisting of 1 to n_ρ members where n_ρ depends on the number of micro-partitions. In most conditions n_ρ can take values from 3 to 10. See section 4.1 for details of micro-partition space.

b) Outlier filtering can be performed easily on the **P** space: Thanks to its capability of unbalanced partitioning, Graclus creates micro-partitions of different sizes. The singletons on the **P** space means the points that does not have enough neighbors, therefore they can be filtered or marked as outliers.

c) Graclus has been used for unbalanced clustering on filtered space: Graclus plays two important roles on our framework. The first role is creating the micro-partition space which is mentioned on item (a). The second role is unbalanced clustering of the filtered space (ΔP) which is denoted by Φ. See section 4.3 for unbalanced clustering of ΔP space.

Graclus yields λ index when applied to the ΔP space. We use CLUSION for visualizing the resulting matrix which is reordered by λ index.

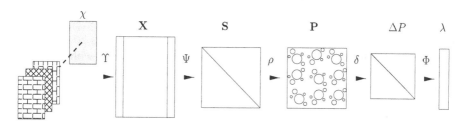

Fig. 3. Our proposed Graclus based framework

4.1 Micropartition Space

Traditional outlier filtering techniques [1] suffer from curse of dimensionality on high dimensional datasets. We propose a better approach independent of dimensionality. In our approach, we apply Graclus to the similarity space **S** to create very small partitions. We call this **P** space. Let k be the number of micro-partitions on the **P** space, then the relation between n and k should be:

$$n > k >> 1 \tag{1}$$

where n is the number of samples. Determining the k value depends on the granularity of the dataset. For the most cases it can be chosen according to the formula (2).

$$4 \geq \frac{n}{k} > 1 \tag{2}$$

Due to the unbalaced partitioning capability of Graclus, resulting tiny partitions can consist of 1,2,3 or 4 points. Therefore we call them micro-partitions.

4.2 Outlier Filtering on Micro-Partition Space

In figure 4, schematic representation of micro-partition space for $n = 7$, $k = 4$ is shown. Let x_j be a sample in dataset with $j \in \{1,..., n\}$ and seven samples be placed as shown in figure 4.a. If similarity s(1, 2) for x_1, x_2 is sufficiently small, Graclus will put them into the same partition. x_4, x_5, x_6 will be another partition accordingly. x_3 and x_7 are singletons discovered by Graclus.

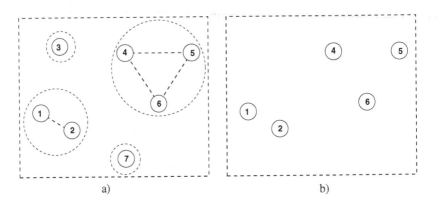

Fig. 4. a) Micro-partition space built by Graclus (n=7,k=4), b) ΔP space (Outliers filtered)

Let C_l denote the set of all objects in the l-th cluster ($l \in \{1,..., k\}$), with $x_j \in C_l$ and n_l is the number of objects within a partition ($n_l = |C_l|$). Then the outliers in P space (P_O) is:

$$P_O = \{\forall x_j, |x_j \in C_l, n_l \le T_O\} \qquad (3)$$

where T_O denotes the outlier threshold value. If a partition contains objects less than the threshold, then it is marked as outlier. Threshold value is usually chosen 1 to filter only singletons.

ΔP space could be easily obtained by subtracting P_O from P.

$$\Delta P = \{\mathbf{P} \setminus P_O\} \qquad (4)$$

4.3 Using Graclus for Clustering the Filtered Space

Our approach employs Graclus twice as it was mentioned in section 4. Here we use Graclus for clustering the filtered space (ΔP). Graclus is a fast kernel based multilevel algorithm which involves coarsening, initial partitioning and refinement phases. Figure 5 shows the graphical overview of the multilevel approach.

ΔP space is input graph for the Graclus algorithm. A graph $G = (V,E,A)$, consisting of a set of vertices V and a set of edges E so that an edge between two vertices

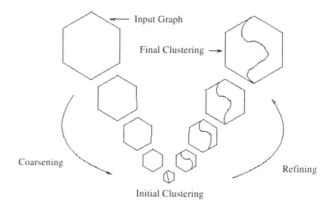

Fig. 5. Overview of the multilevel graph partitioning algorithm for k = 2 [4]

represents their similarity. The affinity matrix A is $|V| \times |V|$ whose entries represent the weights of the edges. An entry of A is 0 if there is no edge between the corresponding vertices.

In the coarsening phase, given the initial graph G_0 is repeatedly transformed into smaller and smaller graphs $G_1, G_2, ..., G_m$ such as :

$$|V_0| > |V_1| > ... > |V_m| \qquad (5)$$

The second phase is the initial clustering of the coarsened graph. A parameter is specified to indicate how small the coarsest graph should be. The parameter is generally chosen between 10k and 20k where k is the number of desired clusters. At this point, an initial clustering is performed by directly clustering the coarsest graph.

The final phase of the algorithm is the refining phase. Given a graph G_i, the graph G_{i-1} is formed (G_{i-1} is the graph used in level i - 1 of the coarsening phase). If a vertex in G_i is in cluster j, then all vertices in G_{i-1} formed from that vertex are in cluster j. This yields an initial clustering for the graph G_{i-1} which is then improved using a refinement algorithm. Graclus uses weighted kernel k-means for the refinement step.

Graclus needs the number of partitions k as input parameter. The k value we use here is different from the one in formula (1). Here k value refers to the exact number of clusters we desire. In formula (1) k refers to the number of micro partitions. From now on we denote the former one by k_1 and the latter one by k_2. On this step, k_2 values between 2 and 6 yield good results in our experiments.

Graclus-based clustering step is denoted by Φ in figure 3. Graclus performs clustering on the ΔP space and produces λ index as output space which is defined as follows:

$$\Delta P \xrightarrow{\Phi} (\lambda \in \mathcal{O}^{n_{\Delta P}} = \{1, ..., k_2\}^{n_{\Delta P}}) \qquad (6)$$

The output space contains $n_{\Delta P}$ objects is denoted by $O^{n_{\Delta P}}$ and the λ index contains cluster id (λ) of the corresponding object.

4.4 Visualization of the Results Using CLUSION Graphs

CLUSION looks at the output of a clustering routine (λ), reorders the ΔP space so that points with same cluster label are contiguous and then visualize the resulting permuted $\Delta P'$.

Since we employ Graclus twice, there are two λ indices produced during clustering process. The first one is created while forming micro-partition space (denoted by ρ) and the second one is created during the clustering of the filtered space ΔP (denoted by Φ) . We only use latter one for CLUSION input, the former one is only used for forming micro-partitions.

5 Experiments

A collection of experiments are performed on two different datasets to compare Strehl & Ghosh's and our framework.

5.1 Datasets

We evaluated our proposed framework on two different real world datasets. The first dataset is BBC news articles dataset constructed by Trinity College Computer Science Department [15]. Dataset consists of 9636 terms from 2225 complete news articles from the BBC News web site, corresponding to the stories in five topical areas (business, entertainment, politics, sport, tech) from 2004-2005. It has five natural classes. The second one is the news dataset of the popular Turkish newspaper Milliyet from the year 2006 [16]. This dataset contains 6223 terms in Turkish from 1455 news articles from three topical areas (politics,economics,sports). This dataset has three natural classes.

These two datasets seem similar at a glance. They both contain terms from news articles. However, Turkish language has very different syntax and characteristics compared to the English language. Consequently the interpretation of the results of two datasets are likely to be different.

5.2 Results

We carried out experiments on the following three frameworks for two different datasets.

- OPOSSUM: Strehl & Ghosh's METIS based original framework
- S&G(Graclus): We replaced METIS by Graclus on Strehl & Ghosh's framework for testing the quality of the clusters produced by Graclus algorithm.
- **P** space + Graclus: Our proposed framework. Only our framework is capable of filtering outliers among these three framework.

We analyzed the purity of the clusters, entropy and mutual information data and visualized the results using CLUSION graphs for visually identifying the results.

In the first experiment we used BBC news articles dataset. There are five natural clusters exists, therefore we set $k = 5$ for OPOSSUM and Graclus-based frameworks.

Table 1. Purity, Entropy and Mutual information values of BBC news articles dataset

	λ	OPOSSUM	S&G(Graclus)	P space+Graclus
Purity	1	0.97534	0.92060	0.98400
	2	0.86261	0.94563	0.97971
	3	0.72135	0.97864	0.92818
	4	0.71910	0.97714	0.94306
	5	0.98652	0.93137	0.93561
Entropy	1	0.08351	0.23415	0.05097
	2	0.29363	0.16363	0.07427
	3	0.58053	0.07658	0.21427
	4	0.55864	0.07710	0.16633
	5	0.05287	0.18694	0.18077
Mutual Info	-	0.68324	0.84951	0.86025

We need two different k values for our framework, one for creating P-space (k_1) and one for clustering (k_2). We set $k_1 = n/3$ and $k_2 = 5$. Table 1 shows the purity, entropy and mutual information of three test systems. λ denotes the cluster labels. OPOSSUM system produces poor purity and entropy values on C_2, C_3 and C_4. Graclus-based framework and our framework produce almost the same values. Although our framework drops out the outliers, purity and entropy of clusters almost remain the same. Figure 6 shows the balanced clusters produced by OPOSSUM system. If we enumerate cluster boxes from upper left to lower right as C_1, ... , C_5 we can see that especially C_3 and C_4 are problematic. They are forced clusters to be balanced. Graclus-based framework is capable of unbalanced clustering, so it correcly identifies them as shown in figure 7. However, neither of these two frameworks deals with outliers. Figure 8 shows the resulting clusters produced by our framework. It is clearly seen that it yields the purest graphics.

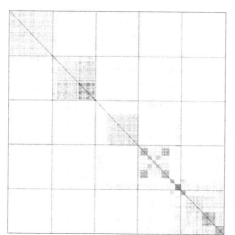
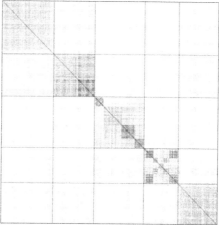

Fig. 6. Strehl & Ghosh's original framework (METIS) with k = 5

Fig.7. Strehl & Ghosh's framework (Graclus used instead METIS) with k =5

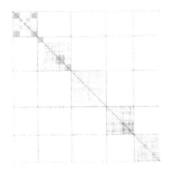

Fig. 8. Our proposed framework on BBC news articles dataset with $k_1 = n/3$ and $k_2 = 5$

Table 2. Purity, Entropy and Mutual information values of Milliyet dataset

	λ	OPOSSUM	S&G(Graclus)	μ space+Graclus
	1	0.90103	0.93064	0.98734
	2	0.90103	1.00000	0.93333
Purity	3	1.00000	0.99339	1.00000
	1	0.29384	0.24292	0.06978
	2	0.29384	0.00000	0.23346
Entropy	3	0.00000	0.04001	0.00000
Mutual Info	-	0.80411	0.90148	0.89623

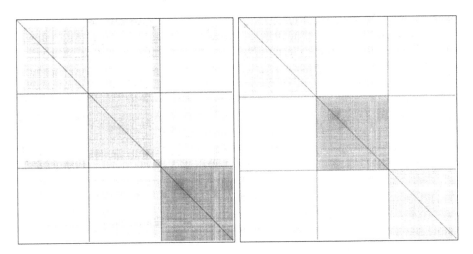

Fig. 9. Strehl & Ghosh's original framework (METIS) with k = 3

Fig. 10. Strehl & Ghosh's framework (Graclus used instead METIS) with k =3.

The second experiment was carried out on the Milliyet dataset. We set $k = 3$ for OPOSSUM and graclus based frameworks, $k_1 = n/3$ and $k_2 = 3$ for our framework. Table 2. summarizes the results of the measures. A purity value of 1 means that OPOSSUM system exactly discovers C_3, however the values for $C1$ and C_2 are not so good. Graclus-based framework exatly discovers C_2 and identifies the others with very high purity rates. Our framework discovers C_3 exactly and almost correctly identifies others.

We can see the CLUSION graph of the OPOSSUM system in figure 9. C_1 and C_2 are forced to be balanced, so they are weird, C_3 is pure and easily identified from the graph. Graclus-based framework correctly identifies the C_2 in figure 10. C_1 and C_2 seem obscure but in contrast to OPOSSUM it discovers the borders almost completely. Figure 11 shows the CLUSION graph of our framework. Our outlier filtering system performs best on this dataset and we get the purest CLUSION graphics. All cluster borders seem correctly positioned and they are easily recognized by human vision system.

On these two text datasets, our proposed framework not only outperformed the existing approaches but also included outlier filtering capability to the existing approaches. It has correctly identified the outliers, removed them and effectively clustered the filtered space.

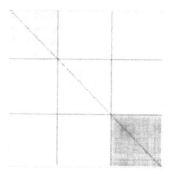

Fig. 11. Our proposed framework on Milliyet dataset with $k_1 = n/3$ and $k_2 = 3$

6   Conclusion

This paper presents a better framework for outlier filtering and unbalanced partitioning on high dimensional datasets. This is achieved by introducing a new space called micro-partition space and using Kernel K-means based Graclus system for clustering. Unlike existing framework, our framework does not limit the clusters to be equal size. We employ Graclus for both outlier filtering and clustering of filtered data.

Using two benchmark datasets, we demonstrated that, filtering the outliers using micro-partition space reduces the data to be clustered, therefore we improve the overall computing performance of the clustering algorithm. In future work, we will try to develop an efficient heuristics for determining optimal size for the micro-partitions.

Acknowledgement. The research for this article was supported by grant number FEN DKR-270306-0055 from Marmara University Scientific Research Projects Committee (BAPKO).

References

1. Bellman, R.: Adaptive Control Processes: A Guided Tour. Princeton University Press (1961)
2. Strehl, A., Ghosh, J.: Relationship-based clustering and visualization for high dimensional data mining. INFORMS Journal on Computing, 208–230 (2003)
3. Karypis, G., Kumar, V.: A fast and high quality multilevel scheme for partitioning irregular graphs. SIAM Journal of Scientic Computing 20(1), 359–392 (1998)
4. Dhillon, I., Guan, Y., Kulis, B.: A Fast Kernel-based Multilevel Algorithm for Graph Clustering. In: Proceedings of The 11th ACM SIGKDD, August 21 - 24, 2005, Chicago, IL (2005)
5. Dhillon, I., Guan, Y., Kulis, B.: Kernel k -means, spectral clustering and normalized cuts. In: Proc. 10th ACM KDD Conference, ACM Press, New York (2004)
6. Jain, A.K., Dubes, R.C.: Algorithms for Clustering Data. Prentice Hall, New Jersey (1988)
7. Kaufman, L., Rousseeuw, P.J.: Finding Groups in Data: An Introduction to Cluster Analysis. John Wiley and Sons, Chichester (1990)
8. Hartigan, J.A.: Clustering Algorithms. Wiley Publishers, New York (1975)
9. Hendrickson, B., Leland, R.: The Chaco users guide -version 2.0. Technical Report SAND94-2692, Sandia National Laboratories (1994)
10. Karypis, G., Kumar, V.: A parallel algorithm for multilevel graph-partitioning and sparse matrix ordering. Journal of Parallel and Distributed Computing 48(1), 71–95 (1998)
11. Kernighan, B.W., Lin, S.: An efficient heuristic procedure for partitioning graphs. Bell Sys. Tech. J. 49(2), 291–308 (1970)
12. Keim, D.A., Kriegel, H.P.: Visualization Techniques for Mining Large Databases: A Comparison. IEEE Trans. Knowledge and Data Eng. 8(6), 923–936 (1996)
13. Gale, N., Halperin, W., Costanzo, C.: Unclassed matrix shading and optimal ordering in hierarchical cluster analy- sis. Journal of Classication 1, 75–92 (1984)
14. Wang, J., Yu, B., Gasser, L.: Concept Tree Based Clustering Visualization with Shaded Similarity Matrices. In: Kumar, V., Tsumoto, S., Zhong, N., Yu, P.S., Wu, X. (eds.) Proceedings of 2002 IEEE International Conference on Data Mining (ICDM'02), Maebashi, Japan, pp. 697–700. IEEE Computer Society, Los Alamitos (2002)
15. BBC news articles dataset from Trinity College Computer Science Department. (Available at: https://www.cs.tcd.ie/Derek.Greene/research/ (downloaded in February 2006)
16. Milliyet dataset is provided by popular Turkish daily newspaper. http://www. milliyet. com.tr

On the Effect of Trajectory Compression in Spatiotemporal Querying

Elias Frentzos and Yannis Theodoridis

Department of Informatics, University of Piraeus,
80 Karaoli-Dimitriou St, GR-18534 Piraeus, Greece
{efrentzo,ytheod}@unipi.gr
Research Academic Computer Technology Institute,
10 Davaki St, GR-11526 Athens, Greece

Abstract. Existing work repeatedly addresses that the ubiquitous positioning devices will start to generate an unprecedented stream of time-stamped positions leading to storage and computation challenges. Hence the need for trajectory compression arises. The goal of this paper is to estimate the effect of compression in spatiotemporal querying; towards this goal, we present an analysis of this effect and provide a model to estimate it in terms of average false hits per query. Then, we propose a method to deal with the model's calculation, by incorporating it in the execution of the compression algorithm. Our experimental study shows that this proposal introduces a small overhead in the execution of trajectory compression algorithms, and also verifies the results of the analysis, confirming that our model can be used to provide a good estimation of the effect of trajectory compression in spatiotemporal querying.

Keywords: Moving Object Databases, Trajectory Compression, Error Estimation.

1 Introduction

The recent advances in the fields of wireless communications and positioning technologies activated the concept of Moving Object Databases (MOD), which has become increasingly important and has posed a great challenge to the database management system (DBMS) technology. During the last decade the database community continuously contributes on developing novel indexing schemes [1, 6, 12, 17, 10] and dedicated query processing techniques [7], in order to handle the excessive amount of data produced by the ubiquitous location-aware devices. However, as addressed by [9], it is expected that all these positioning devices will eventually start to generate an unprecedented data stream of time-stamped positions. Sooner or later, such enormous volumes of data will lead to storage and computation challenges. Hence the need for trajectory compression techniques arises.

The objectives for trajectory compression are [9]: to obtain a lasting reduction in data size, to obtain a data series that still allows various computations at acceptable (low) complexity, and finally, to obtain a data series with known, small margins of error, which are preferably parametrically adjustable. As a consequence, our interest

Y. Ioannidis, B. Novikov, and B. Rachev (Eds.): ADBIS 2007, LNCS 4690, pp. 217–233, 2007.
© Springer-Verlag Berlin Heidelberg 2007

is in lossy compression techniques which eliminate some redundant or unnecessary information under well-defined error bounds. However, existing work in this domain is relatively limited [3, 9, 13, 14], and mainly guided by advances in the field of line simplification, cartographic generalization and time series compression.

Especially on the subject of the error introduced on the produced data by such compression techniques, the single related work [9] provides a formula for estimating the mean error of the approximated trajectory in terms of distance from the original data stream. On the other hand, in this work, we argue that instead of providing a user of a MOD with the mean error in the position of each (compressed) object at each timestamp (which can be also seen as the data (im)precision), he/she would rather prefer to be informed about the mean error *introduced in query results* over compressed data. The challenge thus accepted in this paper is to provide a theoretical model that estimates the error due to compression in the results of spatiotemporal queries. To the best of our knowledge, this is the first analytical model on the effect of compression in query results over trajectory databases.

Outlining the major issues that will be addressed in this paper, our main contributions are as follows:

- We describe two types of errors (namely, *false negatives* and *false positives*) when executing timeslice queries over compressed trajectories, and we prove a lemma that estimates the average number of the above error types. It is proved that the average number of the false hits of both error types depends on the *Synchronous Euclidean Distance* [3, 9, 13] between the original and the compressed trajectory, and the perimeter (rather than the area) of the query window.
- We show how the cost of evaluating the developed formula can be reduced to a small overhead over the employed compression algorithm.
- Finally, we conduct a comprehensive set of experiments over synthetic and real trajectory datasets demonstrating the applicability and accuracy of our analysis.

The model described in this paper can be employed in MODs so as to estimate the average number of false hits in query results when trajectory data are compressed. For example, it could be utilized right after the compression of a trajectory dataset in order to provide the user with the average error introduced in the results of spatiotemporal queries of several sizes; it could be therefore exploited as an additional criterion for the user in order to decide whether compressed data are suitable for his/her needs, and possibly decide on different compression rates, and so on.

The rest of the paper is structured as follows. Related work is discussed in Section 2. Section 3 constitutes the core of the paper presenting our theoretical analysis. Section 4 presents the results of our experimental study, while Section 5 provides the conclusions of the paper and some interesting research directions.

2 Background

In this section we firstly deal with the techniques introduced for compressing trajectories during the last few years, while, we subsequently examine the related work in the field of estimating and handling the error introduced by such compression techniques.

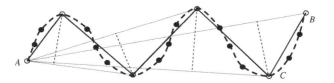

Fig. 1. Top-down Douglas-Peucker algorithm used for trajectory Compression. Original data points are represented by closed circles [9].

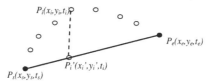

Fig. 2. The Synchronous Euclidean Distance (SED): The distance is calculated between the point under examination (P_i) and the point P_i' which is determined as the point on the line (P_s, P_e) the time instance t_i [9]

2.1 Compressing Trajectories

As already mentioned, existing work in trajectory compression is mainly guided by related work in the field of line simplification and time series compression. Meratnia and By [9] exploit existing algorithms used in the line generalization field, presenting one top-down and one opening window algorithm, which can be directly applied to spatiotemporal trajectories. The *top-down* algorithm, named TD-TR, is based on the well known Douglas-Peucker [5] algorithm (Fig. 1) introduced by geographers in cartography. This algorithm calculates the perpendicular distance of each internal point from the line connecting the first and the last point of the polyline (line *AB* in Fig. 1) and finds the point with the greatest perpendicular distance (point *C*). Then, it creates lines *AC* and *CB* and, recursively, checks these new lines against the remaining points with the same method, and so on. When the distance of all remaining points from the currently examined line is less than a given threshold (e.g., all the points following *C* against line *BC* in Fig. 1) the algorithm stops and returns this line segment as part of the new - compressed - polyline. Being aware of the fact that trajectories are polylines evolving in time, the algorithm presented in [9] replaces the perpendicular distance used in the DP algorithm with the so-called *Synchronous Euclidean Distance* (*SED*), also discussed in [3, 13], which is the distance between the currently examined point (P_i in Fig. 2) and the point of the line (P_s, P_e) where the moving object would lie, supposed it was moving on this line, at time instance t_i determined by the point under examination (P_i' in Fig. 2).

The time complexity of the original Douglas-Peucker algorithm (on which the TD-TR algorithm is based) is $O(N^2)$, with N being the number of the original data points, while it can be reduced to $O(N\log N)$ by applying the proposal presented in [8].

Although the experimental study presented in [9] shows that the TD-TR algorithm is significantly better than the opening window one (presented later in this section) in

Fig. 3. Opening Window algorithm used for trajectory Compression. Original data points are represented by closed circles [9].

terms of both quality and compression (since it globally optimizes the compression process), the TD-TR algorithm has the disadvantage that it is not an on-line algorithm and, therefore, it is not applicable to newcoming trajectory portions as soon as they feed a MOD. On the contrary, it requires the a priori knowledge of the entire moving object trajectory.

On the other hand, under the previously described conditions of on-line operation, the *opening window* (OW) class of algorithms can be easily applied. These algorithms start by anchoring the first trajectory point, and attempt to approximate the subsequent data points with one gradually longer segment (Fig. 3). As long as all distances of the subsequent data points from the segment are below the distance threshold, an attempt is made to move the segment's end point one position up in the data series. When the threshold is going to exceed, two strategies can be applied: either the point causing the violation (*Normal Opening Window, NOPW*) or the point just before it (*Before Opening Window, BOPW*) becomes the end point of the current segment, as well as the anchor of the next segment. If the threshold is not exceeded, the float is moved one position up in the data series (i.e., the window opens further) and the algorithm continuous until the last point of the trajectory is found; then the whole trajectory is transformed into a linear approximation. While in the original OW class of algorithms each distance is calculated from the point perpendicularly to the segment under examination, in the OPW-TR algorithm presented in [9] the *SED* is evaluated. Although OW algorithms are computationally expensive - since their time complexity is $O(N^2)$ - they are very popular. This is because, they are online algorithms, and they can work reasonably well in presence of noise.

Recently, Potamias et al. [13] proposed several techniques based on uniform and spatiotemporal sampling to compress trajectory streams, under different memory availability settings: fixed memory, logarithmically or linearly increasing memory, or memory not known in advance. Their major contributions are two compression algorithms, namely, the *STTrace* and *Thresholds*. The *STTrace* algorithm, utilizes a constant for each trajectory amount of memory *M*. It starts by inserting in the allocated memory the first *M* recorded positions, along with each position's *SED* with respect to its predecessor and successor in the sample. As soon as the allocated memory gets exhausted and a new point is examined for possible insertion, the sample is searched for the item with the lowest *SED*, which represents the least possible loss of information in case it gets discarded. In the sequel, the algorithm checks whether the inserted point has *SED* larger than the minimum one found already in the sample and, if so, the currently processed point is inserted into the sample at the expense of the point with the lowest *SED*. Finally, the *SED* attributes of the neighboring points of the removed one are recalculated, whereas a search is triggered in the sample for the new minimum *SED*. The proposed algorithm may be easily applied in the multiple trajectory case, by simply calculating a global minimum *SED* of all the trajectories stored inside the allocated memory.

It notably arises from the previous discussion that the vast majority of the proposed trajectory compression algorithms base their decision on whether keeping or discarding a point of the original trajectory on the value of *SED* between the original and the compressed trajectory at this particular timestamp. Consequently, a method for calculating the effect of compression in spatiotemporal querying based on the value of *SED* along the original trajectory data points, would not introduce a considerable overhead in the compression algorithm, since it would require only performing additional operations inside the same algorithm.

2.2 Related Work on Error Estimation

To the best of our knowledge, a theoretical study on modeling the error introduced in spatiotemporal query results due to the compression of trajectories is lacking; our work is the first on this topic covering the case of the spatiotemporal timeslice queries. Nevertheless, there are two related subjects: The first is the determination of the error introduced directly in each trajectory by the compression [9], being the average value of the *SED* between a trajectory p and its approximation q (also termed as synchronous error $E(q, p)$). [9] provide a method for calculating this average value as a function of the distance between p and q along each sampled point. The outcome of this analysis turns to a costly formula, which provides the average error (i.e., mean distance between p and q along their lifetime); however, there is no obvious way on how to use it in order to determine the error introduced in query results.

The second related subject is the work conducted on the context of trajectory uncertainty management, such as [4, 11, 16, 19]. This is due to the fact that the error introduced by compression can also be seen as uncertainty, and thus related techniques may be applied in the resulted dataset (e.g., probabilistic queries). However, such methodology cannot be directly used in the presence of compressed trajectory data, since the task of determining the statistical distribution of the location of the compressed trajectory using information from the original one, is by itself a complex task. Moreover, none of the proposed techniques actually deals with our essential proposal, i.e., the determination of the error introduced in query results using information about the compressed (or uncertain) data.

On the other hand, our approach is based only on the fact that the compression algorithm exploits the *SED* in each original trajectory data point and thus, introduces a very small overhead on the compression algorithm.

3   Analysis

The core of our analysis is a lemma that provides the formula used to estimate the average number of false hits per query when executed over a compressed trajectory dataset. In this work, we focus on timeslice queries, which can be used to retrieve the positions of moving objects at a given time point in the past and can be seen as a special case of spatiotemporal range queries, with their temporal extent set to zero

[18, 12]. This type of query can also be seen as the combination of a spatial (i.e., query window W) and a temporal (i.e., timestamp t) component. As it will be discussed in Section 5, the extension of our model to support range queries with non-zero temporal extent is by no means trivial and is left as future work.

It is important to mention that our model supports arbitrarily distributed trajectory data without concerning about their characteristics (e.g., sampling rate, velocity, heading, agility). Therefore, it can be directly employed in MODs without further modifications. The single assumption we make is that timeslice query windows are uniformly distributed inside the data space. Should this assumption be relaxed, one should mathematically model the query distribution using a probability distribution and modify the following analysis, accordingly. Table 1 summarizes the notations used in the rest of the paper.

Table 1. Table of notations

Notation	Description
$S, T^{\dagger}, T$	the unit space, a trajectory dataset and its compressed counterpart.
$T_i^{\dagger}, T_i$	an original trajectory and its compressed counterpart.
T_N, T_P	the set of false negatives and the set of false positives.
$R, R_{a \times b}, W_j$	the set of all timeslice queries over S, its subset with sides of length a and b along the x- and y- axes, and a timeslice query window.
n, m_i	the cardinality of dataset T and the number of sampled points inside $T_i^{\dagger}$.
$SED_i(t)$, $\delta x_i(t), \delta y_i(t)$	the function of the Synchronous Euclidean Distance (SED) between $T_i^{\dagger}$ and its compressed counterpart T_i, and its projection along the x- and y- axes.
$t_{i,k}, SED_{i,k}$ $\delta x_{i,k}, \delta y_{i,k}$	the kth timestamp on which trajectory $T_i^{\dagger}$ sampled its position, its Synchronous Euclidean Distance from its compressed counterpart T_i at the same timestamp, and its projection along the x- and y- axes.
$A_{i,j}$	the area inside which the lower-left corner of W_j has to be found at timestamp t_j in order for it to retrieve T_i as false negative (or false positive).
$AvgP_{i,N}(R_{a \times b})$, $AvgP_{i,P}(R_{a \times b})$	the average probability of all timeslice queries $W_j \in R_{a \times b}$, to retrieve T_i as false negative (or false positives).
$E_N(R_{a \times b})$, $E_P(R_{a \times b})$	the average number of false negatives (or false positives) in the results of a query $W_j \in R_{a \times b}$.

Let us consider the unit $3D$ (i.e., $2D$ spatial and $1D$ temporal) space S containing a set $T^{\dagger}$ of n trajectories $T_i^{\dagger}$ and a set T with their compressed counterparts T_i. Let also R be the uniformly distributed set of all timeslice queries posed against datasets $T^{\dagger}$ and T, and $R_{a \times b}$ be the subset of R containing all timeslice queries having sides of length a and b along the x- and y- axis respectively. Two types of errors are introduced when executing a timeslice query $W_j \in R$ over a dataset with the previously described settings:

- *false negatives* are the trajectories which originally qualified the query but their compressed counterparts were not retrieved; formally, the set of false negatives $T_N \subseteq T$ is defined as $T_N = \left\{ T_i \in T : T_i \notin W_j \mid T_i^{\dagger} \in W_j \right\}$;

- *false positives* are the compressed trajectories retrieved by the query while their original counterparts are not qualifying it; formally, the set of false positives $T_P \subseteq T$ is defined as $T_P = \{T_i \in T : T_i \in W_j \mid T_i^\dagger \notin W_j\}$.

Consider for example Fig. 4 illustrating a set of n uncompressed trajectories $T_i^\dagger$, along with their compressed counterparts T_i. Each uncompressed trajectory $T_i^\dagger$ is composed by a set of m_i time-stamped points, applying linear interpolation in-between them. Fig. 4 also illustrates a timeslice query W; though W retrieves the compressed trajectory T_1, its original counterpart $T_1^\dagger$ does not intersect the query window, encountering a false positive. Conversely, though the original trajectory $T_2^\dagger$ intersects W, its compressed counterpart T_2 is not present in the query results, forming a false negative. Having described the framework of our work, we state the following lemma.

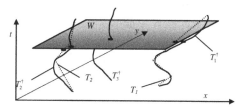

Fig. 4. Problem setting

Lemma 1. *The average number of false negatives $E_N(R_{a \times b})$ and false positives $E_P(R_{a \times b})$ in the results of timeslice queries $W_j \in R_{a \times b}$ uniformly distributed inside the unit space with sides of length a and b along the x- and y- axis respectively, over a compressed trajectory dataset is given by the following formula:*

$$E_N\left(R_{a \times b}\right) = E_P\left(R_{a \times b}\right) =$$

$$\sum_{i=1}^{n} \sum_{k=1}^{m_i-1} \frac{\left(t_{i,k+1} - t_{i,k}\right)}{(1+a)\cdot(1+b)} \cdot \left(\frac{b\left(\left|\delta x_{i,k}\right| + \left|\delta x_{i,k+1}\right|\right)}{2} + \frac{a\left(\left|\delta y_{i,k}\right| + \left|\delta y_{i,k+1}\right|\right)}{2} - \frac{e}{6} \right) \qquad (1)$$

where $e = 2\left|\delta x_{i,k} \delta y_{i,k}\right| + 2\left|\delta x_{i,k+1} \delta y_{i,k+1}\right| + \left|\delta x_{i,k} \delta y_{i,k+1}\right| + \left|\delta x_{i,k+1} \delta y_{i,k}\right|$.

Eq.(1) formulates the fact that the average error in the results of timeslice queries over compressed trajectory data is directly related to the projection of the weighted average *SED* along the x- and y- axis (i.e., $\left(t_{i,k+1} - t_{i,k}\right)$ multiplied by $\left|\delta x_{i,k}\right| + \left|\delta x_{i,k+1}\right|$ or $\left|\delta y_{i,k}\right| + \left|\delta y_{i,k+1}\right|$) multiplied by the respective opposite query dimension (i.e., $b\left(\left|\delta x_{i,k}\right| + \left|\delta x_{i,k+1}\right|\right)$ and $a\left(\left|\delta y_{i,k}\right| + \left|\delta y_{i,k+1}\right|\right)$), while e is a sum of minor importance, since it is the sum of the products between $\left|\delta x_{i,k}\right|, \left|\delta x_{i,k+1}\right|, \left|\delta y_{i,k}\right|, \left|\delta y_{i,k+1}\right|$.

3.1 Proof of Lemma 1

The average number $E_N(R_{a \times b})$ of trajectories being false negatives in the results of a timeslice query $W_j \in R_{a \times b}$, can be obtained by summing up the probabilities

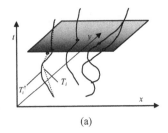 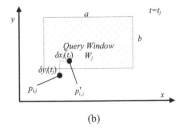

(a)   (b)

Fig. 5. The intersection of a trajectory $T_i^\dagger$ and its compressed counterpart T_i, with the plane of a timeslice query at timestamp t_j

$P\left(T_i \notin W_j \mid T_i^\dagger \in W_j\right)$ of all dataset trajectories T_i $(i=1,\ldots,n)$ to be false negative regarding an arbitrary timeslice query window $W_j \in R_{a \times b}$:

$$E_N\left(R_{a \times b}\right) = \sum_{i=1}^{n} AvgP_{i,N}\left(R_{a \times b}\right) \tag{2}$$

Similarly, the average number $E_P(R_{a \times b})$ of trajectories being false positives can be calculated by the following formula:

$$E_P\left(R_{a \times b}\right) = \sum_{i=1}^{n} AvgP_{i,P}\left(R_{a \times b}\right) \tag{3}$$

Hence, our target is to determine $AvgP_{i,N}(R_{a \times b})$ and $AvgP_{i,P}(R_{a \times b})$. Towards this goal, we formulate the probability of a random trajectory be false negative (or false positive), regarding an arbitrary timeslice query window $W_j \in R_{a \times b}$ invoked at timestamp t_j (i.e., $T_i \notin W_j \mid T_i^\dagger \in W_j$, and $T_i \in W_j \mid T_i^\dagger \notin W_j$, respectively). As also illustrated in Fig. 5(b), the intersection of trajectories T_i, $T_i^\dagger$ with the plane determined by the temporal component of W_j (i.e., timestamp t_j) will be demonstrated as two points (points $p_{i,j}$ and $p_{i,j}^\dagger$, respectively, in Fig. 5(b)) having in-between them, distance $\delta x_{i,j}$ and $\delta y_{i,j}$ along the x- and y- axis, respectively.

In order to calculate the quantity of timeslice query windows that would retrieve trajectory T_i as a false negative (false positive) at the timestamp t_j, we need to distinguish among four cases regarding the signs of $\delta x_{i,j}$ and $\delta y_{i,j}$ as demonstrated in Fig. 6 (Fig. 7, respectively). The shaded (with sided stripes) region in all four cases illustrate the area inside which the lower-left query window corner has to be found in order for it to retrieve trajectory T_i as false negative (or false positive, respectively).

However, as can be easily derived from these figures, the area of the shaded region in all four cases, is equal for both false negatives and false positives, and can be calculated by the following equation:

$$A_{i,j} = a \cdot b - \left(a - \left|\delta x_{i,j}\right|\right) \cdot \left(b - \left|\delta y_{i,j}\right|\right) \tag{4}$$

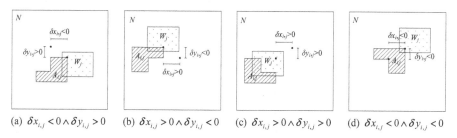

(a) $\delta x_{i,j} < 0 \wedge \delta y_{i,j} > 0$ (b) $\delta x_{i,j} > 0 \wedge \delta y_{i,j} < 0$ (c) $\delta x_{i,j} > 0 \wedge \delta y_{i,j} > 0$ (d) $\delta x_{i,j} < 0 \wedge \delta y_{i,j} < 0$

Fig. 6. Regions inside which the lower-left query window corner has to be found in order to retrieve trajectory T_i as false negative

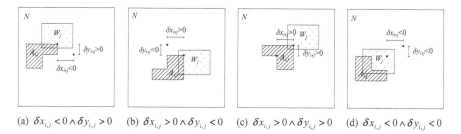

(a) $\delta x_{i,j} < 0 \wedge \delta y_{i,j} > 0$ (b) $\delta x_{i,j} > 0 \wedge \delta y_{i,j} < 0$ (c) $\delta x_{i,j} > 0 \wedge \delta y_{i,j} > 0$ (d) $\delta x_{i,j} < 0 \wedge \delta y_{i,j} < 0$

Fig. 7. Regions inside which the lower-left query window corner has to be found in order to retrieve trajectory T_i as false positive

Given that W_j is valid when it is (either partially or totally) found inside the unit space, the lower-left query window corner must be found inside a space region of area equal to $(1+a)\cdot(1+b)$. Then, since queries are uniformly distributed inside the unit space, the probability of trajectory T_i to be retrieved as a false negative or false positive at timestamp t_j is:

$$P\left(T_i \notin W_j \mid T_i^+ \in W_j\right) = P\left(T_i \in W_j \mid T_i^+ \notin W_j\right) =$$
$$\frac{1}{(1+a)\cdot(1+b)} \cdot A_{i,j} = \frac{1}{(1+a)\cdot(1+b)} \cdot \left(a\cdot b - \left(a - |\delta x_{i,j}|\right)\cdot\left(b - |\delta y_{i,j}|\right)\right) \tag{5}$$

Given also our assumption regarding the distribution of query windows, the average probability of a trajectory T_i to be false negative regarding an arbitrary query window $W_j \in R_{a \times b}$ at any timestamp can be obtained by integrating Eq.(5) over all timestamps inside the unit space. As long as $P\left(T_i \notin W_j \mid T_i^+ \in W_j\right) = P\left(T_i \in W_j \mid T_i^+ \notin W_j\right)$, it follows that:

$$AvgP_{i,N}\left(R_{a \times b}\right) = AvgP_{i,P}\left(R_{a \times b}\right) =$$
$$\int_0^1 P\left(T_i \notin W_j \mid T_i^+ \in W_j\right)dt = \int_0^1 P\left(T_i \in W_j \mid T_i^+ \notin W_j\right)dt \tag{6}$$

However, given that each original trajectory T_i is a set of m_i time-stamped points applying linear interpolation in between them, Eq.(6) is transformed as follows:

$$AvgP_{i,N}\left(R_{a\times b}\right) = AvgP_{i,P}\left(R_{a\times b}\right) =$$

$$\sum_{k=1}^{m_i-1}\frac{1}{t_{i,k+1}-t_{i,k}}\int_{t_k}^{t_{k+1}}P\left(T_i\notin W_j \mid T_i^\dagger \in W_j\right)dt = \sum_{k=1}^{m_i-1}\frac{1}{t_{i,k+1}-t_{i,k}}\int_{t_k}^{t_{k+1}}P\left(T_i\in W_j \mid T_i^\dagger \notin W_j\right)c \qquad (7)$$

and $\delta x_{i,j}$ and $\delta y_{i,j}$ can be trivially formulated as single functions of t when $t_{i,k} \le t \le t_{i,k+1}$, between sampled points:

$$\delta x_i(t) = \delta x_{i,k} + \left(t - t_{i,k}\right) \cdot \frac{\delta x_{i,k+1} - \delta x_{i,k}}{t_{i,k+1} - t_{i,k}}, \qquad \text{and} \qquad (8)$$

$$\delta y_i(t) = \delta y_{i,k} + \left(t - t_{i,k}\right) \cdot \frac{\delta y_{i,k+1} - \delta y_{i,k}}{t_{i,k+1} - t_{i,k}} \qquad (9)$$

Substituting Eq.(8), Eq.(9) and Eq.(5) into Eq.(7) and performing the necessary calculations we result in the following formula:

$$AvgP_{i,N}\left(R_{a\times b}\right) = AvgP_{i,P}\left(R_{a\times b}\right) =$$

$$\sum_{k=1}^{m_i-1}\frac{\left(t_{i,k+1}-t_{i,k}\right)}{(1+a)\cdot(1+b)}\cdot\left(\frac{\frac{b\left(\left|\delta x_{i,k}\right|+\left|\delta x_{i,k+1}\right|\right)}{2}+\frac{a\left(\left|\delta y_{i,k}\right|+\left|\delta y_{i,k+1}\right|\right)}{2}-}{\frac{2\left|\delta x_{i,k}\delta y_{i,k}\right|+2\left|\delta x_{i,k+1}\delta y_{i,k+1}\right|+\left|\delta x_{i,k}\delta y_{i,k+1}\right|+\left|\delta x_{i,k+1}\delta y_{i,k}\right|}{6}}\right) \qquad (10)$$

Finally, by substituting Eq.(10) into Eq.(2) and defining $e = 2\left|\delta x_{i,k}\delta y_{i,k}\right| + 2\left|\delta x_{i,k+1}\delta y_{i,k+1}\right| + \left|\delta x_{i,k}\delta y_{i,k+1}\right| + \left|\delta x_{i,k+1}\delta y_{i,k}\right|$ we haven proven Lemma 1. ∎

3.2 Discussion on Lemma 1

Eq.(1), the main result of Lemma 1, can be straightforwardly used to estimate the average number of false negatives and false positives for timeslice query windows with known size along the x- and y- axes (a and b, respectively). It notably arises from this formula that the average number of false negatives in the results of a timeslice query is equal to the respective average number of false positives, while their values depend mainly on the perimeter of the query window ($a+b$), rather than its area ($a\cdot b$). However, it should be explicitly mentioned that Lemma 1 holds in the case of uniformly distributed query windows only; as such, the estimated average number of false negatives and false positives serves as a metric estimating data losses due to compression, rather than providing an accurate result regarding individual queries.

Obviously, the evaluation of Eq.(1) is a costly operation; given that it involves a double sum, its time complexity is $O(n\cdot m)$ where n is the number of trajectories and m is the (average) number of sampled points per trajectory. In other words, since Eq.(1)

includes the calculation of $\delta x_{i,k}$, $\delta y_{i,k}$, between each tuple of the initial and compressed trajectories on each timestamp the trajectory was originally sampled, it requires to process the entire original dataset along with its compressed counterpart. On the other hand, as already stated in Section 2, the vast majority of the proposed trajectory compression algorithms, base their decision about the point of the original trajectory data to eliminate, on the value of the *SED*; however, since $SED_i(t) = \sqrt{\delta x_i(t)^2 + \delta y_i(t)^2}$, the respective algorithm should first evaluate $\delta x_i(t)$ and $\delta y_i(t)$ at timestamps $t_{i,k}$ producing thus, $\delta x_{i,k}$ and $\delta y_{i,k}$, respectively. Consequently, any trajectory compression algorithm using the *SED* as the criterion to decide which trajectory points to eliminate, also calculates $\delta x_{i,k}$ and $\delta y_{i,k}$. As such, Eq.(1) can be calculated during the algorithm's execution, adding very small overhead in the original algorithm; the above observation is further confirmed in our experimental study presented in the next section.

Moreover, since Eq.(1) involves the query dimensions a and b, it follows that different values of a and b will lead to different calculations for the average error. However, such an approach (i.e., evaluating Eq.(1) from the beginning for every different query size), would lead to high computation cost since it would also require $O(n \cdot m)$ time. In order to overcome this drawback, Eq.(1) can be rewritten as follows:

$$E_N(R_{a \times b}) = E_P(R_{a \times b}) = \frac{A \cdot a + B \cdot b + C}{(1+a) \cdot (1+b)}, \tag{11}$$

where $A = \sum_{i=1}^{n} \sum_{k=1}^{m_i - 1} \left(t_{i,k+1} - t_{i,k} \right) \cdot \frac{\delta y_{i,k} + \delta y_{i,k+1}}{2}$, $B = \sum_{i=1}^{n} \sum_{k=1}^{m_i - 1} \left(t_{i,k+1} - t_{i,k} \right) \cdot \frac{\delta x_{i,k} + \delta x_{i,k+1}}{2}$ and $C = -\sum_{i=1}^{n} \sum_{k=1}^{m_i - 1} \left(t_{i,k+1} - t_{i,k} \right) \cdot \frac{e}{6}$. Therefore, in the case where the average error need to be determined for a variety of query sizes (i.e., different sizes of a and b), rather than directly calculating Eq.(1) for each different query size, the three factors A, B and C could be calculated first, and be subsequently employed in Eq.(11); an approach which dramatically reduces the computation cost to $O(1)$ time.

4   Experiments

In this section, we present several sets of experiments using synthetic and real trajectory datasets. The goal of our experimental study is two-fold:

- first, to present the overhead introduced in the execution of a compression algorithm when calculating during the values of A, B and C factors introduced in Eq.(11), and,
- second, to present the accuracy of the estimation provided by our analytical model regarding the number of false negatives and false positives over synthetic and real trajectory datasets.

Regarding the datasets used, we have exploited on a real-world dataset of a fleet of trucks consisting of 276 trajectories and 112203 entries of trajectory segments [15]. We have also used synthetic datasets produced by a network-based data generator over the San Joaquin road network [2]. The synthetic trajectories generated correspond to 2000 moving objects, each one sampling its position 400 times. All the datasets where normalized in the [0,1] space. In order to test the accuracy of our model and produce compressed datasets, we implemented the TD-TR algorithm proposed by [9]. Then we executed it against all the (real and synthetic) datasets, varying its threshold between 0.001 and 0.02 of the total space, producing thus, the respective compressed datasets. Finally we used the original and compressed datasets and created several 3D R-trees [18] in order to accelerate the querying process used when performing experiments on the quality. Table 2 illustrates summary information about the (original and compressed) datasets used. The experiments were performed in a PC running Microsoft Windows XP with AMD Athlon 64 3GHz processor, 1 GB RAM and several GB of disk size. All structures and algorithms were implemented in Visual Basic.

Table 2. Summary Dataset Information

	Original Datasets		Compressed Datasets (# entries)				
			TD-TR threshold value				
	# trajectories	# entries	0.001	0.005	0.010	0.015	0.020
Trucks	273	112,203	62,067	20,935	12,636	9,274	7,571
Synthetic	2,000	800,000	229,167	120,437	88,565	74,638	65,410

4.1 Experiments on the Performance

In order to demonstrate the applicability of our proposal in trajectory data and estimate the overhead introduced in a trajectory compression algorithm when calculating the values of A, B and C factors introduced in Eq.(11), we run the TD-TR compression algorithm over the real data and measured the average execution time required for each trajectory, scaling also the threshold of the algorithm. We then modified the algorithm in order to calculate the model parameters (i.e., the values of A, B and C in Eq.(11)) within its execution and also run it against the same dataset with the same parameters. The respective results are illustrated in Fig. 8.

In particular, Fig. 8(a) and Fig. 8(b) illustrate the execution time of the TD-TR algorithm per compressed trajectory (in milliseconds), with and without the evaluation of the model parameters, against the trucks, and the synthetic datasets, respectively. A first conclusion is that the algorithm's execution time reduces as the value of the TD-TR threshold increases; this is an expected result, since typically, the number of the algorithm's iterations increase, as the value of the threshold decreases.

However, the main result gathered from Fig. 8 is that the overhead introduced in the algorithm's execution, is typically small (i.e., the difference between the two bars). In all cases, the overhead introduced in the algorithm is between 7% and 19% of the originally required execution time; furthermore, in absolute times, the overhead introduced never exceeds 0.2 milliseconds per trajectory. As a consequence, the discussion presented in Section 3.2 is further confirmed, and our model can be evaluated as an extension of the compression algorithm's execution, introducing a small / perhaps negligible overhead.

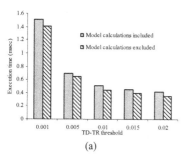
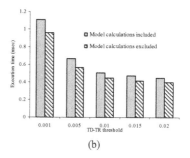

(a)   (b)

Fig. 8. Execution time for the TD-TR algorithm with and without the calculation of the model parameters over (a) the trucks, and, (b) the synthetic datasets, scaling the value of the TD-TR threshold

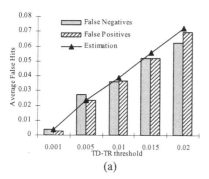
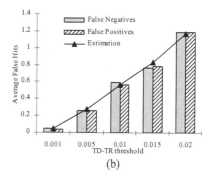

(a)   (b)

Fig. 9. Accuracy of the model scaling the value of the TD-TR threshold over (a) the trucks, and, (b) the synthetic datasets

4.2   Experiments on the Quality

The statistical measure employed in order to demonstrate the quality of our estimation, are the reported *average number of false negatives and false positives*, $\overline{E_N}$ and $\overline{E_P}$, respectively. Formally, these measures are defined as:

$$\overline{E_N} = \frac{1}{n}\sum_{i=1..n} E_{N,i}, \quad \overline{E_P} = \frac{1}{n}\sum_{i=1..n} E_{P,i} \qquad (12)$$

where, n is the number of executed queries and $E_{N,i}$ ($E_{P,i}$) the actual number of false negatives (false positives, respectively) in the ith query. In the next experiments, n is set to 10000 timeslice queries.

Our first set of experiments was performed against both the real and the synthetic datasets. Specifically, we executed 10000 rectangular timeslice queries of 0.10×0.10 size (i.e., covering 1% of unit space) randomly distributed inside the unit space, over both the original and the compressed datasets (each one stored in separate 3D R-trees), and then, utilizing the results of each particular query over the two datasets, we

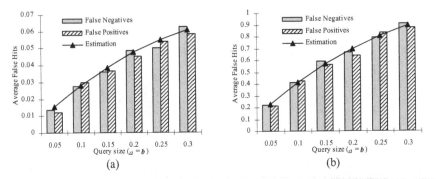

Fig. 10. Accuracy of the model scaling the square query size over (a) the trucks, and, (b) the synthetic datasets

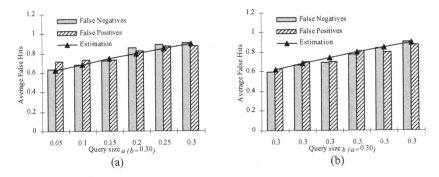

Fig. 11. Accuracy of the model scaling the non-square query size towards (a) the x- axis, and (b) the y- axis, against the synthetic datasets

counted *the actual number of false negatives and false positives*, $E_{N,i}$ and $E_{P,i}$, respectively and then calculated their average values over all the executed queries (termed as *average false hits* – negatives and positives - in all figures describing the experimental evaluation). Fig. 9 illustrates the results of this experiment scaling the value of the compression threshold over the trucks and the synthetic dataset. A first conclusion is that the average number of false hits (negatives and positives) is linear with the value of the TD-TR compression threshold.

Moreover, the estimations, $\overline{E_N}$ and $\overline{E_P}$, of our model are very close to the actual values of average false negatives and false positives reported by the experiments, regardless of the value of the compression threshold. In particular, the average error in the estimation (i.e., the difference between the bars describing the reported by the experiment average number of false negatives and positives, and our model estimation drawn by a solid line) for the synthetic dataset is around 6%, varying between 0.2% and 14%; regarding the trucks dataset (i.e., Fig. 9(a)), the average error increases around 10.6%, mainly due to the error introduced in small values of TD-TR threshold.

In our second experiment we used the same experimental settings (i.e., datasets, number of queries), but we fixed the TD-TR threshold to 0.01 and scaled the size of the timeslice query window between 0.05×0.05 and 0.30×0.30 (resulting in 0.25% and 9% of unit space, respectively). The corresponding results are illustrated in Fig. 10(a) and Fig. 10(b) against the trucks and the synthetic datasets, respectively. Again, it is clear that our model is highly accurate, producing estimates $\overline{E_N}$ and $\overline{E_P}$ with errors (i.e., the difference between the average values reported by the experiment and our model estimation) for the synthetic dataset between 0.2% and 8.7% and the average error being around 2.9% (while the respective average error for the trucks dataset is 7.5%). Another notable conclusion is that the average number of false positives and false negatives are sub-linear with the query size; an expected result gathered directly from the way that Eq.(11) involves the lengths a and b of the query sides.

In the last experiment we verified the effect of using non-square timeslice queries (i.e., $a \neq b$) over the synthetic datasets (while the experiments with the trucks dataset produced similar results). Specifically, we used timeslice query windows with sizes varying from 0.05×0.30 (where $a \ll b$) to 0.30×0.30 (where $a=b$); we also scaled the query size towards the other direction (from 0.30×0.05 to 0.30×0.30). The results of this experiment, illustrated in Fig. 11(a) and (b) respectively, resulted in similar outcomes as the ones presented in the previous paragraph regarding square (i.e, $a=b$) timeslice queries. Specifically, our model is once again very accurate, producing estimates with error between 0.6% and 7.2%, while the average error is 3.5%.

5 Conclusions

Related work on the subject of trajectory compression has focused on the development of compression algorithms also emphasizing on the error introduced in the position of each object from the compression. In this work, acknowledging that users are more likely concerned about the error introduced by the compression in spatiotemporal *query results*, we presented the first theoretical model that estimates this error in the results of timeslice queries. We provided a closed formula of the average number of false hits (false negatives and false positives) covering the case of uniformly distributed query windows and arbitrarily distributed trajectory data with various speeds, headings etc. Under various synthetic and real trajectory datasets, we first illustrated the applicability of our model under real-life requirements – it turns out that the estimation of the model parameters introduce only a small overhead in the trajectory compression algorithm - and then presented the accuracy of our estimations, with an average error being around 6%.

There are numerous interesting research directions arising from this work, including the development of the model's counterparts for nearest neighbor queries, or even more, general spatiotemporal range queries (i.e., with temporal extent $\neq 0$). More specifically, the extension of our approach towards the second direction, would require to determine the shape of the spatiotemporal space inside which the lower left range query corner (i.e., the minimum point of the range query) has to be found in order for the compressed trajectory to be retrieved as a false hit (negative of positive), in accordance with Fig. 6, Fig. 7, and subsequently to determine its volume in accordance with Eq.(4). Although this volume can be calculated when δx_i and δy_i are expressed as single functions (i.e., between consecutive timestamps), in the general case where δx_i and δy_i are expressed as multi-functions (i.e., different functions in different original trajectory line segments), the respective volume is very hard to be determined. Nevertheless, it is a great challenge for future work.

Finally, we plan to examine the application of our model to trajectory data warehouse environments, which manage aggregate data. Considering for example a trajectory data warehouse with population measurements (i.e., the number of trajectories located in each cell of the partitioned space), our model could be utilized in order to estimate the number of false hits introduced in the number of objects contained within each cell.

Acknowledgements

Research partially supported by FP6/IST Programme of the European Union under the GeoPKDD project (2005-08) [www.geopkdd.eu].

References

1. Almeida, V.T., Guting, R.H.: Indexing the Trajectories of Moving Objects in Networks. GeoInformatica 9(1), 33–60 (2005)
2. Brinkhoff, T.: A Framework for Generating Network-Based Moving Objects. Geoinformatica 6(2) (2002)
3. Cao, H., Wolfson, O., Trajcevski, G.: Spatio-temporal Data Reduction with Deterministic Error Bounds. In: Proceeding of DIALM–POMC (2003)
4. Cheng, R., Kalashnikov, D., Prabhakar, S.: Querying Imprecise Data in Moving Object Environments. IEEE TKDE 16(9) (2004)
5. Douglas, D.H., Peucker, T.K.: Algorithms for the reduction of the number of points required to represent a digitized line or its caricature. The Canadian Cartographer 10, 112–122 (1973)
6. Frentzos, E.: Indexing Objects Moving on Fixed Networks. In: Proceedings of SSTD (2003)
7. Frentzos, E., Gratsias, K., Pelekis, N., Theodoridis, Y.: Algorithms for Nearest Neighbor Search on Moving Object Trajectories. Geoinformatica 11(2) (2007)
8. Hershberger, J., Snoeyink, J.: Speeding up the Douglas-Peucker line-simplification algorithm. In: Proceeedings of SDH (1992)
9. Meratnia, N., By, R.: Spatiotemporal Compression Techniques for Moving Point Objects. In: Proceedings of EDBT (2004)
10. Ni, J., Ravishankar, C.: Indexing Spatiotemporal Trajectories with Efficient Polynomial Approximation. IEEE TKDE 19(5) (2007)
11. Pfoser, D., Jensen, C.S.: Capturing the uncertainty of moving-object representations. In: Güting, R.H., Papadias, D., Lochovsky, F.H. (eds.) SSD 1999. LNCS, vol. 1651, Springer, Heidelberg (1999)
12. Pfoser, D., Jensen, C.S., Theodoridis, Y.: Novel Approaches to the Indexing of Moving Object Trajectories. In: Proceedings of VLDB (2000)

13. Potamias, M., Patroumpas, K., Sellis, T.: Sampling Trajectory Streams with Spatiotemporal Criteria. In: Proceedings of SSDBM (2006)
14. Potamias, M., Patroumpas, K., Sellis, T.: Amnesic online synopses for moving objects. In: Proceedings of CIKM (2006)
15. Theodoridis, Y. (ed.): The R-tree Portal (accessed 5 March, 2007) URL: www. rtreeportal. org
16. Trajcevski, G.: Probabilistic Range Queries in Moving Objects Databases with Uncertainty. In: Proceedings of MobiDE (2003)
17. Tao, Y., Papadias, D.: MV3R-Tree: A Spatio-Temporal Access Method for Timestamp and Interval Queries. In: Proceedings of VLDB (2001)
18. Theodoridis, Y., Vazirgiannis, M., Sellis, T.: Spatio-temporal Indexing for Large Multimedia Applications. In: Proceedings of ICMCS (1996)
19. Trajcevski, G., Wolfson, O., Hinrichs, K., Chamberlain, S.: Managing uncertainty in moving objects databases. ACM TODS 29(3) (2004)

Prediction of Bus Motion and Continuous Query Processing for Traveler Information Services

Bratislav Predic, Dragan Stojanovic, Slobodanka Djordjevic-Kajan,
Aleksandar Milosavljevic, and Dejan Rancic

Faculty of Electronic Engineering, University of Nis
Aleksandra Medvedeva 14, 18000 Nis, Serbia
{bpredic,dragans,sdjordjevic,alexm,ranca}@elfak.ni.ac.yu

Abstract. The paper presents the methods for prediction of bus arrival times and continuous query processing as foundations of traveler information services. The time series of data from automatic vehicle location (AVL) system, consisting of time, location and speed data, is used with historical statistics and bus schedule information to predict future arrivals and motion. Based on predicted and AVL data, continuous query processing technique is proposed to extend traveler information service with notification/alarm features. Extensive experiments have shown that the proposed algorithm for bus motion prediction is efficient enough to function in real conditions and that augmented with continuous query processing techniques can produce services that useful to the travelers.

Keywords: Automatic Vehicle Location (AVL), prediction of arrival/travel time, continuous query processing, information services, intelligent transportation systems (ITS).

1 Introduction

The recent advance of automatic vehicle location (AVL) systems based on global positioning system (GPS) has provided the transit industry and public transport enterprises with tools to monitor and control operations of their vehicles and manage their fleet in an efficient and cost effective way. It has also provided the opportunity to provide customers with reliable, up-to-date information on the transit status through traveler information services (TIS). A major component of such service is travel time/location information, i.e. the time when a vehicle will reach desired location, or the location where a vehicle will be at specific time. The provision of timely and accurate transit travel time/location information is important because it attracts additional users and increases the satisfaction and convenience of transit users. Such reliable real-time information assists the customer in making their transit and multimodal trip plans and enables them to make better pre-trip and en-route decisions.

Y. Ioannidis, B. Novikov, and B. Rachev (Eds.): ADBIS 2007, LNCS 4690, pp. 234–249, 2007.

In our work we develop and apply a model to predict bus motion and bus arrival time using real-time AVL data and historical data on previous bus motion along the same route. Our travel time prediction model includes:

- Static bus schedule data and schedule adherence,
- Bus motion and traffic flow histories divided in classes depending on time of the day/week (rush hour, morning, evening, etc.),
- Real-time traffic events such as traffic congestions, jams and bus breakdowns that have influence on bus motion and arrival times.

The test bed represents bus routes running in the city of Nis, Serbia, with AVL data obtained periodically from each bus and collected at the server. The model and algorithms for prediction of bus motion are also based on data representing bus routes and bus stops, and a road network data with segments and intersections. The results of extensive experiments performed on real AVL data shown that the model and algorithms offer acceptable performance and accuracy in prediction of bus motion and bus arrival times. Based on techniques and algorithms for continuous query processing, a "notification/alarm" feature has been added to the TIS.

The rest of the paper is structured as follows. The next section presents the problem statement and review the related work in the area. Section 3 describes in detail the proposed model of bus motion and algorithms for prediction of arrival times. Section 4 presents the experimental evaluation results, whereas Section 5 describes implementation of continuous query strategy for reactive behavior of TIS. Finally, Section 6 concludes the article and shortly suggests future research work.

2 Problem Statement and Related Work

2.1 AVL System in Public Transport Monitoring

Public transport monitoring systems are increasingly based on Automatic Vehicle Location (AVL) systems, which provide the means for determining the geographic location of a vehicle and transmitting this information to a point where it can be used (transport monitoring centre). In such systems the vehicles are mostly equipped with tracking devices which contain GPS systems and are connected through dedicated wireless networks (satellite, terrestrial radio or cellular networks) to transportation control centers. Several small and big bus systems and operators worldwide use or plan to use AVL technology for mass transit tracking and for real-time traveler information services [5], [6], [9], [10], [19]. The city of Nis in Serbia has also started to deploy the AVL system for control of the public bus transport using satellite based (GPS) vehicle location method. The primary objective of the system is to monitor and track the buses in the real-time for the purpose of analysis and management of public bus transport. The second objective is to provide traveler information services and travel time/location information in variety of ways.

The high-level system architecture is shown in figure 1. The tracking devices installed in each bus include GPS receiver, GPRS modem, a microcontroller and a local memory for storage of positional/time/speed updates. Bus positions are detected periodically (every 15 seconds) and position/time/speed data is transferred to the

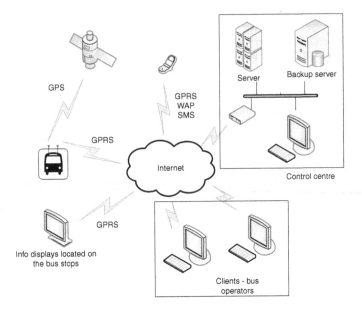

Fig. 1. The architecture of the public transport monitoring system

server located in the control centre using GSM phone network and GPRS packet data transfer service. This local memory is used in the case of restricted GPRS network coverage preventing any positional data loss. In this case, real time bus visualization is unavailable, but no data is lost. The traveler information service generally uses the current vehicle location to compute the estimated arrival time at the upstream stops using data that may include vehicle speed, distance, travel time history, bus schedule, and traffic flow history.

2.2   Related Work

Some research on this topic has been conducted and presented in the literature, but more research is required and has to be done to further improve reliability and effectiveness of traveler information services. At the heart of such services are methodology, models and techniques to predict bus arrival time, bus travel time/location, as well as continuous processing of queries on real-time AVL and predicted data [1], [2], [3], [4], [13], [16], [20]. The accurate prediction of link travel time is critical to ITS transit applications especially with the development of advanced traveler information systems and services. A number of prediction models, including historical data based models, Kalman filtering models and artificial neural network models (ANN) have been developed over the years by various transit agencies [7].

Historical data based models predict travel time for a given time period using the average travel time for the same time period obtained from a historical data base. These models are based on statistical processing of data as well as Kalman filtering technique [14]. Chien and Kuchipudi propose prediction algorithm that combines the real-time model and the historical data model [3]. First the historical data is used to obtain estimated travel time. This time is subsequently adjusted as real-time location data are obtained. The Kalman filtering algorithm is applied for travel time prediction because of its significance in continuously updating the state variable as new observations arrive. They assume that traffic patterns are cyclical and the ratio of the historical travel time on a specific link to the current travel time reported in real-time will remain constant. The Los Angeles Department of Transportation develops the Bus Arrival Information System that contains a prediction model based on historical data [13]. Such model operates using bus detectors at each stop and estimates bus travel time using information from last bus to traverse the same section. Estimated time of arrival of the next bus is calculated based on the previous bus travel time under the assumption that the current bus would experience the same or similar traffic conditions in the same segment of the corridor. Dailey, et al. in [4] and Cathey and Dailey in [1] propose the prediction algorithm in which the time series of vehicle location data is processed by the Kalman filter in their Tracking and Filter components. The third component, Predictor, is driven by reports from the Filter component and produces arrival/departure predictions as a sequence of scheduled time-points ahead of the vehicle. The Predictor uses historical statistics and is based on information obtained from the Tracking/Filter: the current time, the current estimate of the distance-until-destination, and a travel-time function for the route that the transit vehicle is on. Shalaby and Farhan develop a bus travel time prediction model also using the Kalman filtering technique [16]. They use downtown Toronto data collected with four buses equipped with AVL and automatic passenger counter (APC). They find that Kalman filtering techniques outperform the historical models, regression models, and time lag recurrent neural network models. They develop two Kalman filtering algorithms to predict running times and dwell times separately.

In [2], the authors propose an artificial neural network model to predict dynamic bus arrival time in New Jersey. They state that the back-propagation algorithm, which is the most used algorithm for transportation problems, is unsuitable for on-line application because of its time consuming learning process. Consequently they develop an adjustment factor to modify their travel time prediction using recent observed real-time data. They use generated data to predict bus arrival time, and they do not consider dwell time and scheduled data. The bus arrival times predicted by the ANNs are assessed with the microscopic simulation model CORSIM, which is calibrated and validated with real-world data collected from route number 39 of the New Jersey Transit Corporation. In [11] the authors propose an algorithm for estimating transit vehicle's link travel time from data transmitted through wireless communication channel based on a neural network algorithm tailored for each period of the day under mobile environment. In this model, link weights are updated every 7 days in order to provide feedback control based on bias between estimate and real measurement. Jeong in [7] implement three main modeling techniques that are used for prediction of bus motion and arrival time: a simple statistical model (historical data based model), a multi linear regression model and an artificial neural network model. The input variables to these models are arrival time, dwell time, and schedule adherence at each stop. After evaluation and statistical testing of the three prediction models, the artificial neural network models give superior results than historical data based models and regression models in term of prediction accuracy. The related research shows that the Kalman filtering algorithms that utilize real-time data perform well for short prediction, while ANN techniques that utilize historical data are computationally expensive and are not suitable for real-time applications [18].

3   Prediction of Bus Motion and Bus Arrival Times

Before development of algorithm for prediction of bus motion and bus arrival times it is necessary to analyze characteristics of bus motion in city traffic and to detect certain regularities that inevitably influence the prediction process. Furthermore, the vehicles in public bus transportation system follow certain rules and motion restrictions. In order to perform efficient and precise enough prediction these facts should be carefully considered and incorporated in the motion model that will be used in the prediction algorithm.

3.1   Model of Motion and Prediction

The source of vehicles positions data in the typical public transportation tracking and control system is AVL system. The AVL output represents an array of points in time and two dimensional space, as well as speed of vehicle at certain time instants. This form of location data is well suited for vehicle location visualization on the map or analysis whether a vehicle follows the assigned route. Such data also enables map matching: a technique that positions an object on a network segment, at some distance from the start of that segment, based on location information from a GPS device. The problem of bus travel/arrival time prediction is essentially two dimensional (time and one spatial dimension) (figure 2) [7].

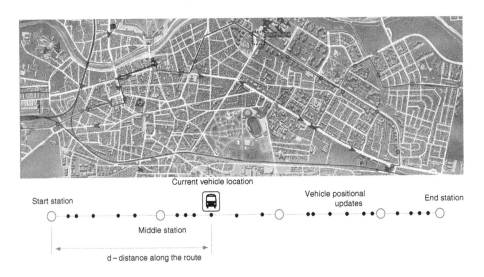

Fig. 2. Tracking the position of busses in 2D/time and 1D/time

Bus locations streamed from AVL system for each vehicle are firstly converted into distance along its assigned route using the output from the map matching component. Location updates that are available at regular time intervals are usually spaced apart with insufficient frequency to completely model every aspect of the real motion of the vehicle. The principle is shown in figure 2. All city bus routes are cyclic in nature and can be categorized as unidirectional and bidirectional. Bidirectional bus routes can be generalized as unidirectional with identical start and end stops. They allow that some (or even all) of the intermediate stops appear more than once in this "unfolded" route representation since these stops belong to both route directions.

Bus arrival/travel time prediction based solely on current motion parameters detected by AVL (speed, direction) is unsuitable for city traffic conditions and can produce only very approximate results [4]. Therefore it is necessary to use data streamed from the AVL system and matched to the routes to record the characteristics of bus motion along the route over time during different (representative) time frames (the whole day, rush hour, weekends, holydays, vacations etc.). The example of recorded location/time data along the specified route is shown in figure 3. Apart from certain irregularities it is possible to note repeating patterns in distribution of recorded positional data during certain periods of the day.

Recorded characteristics of bus motion along the route are classified according to the period it was recorded in. This information is stored as a bus motion profile. Statistical analysis of current bus motion and matching against previously formed profiles allows for much more precise estimation. More importantly, this estimation shows much more stability over longer periods of time.

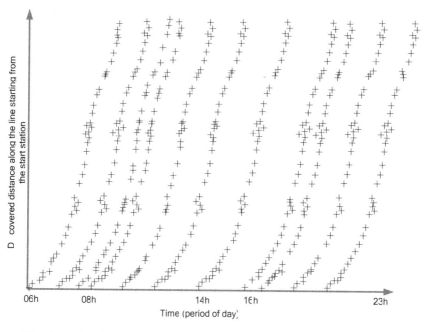

Fig. 3. Bus locations represented as distances along the route in function of time

3.2 Algorithms for Prediction of Arrival Times

As noted in the previous section, the raw location/time data streamed from the AVL system is firstly processed by the map and route matching components. Each bus is assigned with the route it is operating on. This assignment is usually fixed and predefined by the operators. But, the policy adopted by the bus operators in the city of Nis is to dynamically allocate busses to different routes during different periods of the day. Then, the filter component has to be able to dynamically assign the bus to the route in real time, based on immediate history of the bus motion. For each of the operating busses the system keeps track of last N stops along the corresponding routes. The parameter N is set to 100. Arrival of the vehicle at the bus stop is detected if at least one positional update for that vehicle is detected inside predefined circular proximity of the stop and the speed at that position(s) is lower than the predefined upper threshold. The more natural and precise condition would be at least two consecutive detected positional updates within defined proximity of the station with speed 0km/h. Unfortunately, this method would require much more frequent positional updates unnecessarily straining the system and making GPRS costs unacceptably high. If multiple bus stops fulfill the defined condition (the same stop on the route used in different directions) the stop that is closer is selected. The formed array of last 100 detected stops is matched against stop arrays for both directions for the routes that operator held. Since GPS positioning system is less reliable and precise in "urban canyon" conditions, the matching algorithm allow for up to one station to be skipped during matching [16], which is in accordance with the practice of some of the operators to have stop on demand. If no passengers are waiting on that stop and no passenger on the bus is getting of at that stop it is skipped. For each route that is matched against maintained array of last N stops, started from the last detected stop backwards, a probability is given depending on the number and order of matched stops. The route detected with the highest probability is pronounced the current route the bus is operating on. There is also a watchdog feature that designates the bus "in the garage" if no bus stops is detected for 30 minutes of bus' current ride. When the filter component assigns vehicles to the routes it is possible to proceed with bus station arrival time estimation.

In order to perform bus arrival/travel time predictions it is necessary to extract instances of route rides from the filtered positional data. The single instance of the route ride consists of all positional updates for a single vehicle received after bus departure from the starting stop to the last stop in direction A and all updates in direction B back to the starting stop. After that, a new ride for that route is initiated. In order to compensate for different traffic conditions during different periods of the day, the day period the ride was recorded is assigned to each ride recording. This information is important and used later during arrival estimation in order to assign a relevance weight factor to the each estimation. Naturally, rides recorded in similar period of the day are more relevant than others [20]. After each instance of the ride is detected and extracted it is stored in an array of rides that is associated with each of the routes. Last 10 rides for each of the routes are always available to be used by the estimation algorithm. This step of the algorithm completes all necessary data for the estimation.

From here, the estimation algorithm proceeds as follows (figure 4). For each of the bus stops estimation is performed for (S_{pr}), the list of routes that stop belongs to is extracted. Then, for each of the extracted routes a list of vehicles currently operating on that route is formed. For each of these vehicles the last stop (S_{pz}) it has stopped on and time (t_{pz}) is known. Knowing these two pieces of information it is possible to find bus stop (S_{pr}) in every one (out of 10) rides associated with that route. The algorithm notes the time (t_{pr}) when the vehicle which recorded that ride was on the station (S_{pr}) in question. In all of 10 rides the algorithm also notes the time (t_{pzv}) when the vehicle which recorded the ride arrived at the station (S_{pz}). After that the arrival time for that vehicle on the station (S_{pr}) is given as:

$$t_p = (t_{pr} - t_{pzv}) - (< current\_time > -t_{pz})$$

Based on the time period of the day the estimation is calculated and the time period for each of the estimation is included in the final estimation with the certain weight factor. As it was previously noted this weight factor models the relevance of the estimation.

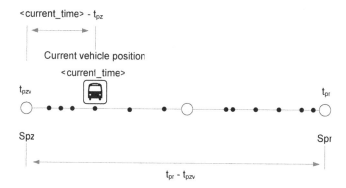

Fig. 4. Graphical representation of arrival time prediction algorithm based on one instance of bus ride along the route

The estimated arrival time is displayed at the information display located at the bus stops and other key locations in city on electronic signs, as well as other public information devices, such as interactive kiosks. Once estimated arrival/travel times are available in the system, various forms of advanced traveler information services (TIS) can be distributed to the passengers (customers). Customers with access to the Internet is be able to obtain real-time transit information, get notifications, as well as static and general transit information (e.g., schedule and fare) from a city portal or transit agency's web site. Customers with a mobile, Internet-enabled device is able to acquire real-time arrival/travel time information, either in the bus during the ride ("When the bus will arrive to the specified stop, or where the bus will be for the T minutes from now?") or at or near the bus stop ("When the bus will arrive to this stop?") using SMS or by accessing the city WEB/WAP portal [8], [15]. More

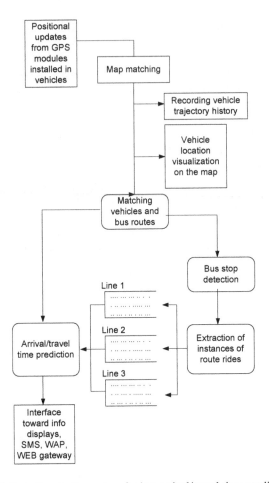

Fig. 5. Data flow in the system for bus arrival/travel time prediction

sophisticated TISs can be provided to passengers by use of continuous query techniques in the form of notification/alarms service regarding the current and predicted bus motion.

Data flow in the described system for city bus public transportation monitoring is shown in figure 5. City public transportation service is more interested in daily and monthly reports on mileage, bus stop schedule accomplishment, number of active vehicles etc., than real-time bus tracking.

4 Experiments

In this section, we present the results of some experiments to analyze the performance of our methodology and algorithms for prediction of bus arrival times.

4.1 Experimental Setup

In order to verify assumptions presented in previous sections, an experiment was performed. The experimental testbed consists of a server running in city bus service monitoring center which was tasked with receiving positional updates from the vehicles, storing this data to the database and running analysis algorithms (matching vehicles to routes, detecting bus stops and making arrival/travel time predictions). In this stage of system implementation 150 busses had tracking devices installed. For the purpose of bus arrival time prediction a bus station number 45 was selected and equipped with information display. The information display is a LED matrix capable of scrolling information that is fed to it. It contains microcontroller logic and GPRS modem that is used to connect it to the central server and transfer estimated times to be displayed. This bus station is located in the city center and a majority of routes stop at this bus stop making it a very frequent one. It was expected that the prediction algorithm applied to this bus stop would give relevant results. There are 12 bus routes stopping at the selected stop 045 (routes 6, 13, 1, 2, 5, 38, 18, 21, 20, 22, 23, 37)

The experimental setup was installed in monitoring center and fed with real life data acquired from the vehicles operating in the city. Since the prediction algorithm relies on a number of data structures to be filled the server was left running for 24 hours before the log files were extracted for analysis.

4.2 Experimental Results

Predicted arrival times are stored and latter compared to the real arrival times to assess the algorithm performance. Arrival times for various busses were recorded for bus stops 011 through 045 (011, 164, 165, 067, 070, 072, 091, and 045). Without loss of generality this section presents data concerning route 5. Table 1 shows recorded bus stop times along route 5 for the selected bus stops being analyzed. These stops on different rides were performed by different busses; hence they are overlapping in time.

Since the proposed algorithm periodically outputs estimated times of the next bus arrival on station 045 for the purpose of rating the quality of the estimation full history of arrival times was calculated. This was possible because during operation in the debug mode the algorithm created full log of the events and times of events

Table 1. Recorded buses stops during analysis period

	Stop 011	Stop 164	Stop 165	Stop 067	Stop 070	Stop 072	Stop 091	Stop 045
Ride1	7:01:37	7:02:47	7:04:53	7:06:33	7:09:07	7:10:45	7:13:07	7:14:17
Ride2	7:37:40	7:38:47	7:39:57	7:42:45	7:45:38	7:46:59	7:48:09	7:50:57
Ride3	7:45:43	7:47:07	7:48:59	7:50:37	7:52:17	7:54:23	7:57:13	7:58:37
Ride4	8:37:51	8:39:01	8:41:09	8:43:01	8:45:21	8:48:39	8:49:35	8:50:45
Ride5	9:24:59	9:25:55	9:27:35	9:30:37	9:33:13	9:35:47	9:38:07	9:39:33
Ride6	10:13:05	10:14:01	10:15:53	10:17:31	10:19:53	10:24:47	10:26:25	10:27:37
Ride7	11:02:05	11:03:15	11:05:07	11:07:13	11:09:35	11:14:03	11:15:13	11:16:51
Ride8	11:29:50	11:31:14	11:34:18	11:36:38	11:38:30	11:40:52	11:43:26	11:44:38

occurrences. The graph in figure 6 shows times needed for the bus to arrive from different consecutive stations to the station 045. The graph shows values for 8 consecutive recorded rides performed by different vehicles.

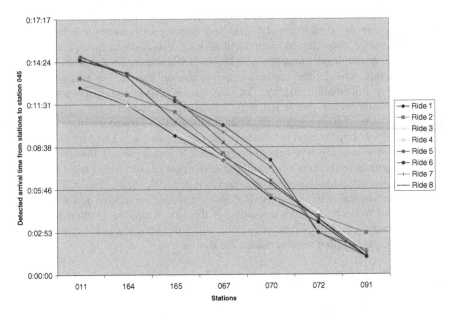

Fig. 6. Recorded times needed for busses to travel from selected stations to station 045

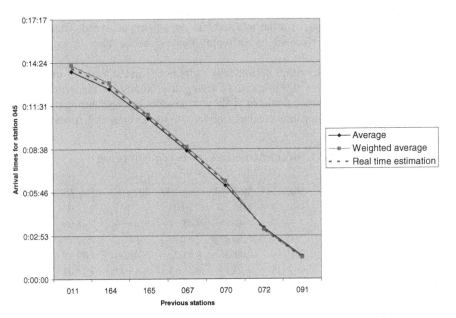

Fig. 7. Average, weighted average and real-time arrival time estimation

During recording of the data shown in figure 6, the algorithm gave estimates of arrival time of a next bus on station 045 in a real time. For easier algorithm performance evaluation the figure 7 shows graphs for averages, weighted averages of recorded arrival times and outputted estimates during recording.

4.3 Summary of the Experiments

The trend visible in the data shown in the previous section was also detected on other routes stopping at bus stop 045. Therefore, the presented results are representative and show that proposed algorithm for arrival time estimation in city public transportation system is precise enough to be used in traveler information services.

5 Processing of Continuous Queries for Notification Services

Along with standard TIS features regarding arrival/travel time prediction, an additional feature is added in the form of "alarm feature". A user can request the notification (set the alarm) to tell the service to alert when it is time to leave to catch the bus on time. The user chooses specific location/bus stop at the bus routes and sets the alarm to notify him/her when the bus will reach the selected location in defined time period ("Notify me when bus will be 5 minutes from this bus stop, and when it is time to leave the home"). The use of instantaneous queries for such service ("What is the time period until the bus arrives to the bus stop?") will not provide accurate results. Such query is evaluated immediately and the answer is transmitted to the user. To get desired time for leaving home, either user must issue the query periodically until the answer represents the time he/she specified in the query, or to rely on uncertain information given in a query answer because the bus motion onward may be slowing down due to traffic jams, accidents etc.

Thus, the solution is in application of continuous query processing methodology. The continuous query remains active over period of time and periodically examines the satisfaction of the query condition given in the form of: "Notify me when the bus will be at specific location, or will be N minutes from the specified location/bus stop". The accuracy of continuous query processing answers and notifications sent to the user depend also on predicted bus travel and arrival times at specific locations/bus stops.

The methodology for processing continuous range queries in a mobile environment is developed as a part of ARGONAUT, a service framework for mobile object data management [17]. We base our approach on the application scenario appropriate in LBS for monitoring and tracking mobile objects. In this scenario, users have wireless devices (e.g., mobile phones, PDA, in-car tracking device) that are on-line via some form of wireless communication network and can obtain their positions using Global Positioning System (GPS) technology, like in AVL system for bus transport. We successfully applied ARGONAUT continuous query processing to notification request/respond service in TIS.

The ARGONAUT methodology employs an incremental continuous query evaluation paradigm. Thus the server reports to the clients only the changes of the answer from the last evaluation time of their continuous queries. This significantly saves the network bandwidth by limiting the amount of transmitted data to the updates of the answer only rather than the whole query answer. Two types of updates are distinguished: positive updates and negative updates. The positive/negative update indicates that a certain object needs to be added/removed to/from the query answer. In the notification query/service the user is interested in a positive update of the underlying query, i.e. when the bus he/she is expecting is within the time period from the selected location/bus stop. As such, the ARGONAUT continuous query processing service is suitable for a wide range of location based notification, sense-and-respond, publish-subscribe and geofencing services.

Since the vehicle motion is constrained by an underlying transportation (road, rail, etc.) network, the methodology maintains two representations of the network data (figure 8). Both structures support the filter step of the query processing algorithm and the map matching procedure. The first representation organizes network segments according to Euclidean distance in a main memory R*-tree index structure and is named SR*-tree (Segment R*-tree). The graph representation of a spatial network is maintained by the Network Connectivity Table (NCT), a data structure that stores information about the connectivity of network segments.

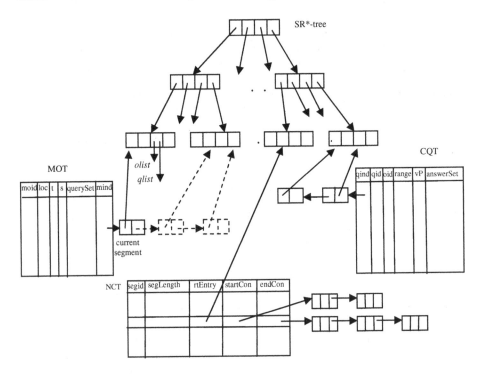

Fig. 8. Index/Data structures in Argonaut query processing framework

Both network representations are interconnected, i.e. there is a reference from SR*-tree segment representation to the corresponding NCT segment representation and vise versa. Both mobile objects (buses) and static/mobile queries are indexed by these index/data structures, according to their location/route along the road network segments and stored in corresponding list of objects/queries in leaf node entries of the SR*-tree. In addition, two additional main-memory data structures are defined (figure 8). The Continuous Query Table (CQT) is used to store information about continuous range queries (query identifier, identifier of the focal object of the query, range/distance, and valid period). For every query in CQT, a temporal query answer set is maintained, i.e. the set of objects and the periods in which they satisfy the query condition. The start of the period indicates the time when an object becomes the part of the continuous query answer (enter the query region, or approach to the defined location/bus stop on specified distance). Analogously, the end of the period indicates that the object is not a part of the query answer any more (leave the specified region or is far from the defined location by specified distance). For each mobile object in the system, an in-memory Mobile Object Table (MOT) is created and maintained. Besides the basic attributes (identifier, location, time and speed) for each mobile object the list of queries in which such object participates, either in a query answer, or as a reference object of a query is maintained. To support fast and efficient updates of leaf node entries (object and query lists), which do not require changes in the SR*-tree non-leaf node entries and the structure of the SR*-tree, we extend these two main-memory data structures to contains the pointers to the leaf node entries of the SR*-tree where the particular mobile object/query currently resides.

We introduce the pre-refinement step performed after the filter step to further refine the answer obtained by the filter step, to build main memory data structures CQT and MOT and to support periodical and incremental refinement steps. In the pre-refinement step the temporal query answer is generated for each continuous query in the system. The usage of the pre-refinement step is proven to be especially useful for processing of continuous queries over objects with known route, like busses in the system for monitoring public bus transport. For the purpose of notification/alarm feature of TIS, the ARGONAUT continuous query processing services for each continuous query (notification request) maintains the list of mobile objects/busses which represents their temporal answer sorted according to the start time of their answer periods. When a mobile object (bus) change its motion parameters (schedule) and the new bus travel time is predicted, the service updates the temporal query answer of the continuous queries (notification requests) whose answer includes this bus, accordingly. When the start time of the answer period for particular bus is reached, the service generates and delivers the notification according to the preferences of the user who requests the notification (WAP push, SMS, MMS, etc.)

The detailed algorithms for ARGONAUT continuous query processing methodology are given in [17]. With respect to the experimental results presented in [17], the ARGONAUT methodology offers acceptable performance in updating the answering periods of the continuous query answers, when receiving new values for the location or speed of a mobile object (busses), and thus is useful in providing notification in a traveler information service.

6   Conclusion

This paper presents an algorithm to predict the arrival time of public buses as a central component of the traveler information services. The algorithm combines real-time AVL data with a historical data source to produce an estimation of travel time based on the current time. Using this estimation, the algorithm calculates the expected travel-time as the arithmetic mean of the weighted estimations of travel-time. Empirical results have shown that the proposed algorithm is flexible enough to function in real life conditions and is able to produce predictions that are useful to the travelers. Based on prediction of bus motion and bus arrival time, as well as spatial and spatio-temporal data about routes and schedules, we implement continuous query processing techniques to further improve traveler information service with notification/alarm feature.

Further work can improve the prediction model and services presented here in several ways. The possible direction for future research includes adaptation and implementation of advanced techniques based on trajectory clustering and mining and Kalman filtering for better predictions based both on historical and real-time data. Furthermore, data from other sources are to be incorporated into the prediction model and algorithms. Example sources are real-time traffic condition and performance of previous predictions. Besides advanced traveler information services, we plan to use real-time and off-line analysis of AVL data and predicted bus motion for traffic monitoring and detection of traffic queues, jams and bottlenecks in the traffic flow which are of interest to other drivers.

References

1. Cathey, F.W., Dailey, D.J.: A Prescription for Transit Arrival/Departure Prediction using AVL Data. Transportation Research 11(3-4), 241–264 (2003)
2. Chien, S.I.J., Ding, Y., Wei., C.: Dynamic Bus Arrival Time Prediction with Artificial Neural Networks. Journal of Transportation Engineering, 429–438 (2002)
3. Chien, S.I.J., Kuchipudi, C.M.: Dynamic Travel Time Prediction with Real-Time and Historic Data. Journal on Transportation Engineering 129(6), 608–616 (2003)
4. Dailey, D.J., Wall, Z.R., Maclean, S.D., Cathey, F.W.: An Algorithm and Implementation to Predict the Arrival of Transit Vehicles. In: ITSC 2000 (last accessed March 2007), Available from http://www.its.washington.edu/its_pubs.html
5. Dublin Bus, Automatic Vehicle Location (AVL) and Control (AVLC) and Real Time Passenger Information System (last accessed March 2007) http://www.dublinbus.ie/projects/real_time_passenger_information_system.asp
6. iBus - Passenger information System, London Buses (last accessed March 2007) http://www.tfl.gov.uk/buses/ini-ibus.asp
7. Jeong, R.H.: The Prediction of Bus Arrival time Using Automatic Vehicle LocationSystems Data. A Ph.D. Dissertation at Texas A&M University (2004)
8. Maclean, S.D., Dailey, D.J.: Real-time bus information on mobile devices. In: Proceeding of the Intelligent Transportation Systems, pp. 988–993 (2001)
9. Maclean, S., Dailey, D.: MyBus: helping bus riders make informed decisions. IEEE Intelligent Systems 16(1), 84–87 (2001)
10. NextBus, http://www.nextbus.com/ (last accessed March 2007)
11. Park, T., Lee, S., Moon, Y.-J.: Real Time Estimation of Bus Arrival Time under Mobile Environment. In: Laganà, A., Gavrilova, M., Kumar, V., Mun, Y., Tan, C.J.K., Gervasi, O. (eds.) ICCSA 2004. LNCS, vol. 3043, pp. 1088–1096. Springer, Heidelberg (2004)
12. Philipp, M.: Arrival Time Prediction in BTAPS. BTAPS project technical guide (last accessed March 2007), http://hal.trinhall.cam.ac.uk/ bravo/ bravo/ public_html/ UserManual.html

13. Raman, M., Schweiger, C., Shammout, K., Williams, D.: Guidance for Developing and Deploying Real-time Traveler Information Systems for Transit. Federal Transit Administration, U.S. Department of Transportation, National Technical Information Service/NTIS, Springfield, Virginia (2003)
14. Ramjattan, A.N., Cross, P.A.: A Kalman Filter Model for an Integrated Land Vehicle Navigation System. Journal of Navigation 48(2), 293–302 (1995)
15. Repenning, A., Ioannidou, A.: Mobility agents: guiding and tracking public transportation users. In: Proceedings of the Working Conference on Advanced Visual interfaces, pp. 127–134. ACM Press, New York, NY (2006)
16. Shalaby, A., Farhan, A.: Prediction Model of Bus Arrival and Departure Times Using AVL and APC Data. Journal of Public Transportation 7(1), 41–61 (2004)
17. Stojanovic, D., Djordjevic-Kajan, S., Papadopoulos, A.N., Nanopoulos, A.: Continuous Range Query Processing for Network Constrained Mobile Objects. In: International Conference on Enterprise Information Systems (ICEIS), pp. 63–70 (2006)
18. Tiesyte, D., Jensen, C.: Challenges in the Tracking and Prediction of Scheduled-Vehicle Journeys. In: Proc. Of the 5th IEEE International Conference on Pervasive Computing and Communications Workshops, pp. 407–412. IEEE Computer Society Press, Los Alamitos (2007)
19. TransDB: GPS Data Management with Applications in Collective Transport (last accessed March 2007), http://www.cs.aau.dk/TransDB/
20. Wall, Z., Dailey, D.J.: An Algorithm for Predicting the Arrival Time of Mass Transit Vehicles Using Automatic Vehicle Location Data. In: Transportation Research Board 78th Annual Meeting Washington, D.C. (1999)

Optimal Query Mapping in Mobile OLAP

Ilias Michalarias and Hans-J. Lenz

Freie Universität Berlin
Garystr. 21, 14195 Berlin, Germany
{ilmich,hjlenz}@wiwiss.fu-berlin.de

Abstract. Query mapping to aggregation lattices is used in order to exploit sub-cube dependencies in multidimensional databases. It is employed in mobile OLAP dissemination systems, in order to reduce the number of handled data items and thus optimize their scheduling and dissemination process. This paper analyzes the impact of choosing between mapping to the data cube lattice or alternatively to the respective hierarchical data cube lattice. We analyze the involved tradeoffs and identify the exploitation degree of sub-cube derivability as the deciding factor. We therefore introduce an analytical framework which computes derivability related probabilities and thus facilitates the quantification of this degree. The information provided by the framework is consistent with experimental results of state of the art mobile OLAP dissemination systems.

1 Introduction

Wireless information systems become ever more present due to great advances, both in wireless network technology and mobile, portable devices. These devices are continuously getting smaller, more powerful, running more sophisticated applications. Consequently, organizations are adopting mobile applications because substantial business benefits can be safely assumed.

Increasing data volumes and accelerating update speed are fundamentally changing the role of data warehouses in modern business. More data, coming in faster, and requiring immediate conversion into decisions means that organizations are being confronted with the need for real-time data warehousing. Evidently, enabling mobile, portable devices to participate in such systems is a fundamental requirement.

As a consequence, the research field of mobile OLAP (*mOLAP*) has emerged. As an application scenario, we refer to an example described in [1], which considers the case of brokers accessing a stock market gallery data mart. At opening and closing times, different stocks in different financial dimensions are analyzed by many traders using some mobile device, typically laptops. Some of these stocks are more popular than other, similarly, some analytical dimensions are more important than other. In such scenarios, a data mart equipped with a broadcast gateway is responsible for serving the incoming requests.

Data broadcast technology plays a fundamental role on wireless dissemination systems, since it is a 1-to-n process, enabling enhanced scalability. Nevertheless, early data broadcast systems were designed under the principle, that the number of handled data items is not too high, that the data items occupy relatively small size and do not possess

Y. Ioannidis, B. Novikov, and B. Rachev (Eds.): ADBIS 2007, LNCS 4690, pp. 250–266, 2007.

any semantic connection with each other (e.g. web pages). Content or characteristics of data items were practically ignored. In this context, the dissemination of multidimensional cubes, which are order of magnitude bigger than web pages, and between which semantic connections exist, has to be tackled differently.

To cope with this problem, *mOLAP* dissemination systems employ *query mapping* to aggregation lattices in order to reduce the number of handled data items. Moreover, they exploit *sub-cube derivability* to serve multiple requests with one transmission and thus increase scalability. Since the number of different multidimensional queries, which can be issued by the mobile clients is unlimited, the reduction of handled data items is achieved by mapping them to an aggregation lattice, which is a graph representing different views of the data cube. Having mapped queries to lattice nodes, two different lattice nodes for which a dependency exists, do not have to be served by two separate connections, but from a single broadcast instead. All transmissions at the physical layer of the wireless network are anyway broadcasts.

There are two types of aggregation lattices, depending on the inclusion of hierarchical levels of dimensions or not. The first discussed type is the Data Cube Lattice (*DCL*), which consists of nodes that represent dimensions at their lowest level (fact table data) ignoring hierarchies, leading to *coarse grained* representations. The second type is the hierarchical Data Cube Lattice (*hDCL*) which includes hierarchies leading to *fine grained* representations.

Previous work regarding query mapping does not provide a clear answer regarding which aggregation lattice is more appropriate in mobile data warehouses. The main reason is that this problem strongly depends on the observed domain. For example, in [2], *hDCL* mapping is used for cache management. In contrast, *mOLAP* dissemination systems in [3,1] operate on *DCL* query mapping.

In this paper, we deal with optimal mapping of queries in *mOLAP* systems. We analyze the tradeoffs that arise when choosing between them and identify the exploitation degree of sub-cube derivability as the deciding factor. Therefore, we propose an analytical model to quantify this degree in both query mapping modes. On the one hand, our analytical model provides a general framework for computation of sub-cube derivability probabilities. On the other hand, it computes probabilities explicitly related to the *mOLAP* domain. Its main feature is, that it does not presume a specific query distribution, making it applicable for any query workload. We believe that the information provided by the framework should be useful for any future *mOLAP* system. We use the information provided by the model for the evaluation of state of the art *mOLAP* dissemination systems with respect to query mapping. Both analytical and experimental results reveal that coarse grained query mapping is more suitable for *mOLAP*. In short the major contributions of this paper are:

- Presenting the case of query mapping in multidimensional databases (*MDDBs*)
- Analyzing the tradeoffs of different query mapping modes in mobile OLAP systems
- Introducing an analytical framework for computing sub-cube derivability probabilities
- Evaluating state of the art *mOLAP* systems with respect to query mapping.

The remainder of the paper is structured as follows: In Section 2, we describe aggregation lattices and the way in which multidimensional queries are mapped to them.

Section 3 explains why query mapping is critical in *mOLAP* systems. In Section 4, we introduce an analytical framework for evaluating the exploitation degree of derivation semantics in *mOLAP* systems. In Section 5, we compare state of the art *mOLAP* schedulers with respect to query mapping and provide experimental results. Finally, Section 7 concludes our results and presents future work topics.

2   Derivability in Aggregation Lattices

Consider two queries q_1 and q_2. q_1 is dependent on q_2 ($q_2 \succeq q_1$) when q_1 can be answered using the result of q_2. This property is known as *query dependency*. The reuse of queries in *MDDBs* is mainly related to the *data cube* operator, [4,5]. The data cube operator is the union of all possible group-by operators applied on a fact table. [6] notes that some of the group-by queries in the data cube query can be answered using the results of other. In *MDDBs* there are two types of query dependencies. *Dimension dependency* is caused by the interaction of the different dimensions with one another. *Attribute dependency* is caused within a dimension by attribute hierarchies.

Data cubes are created from group-by queries, for which dependencies exist. Therefore, dependencies also exist between the produced sub-cubes. An aggregation *lattice* is a graph, whose nodes represent different views of the data cube and whose arcs the *derivability of sub-cubes*. As to be seen in the following paragraphs, there are two kinds of aggregation lattices, depending on whether the hierarchical levels of the dimensional attributes are considered or not.

Derivability is not a new research area. It was introduced it in the context of statistical databases, when checking derivability of summary data under different classifications. In the following years, derivability became very important, both in relational and multidimensional databases (*MDDBs*), in the context of materialized views. Derivability is exploited in finding the optimal set of materialized views, [7].

It is important to underline, that our work observes derivability from a completely different perspective. We do not use derivability to optimize query execution by a database server, but to optimize the scheduling and dissemination process of a *mOLAP* application server instead.

2.1   Data Cube Lattice

A data cube stemming from a schema with D dimensional attributes has 2^D possible sub-cubes. Assume a 3-dimensional cube. Let the three dimensional attributes be *Product, Store* and *Time*. Table 1 contains the declared hierarchical levels of this schema. Given multidimensional data, the *Data Cube Lattice* is the lattice of the set of all possible grouping queries that can be defined on the foreign keys of the fact table, [8]. It is a directed, acyclic graph, that depicts the relationships between all 2^D sub-cubes. In Fig.1, one 3-dimensional cube, three 2-dimensional sub-cubes and three 1-dimensional sub-cubes are shown. Each of every possible sub-cube is represented in the lattice by one node.

DCL nodes can be labeled by a sequence of bits (bitmap), as depicted in Fig.1. The number of necessary bits is equal to the dimensionality of the cube. Each bit represents

Table 1. Declared hierarchies of a 3-dimensional data cube

Hierarchies						
Product		**Store**		**Time**		
ALL	P_0			ALL	T_0	
↑	↑	ALL	S_0	↑	↑	
Category	P_1	↑	↑	Year	T_1	
↑	↑	StoreId	S_1	↑	↑	
Code	P_2			Day	T_2	

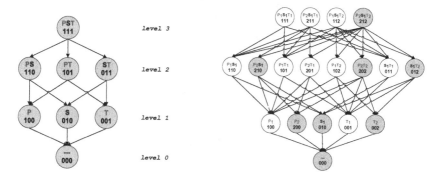

Fig. 1. The DCL of a 3-dimensional data cube   **Fig. 2.** The hDCL of a 3-dimensional data cube

one dimension. If the dimension exists in the node then the bit is set to 1, otherwise it is set to 0.

Note that the hierarchical levels of each dimension are completely ignored in *DCLs*. This is how the two lattices discussed in this paper differ from one another.

2.2   Hierarchical Data Cube Lattice

Hierarchies on aggregation lattices were introduced in [6,9]. A *hierarchical Data Cube Lattice*, is a directed, acyclic graph, that depicts the relationships between all $\prod_{n=1}^{D}$ $(gr_i + 1)$ sub-cubes, given a D-dimensional cube and the number of grouping attributes (levels) gr_i of dimension i.

Note that this definition considers a limited set of grouping attributes. It considers only the dimension's key attribute and the non key attributes that are functionally dependent on it (practically an attribute hierarchy). If the set of grouping attributes additionally includes attributes not functionally dependent on the key attribute, the produced lattice is called *MD-lattice* (Multidimensional lattice), as defined in [8]. *MD-lattices* are not considered in this paper since the number of its nodes is so high, that the fundamental objective behind query mapping of reducing the handled data items, can not be fulfilled. Moreover the hierarchies are *strict*, namely each object at a lower level belongs to only one value at a higher level.

It is important to underline, that the only key difference between *DCL* and *hDCL* is in the degree of detail. Figure 2 contains the *hDCL* that corresponds to the schema of Table 1. There are $(gr_P+1) \times (gr_S+1) \times (gr_T+1)=3 \times 2 \times 3=18$ possible views or sub-cubes. Similarly to the bitmap of the *DCL*, we notate *hDCL* nodes with a sequence of digits. Each digit represents the dimension and the hierarchical level. If the dimension does not exist in the node then the digit is set to 0, otherwise it is set to the number of hierarchical level. For example the sub-cube P_2S_1 in Fig.2 is marked with *210*. The first digit is 2, and indicates the second hierarchical level of dimension *Product*, the second digit is 1 indicating the first hierarchical level of dimension *Store* and the last digit is 0, indicating that dimension *Time* has been projected.

A sub-cube can be derived by another sub-cube, when there is a path in the lattice that connects the nodes that represent these sub-cubes. This is known as *subsumption*. This derivation is feasible for distributive SQL aggregation functions such as *sum, min, max* or *count*, but is neither allowed for algebraic functions such as *average* or *covariance*, nor for holistic functions like *median*, [10]. Aggregating over a time dimension is allowed if the fact (measure) is of type *flow*.

The *ancestor* and *descendant* operators, as defined in [8,11], reveal query dependencies. The result of the ancestor operator $\oplus$ on two queries is the smallest query containing all the necessary information to answer both queries, whereas the descendant operator $\ominus$ computes the greatest among the set of attributes characterizing the queries that can be computed by the two queries. In this paper though, we do not use the term ancestor as an operator, but as a property to representing sub-cube derivability. In this context, a lattice node n_a is an ancestor of a lattice node n_b when there is a downward path from n_a to n_b in the lattice.

hDCL arcs represent either dimension or attribute dependencies, whereas *DCL* arcs represent exclusively dimension dependencies.

2.3 Query Mapping

Query mapping is the process of mapping a query to its respective node in the lattice. The question which immediately arises is which of the two lattices is more appropriate for the mapping of the queries. Before dealing with this issue, we provide an example of how a query would be mapped in the two lattices. Assume the following query targeting the schema of Table 1 and one measure attribute. Without loss of generality, we use Multidimensional Expressions (MDX) as the query language.

```
SELECT
  { [Product].[Category].[Drinks] } ON COLUMNS,
  { [Time].[Year].AllMembers } ON ROWS
FROM [TestCube]
```

This query is mapped to node *PT* of the *DCL*, since it only matters that dimensions *Product* and *Time* are involved. Considering the *hDCL*, the query is mapped to node P_1T_1, since the member *Drinks* belongs to the hierarchical level *Product.Category* (P_1) and all members of the hierarchical level *Time.Year* must be retrieved (T_1).

3 Query Mapping in Mobile OLAP Dissemination Systems

mOLAP dissemination systems serve multidimensional queries issued from mobile clients. They are responsible for the scheduling of queries and the dissemination of results. In a typical *mOLAP* scenario, mobile clients issue queries to one or more data cubes. The server is able to respond to any incoming query referring to a data cube, either by having already stored all possible sub-cubes, or by retrieving them from the backend data warehouse, when necessary. There is a single broadcast channel that is monitored by all clients and an uplink channel for issuing requests. Clients continuously monitor the downlink channel after making a request, to check for requested data.

The entire concept of *mOLAP* systems is founded on providing offline functionality. This is crucial when considering that portable devices are not permanently connected to a network. Therefore, clients are aware of the schema metadata and can consequently *locally* store received data and perform local processing in order to enable subsequent analysis.

FCLOS, [3], and *STOBS*, [1], are scheduling algorithms, explicitly designed for dissemination of multidimensional data in wireless networks. Despite many differences in their scheduling approach, both of them build on *subsumption*. In this way, multiple benefits are gained. Not only can the server answer the requests faster, but the clients experience improved access time as well. Moreover, energy and generated traffic are reduced.

mOLAP systems map the incoming queries to respective nodes of the lattice. This is justified by the fact that the point-to-point communication model is inefficient in a wireless network, especially for an OLAP application. In other words, serving each request individually, assuming that the client is not able to perform local processing, not only exhibits poor performance in terms of access time, but it does not scale with the number of requests as well. Query mapping substantially reduces the number of data items that the scheduler has to handle, which is a desired property. Figure 3 highlights the question which naturally arises: which of the two lattices should be used for the mapping of the queries? The *DCL* or the *hDCL*? How does this decision influence the system's behavior?

Both *FCLOS* and *STOBS* operate on *DCL* mapping. This results in a *coarse grained* query mapping. The absence of hierarchies, practically imposes that the clients have to locally aggregate fact table tuples in order to compose the hierarchical level aggregations.

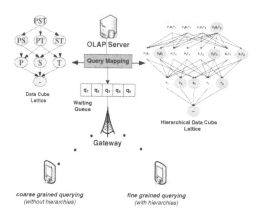

Fig. 3. A mOLAP dissemination system and two possible query mappings

Table 2. Query mapping tradeoff in mOLAP

Querying/Property	DCL	hDCL
Generated Traffic for a given query	-	+
Client Side Processing	-	+
Number of Data Items	+	-
OLAP Operations	+	-

Although previous work in this area mainly assumes coarse grained querying, it is worthwhile investigating the impact of *fine grained* query mapping, which can be achieved when queries are mapped to *hDCLs*. In fine grained querying scenarios, transmitted structures include aggregated values according to the dimension hierarchy levels. Such querying imposes that end users receive datasets, which require less local processing. Moreover, data transfer in preferred granularities is supported.

Table 2 summarizes the tradeoff which arises: On the one hand, using fine grained query mapping, generated traffic for a given query is reduced (or is equal in worst case). Moreover, clients are forced to perform intensive local processing, which might not always be feasible. On the other hand though, sub-cube derivability is more straightforward in *DCL*. A *DCL* consist of less nodes than its respective *hDCL*, so that the number of handled data items is reduced. In addition to that, *DCL* mapping enables better offline functionality and OLAP operations like roll-up or drill-down. Consider the query of Section 2.3. A client having received the sub-cube PT can answer locally any query that involves only these two dimensions, namely it is able to locally perform any roll-up, drill-down or projection. A client having received the sub-cube P_1T_1 instead, is not able to locally perform any drill-down. A drill-down would inevitably have to invoke a new query to the server.

4   An Analytical Framework for Computation of Query Dependency Probabilities

4.1   Model

In the previous Section we considered a *mOLAP* system and described the tradeoffs imposed by selecting between coarse and fine query mapping modes. We suggest that the exploitation of the derivation semantics is the key factor influencing the query mapping mode selection. However, evaluating this exploitation degree is not a straightforward task. This section presents an analytical model which facilitates this evaluation. This model provides the basis for computing relevant subsumption probabilities, that *mOLAP* schedulers has to consider when scheduling and disseminating requests.

mOLAP schedulers, regardless of their individual approach, are keen to exploit the subsumption property among nodes of the lattice, but naturally, there are fundamental differences in the way that this is pursued. Therefore, the usefulness of a probability strongly depends on the scheduler itself. Considering that our objective is to evaluate the appropriateness of the two different query mapping modes, we do not restrict our

Table 3. Notation

Notation	Definition
S	Set which contains every lattice node
Q	Multiset which contains every queue element
D	Number of dimensions
n_j	A lattice node ($n \in S$)
n_j^i	The i_{th} bit of the binary representation of n_j where i $\in [0, D-1]$
$n_a \succeq n_b$	n_a is an ancestor of n_b
$n_j \succeq MS$	n_j is an ancestor of every element of the multiset MS (ancestor of $|MS|$ elements)
e_j	An element of the queue ($e \in Q$)
gr_d	Number of hierarchical attributes of dimension d
D_n	Set of all dimensions existing in node n
$lev(n, d)$	The hierarchical level of node n at dimension d ($lev(n, d) \in (0, gr_d]$)
P_d	The probability that dimension d exists in node n
$P_{d,l}$	Given that dimension d exists in node n, the probability that the hierarchical level l is selected
p_n	The probability that a node n is selected from a multiset Q of lattice nodes

model to a specific scheduler. Nonetheless, we will show in following paragraphs which of the computed probabilities are directly applicable to *STOBS* and to *FCLOS*.

Consider the architecture of Fig.3, where mobile clients issue queries targeting a given data cube. These queries are mapped by the scheduler to the corresponding lattice node and subsequently inserted in the waiting queue. Naturally, one node of the lattice might have been requested by more than one client. Therefore, our model considers the queue as a *multiset Q* and thus $|Q|$ is the size of the queue. Table 3 provides a notation overview.

We employ the following evaluation metrics, which can be used by any scheduler, when trying to exploit dependencies:

1. $P(e_a \succeq e_b)$: In a multiset Q of lattice nodes, the probability that a selected element $e_a \in Q$ is an ancestor of another selected element $e_b \in Q$
2. $P(e_a \succeq Q)$: In a multiset Q, the probability that a selected element $e_a \in Q$ is an ancestor of every element in Q
3. $P(\exists\, e : e \succeq Q)$: In a multiset Q, the probability that there exists one element, which is an ancestor of every element in Q
4. $P(e_a \succeq q^+ \subseteq Q)$: In a multiset Q, the probability that a selected element $e_a \in Q$ is an ancestor of at least $|q|$ ($|q| - 1 + itself$) elements of Q
5. $P(\exists\, e : e \succeq q \subseteq Q)$: In a multiset Q, the probability that there exists at least one element, which is ancestor of exactly $|q|$ ($|q| - 1 + itself$) elements of Q

$P(e_a \succeq e_b)$ for example provides a general evaluation of sub-cube derivability and is not directly to *mOLAP*. $P(e_a \succeq q^+ \subseteq Q)$ is intended to be useful for any *mOLAP* scheduler that checks subsumption after having determined the element to be transmitted, as in the case of *STOBS*. $P(\exists\, e : e \succeq q \subseteq Q)$ on the contrary targets schedulers that consider every element as candidate for transmission and decide according to its subsumption, as in the case of *FCLOS*.

A fundamental objective of this work is to be applicable regardless of the query distribution. Assuming a specific distribution would severely restrict the applicability of our approach. Nevertheless, the discussed probabilities are obviously dependent on the query distribution. We overcome this by forcing the server to collect simple statistics about the incoming queries. Particularly, the server measures the probability P_d that a dimension d exists in an incoming query. For example, using the data cube of Fig.1 the probability $P_{Product}$ that dimension *Product* appears in an incoming query can be computed as $P_{Product} = p_{PST} + p_{PS} + p_{PT} + p_P$. Additionally, since every dimension has hierarchical levels, for every dimension d, the probability $P_{d,l}$ that the hierarchical attribute l is requested, is also computed by the server. In this way, our model can be applied to any possible query distribution. Naturally, in order to be able to react to fluctuations in the incoming workload the server will have to consider each time the more recent statistics.

It should be clear that if we consider every node of the lattice: $\sum_{n \in S} p_n = 1$, and that for every dimension d:

$$\sum_{l=0}^{gr_d} P_{d,l} = 1.$$

4.2 DCL Mapping

A quick way to find out whether a node n_b can be subsumpted by n_a is to apply the binary AND operator on their binary representations:

- If $\left(n_a{}^{bin}\ AND\ n_b{}^{bin}\right) = n_b{}^{bin}$ then $n_a \succeq n_b$
- If $\left(n_a{}^{bin}\ AND\ n_b{}^{bin}\right) \neq n_b{}^{bin}$ then $n_a \not\succeq n_b$

In Fig.1 for example, it is $\left(n_{PS}{}^{bin}\ AND\ n_S{}^{bin}\right) = (110\ AND\ 010) = 010 = n_S{}^{bin}$, which confirms that sub-cube S can be subsumpted by sub-cube PS.

The *DCL* graph is separated into distinct levels according to the dimensionality of the nodes. We enumerate the levels of the *DCL* as follows: Nodes with dimensionality l appear in the l_{th} level of the graph. For example the root node with dimensionality D appears at the D_{th} level. Of course $l \in [0, D]$. Every level of the *DCL* consists of $\binom{D}{l}$ nodes. The number of nodes in the *DCL* is $\sum_{l=0}^{l=D} \binom{D}{l} = 2^D$. A node n_a belonging to the l_{th} level of the graph has $2^l - 1$ successors.

Proposition 4.2.1. *In a multiset Q of DCL elements, the probability that a selected element $e_a \in Q$ is an ancestor of another selected element $e_b \in Q$ is :*

$$P(e_a \succeq e_b) = \prod_{d=0}^{D-1} \left(P_d^2 - P_d + 1\right)$$

Proof. We isolate one bit i of the bitmap (which represents one dimension): $P(e_a^i \succeq e_b^i) = P(e_a^i = 0) \cdot P(e_b^i = 0) + P(e_a^i = 1) \cdot P(e_b^i = 0 \vee e_b^i = 1) = (1 - P_i) \cdot (1 - P_i) + P_i \cdot 1 = P_i^2 - P_i + 1$

For the subsumption property to be valid this must be valid for every dimension:

$$P(e_a \succeq e_b) = P((e_a^0 \succeq e_b^0) \wedge \cdots \wedge (e_a^{D-1} \succeq e_b^{D-1})) = \prod_{d=0}^{D-1} \left(P_d^2 - P_d + 1 \right)$$

Proposition 4.2.2. *In a multiset Q of DCL elements, the probability that a selected element $e_a \in Q$ is an ancestor of every element in Q is:*

$$P(e_a \succeq Q) = \sum_{n \in S} p_n \psi^{|Q|-1} \quad where \; \psi = \prod_{d \notin D_n} (1 - P_d)$$

Proof. For the remaining $(|Q| - 1)$ elements of Q, it suffices that dimensions $d \notin D_{e_a}$ also do not exist in these elements. This is represented by ψ. Thus:

$$P(e_a \succeq Q) = P((e_a \succeq e_0) \wedge \cdots \wedge (e_a \succeq e_{|Q|-1})) = \sum_{n \in S} p_n \psi^{|Q|-1}$$

Proposition 4.2.3. *In a multiset Q of DCL elements, the probability that there exists one element, which is an ancestor of every element in Q is:*

$$P(\exists \, e : e \succeq Q) = \sum_{k=1}^{|Q|} \left\{ (-1)^{k+1} \binom{|Q|}{k} \sum_{n \in S} p_n^k \psi^{|Q|-k} \right\}$$

Proof

$$P(\exists \, e : e \succeq Q) = P(e_0 \succeq Q \vee \cdots \vee e_{|Q|-1} \succeq Q)$$
$$= \binom{|Q|}{1} \cdot P(e_a \succeq Q) - \binom{|Q|}{2} \cdot P(e_a \succeq Q \wedge e_b \succeq Q) + \cdots \pm$$
$$\binom{|Q|}{k} P(e_0 \succeq Q \wedge \cdots \wedge e_{k-1} \succeq Q)$$

Proposition 4.2.4. *In a multiset Q of DCL elements, the probability that a selected element $e_a \in Q$ is an ancestor of at least $|q|$ $(|q| - 1 + itself)$ elements of Q is:*

$$P(e_a \succeq q^+ \subseteq Q) = \sum_{j=|q|}^{|Q|} \left\{ \binom{|Q|-1}{j-1} \cdot \sum_{n \in S} \left(p_n \psi^{j-1} (1 - \psi)^{|Q|-j} \right) \right\}$$

Proof. It suffices to compute the probability for exactly $|q|$ elements: $P(e_a \succeq q^+ \subseteq Q) = \sum_{j=|q|}^{|Q|} P(e_a \succeq j \subseteq Q)$ If $j = |q|$, $|q|$ elements of Q must be chosen from the 2^l successors and the rest $|Q| - |q|$ from the $2^D - 2^l$ non-successors. There are $\binom{|Q|-1}{|q|-1}$ combinations of places in the queue where the $|q| - 1$ successors can be. For $|q| - 1$ elements, we demand that every dimension $d \notin D_n$ does not exist in the examined node. For the rest $|Q| - |q|$ non-successors, we demand that at least one dimension $d \notin D_n$ does exist in the examined node.

Proposition 4.2.5. *In a multiset Q of DCL elements, the probability that there exists at least one element, which is ancestor of exactly $|q|$ ($|q| - 1 + itself$) elements of Q is:*

$$P(\exists\, e : e \succeq q \subseteq Q) = 1 - (1 - P(e_a \succeq q \subseteq Q))^{|Q|}$$

Proof. $P(\exists\, e : e \succeq q \subseteq Q) = 1 - P(\overline{\exists\, e : e \succeq q \subseteq Q}) = 1 - P(\overline{e_a \succeq q \subseteq Q})^{|Q|} = 1 - (1 - P(e_a \succeq q \subseteq Q))^{|Q|}.$

4.3 hDCL Mapping

In this Section, we compute the same probabilities as in Section 2.1, but for *hDCL* mapping.

Proposition 4.3.1. *In a multiset Q of hDCL elements, the probability that a selected element $e_a \in Q$ is an ancestor of another selected element $e_b \in Q$:*

$$P(e_a \succeq e_b) = \sum_{n \in S} p_n \prod_{d \in D_n} \left(P_d \sum_{l=lev(n,d)}^{gr_d} P_{d,l} + (1 - P_d) \right) \cdot \prod_{d \notin D_n} (1 - P_d)$$

Proof. Analogously to the proof of Proposition 4.2.1 we isolate one digit i of the representation. For the subsumption property to be valid for one specific dimension, we demand that in the examined element either this dimensions appear but in a higher, less detailed, level or that it does not exist at all.

$$P(e_a^i \succeq e_b^i) = P(e_a^i \geq e_b^i | e_a^i \geq 1) + P(e_b^i = 0 | e_a^i = 0)$$

For the subsumption property to be applied for the element this must be valid for every dimension. We differentiate between dimensions $d \in D_{e_a}$ and dimensions $d \notin D_{e_a}$:

$$P(e_a \succeq e_b) = \prod_{d \in e_a} P_d P((lev(d, e_a) \geq lev(d, e_b > 0)) \vee d \notin D_{e_b}) \cdot$$

$$\cdot \prod_{d \notin e_a} (1 - P_d) P(d \notin D_{e_b}) = \cdots$$

Proposition 4.3.2. *In a multiset Q of hDCL elements, the probability that a selected element $e_a \in Q$ is an ancestor of every element in Q is:*

$$P(e_a \succeq Q) = \sum_{n \in S} p_n \omega^{|Q|-1} \ where$$

$$\omega = \prod_{d \in D_n} \left(P_d \sum_{l=lev(n,d)}^{gr_d} P_{d,l} + (1 - P_d) \right) \prod_{d \notin D_n} (1 - P_d)$$

This can be proven using Proposition 4.3.1 for the remaining ($|Q| - 1$) elements of Q.

Proposition 4.3.3. *In a multiset Q of hDCL elements, the probability that there exists one element, which is an ancestor of every element in Q is:*

$$P(\exists\, e : e \succeq Q) = \sum_{k=1}^{|Q|} \left\{ (-1)^{k+1} \binom{|Q|}{k} \sum_{n \in S} p_n^k \omega^{|Q|-k} \right\}$$

The proof is analogous to the proof of Proposition 4.2.3.

Proposition 4.3.4. *In a multiset Q of hDCL elements, the probability that a selected element $e_a \in Q$ is an ancestor of at least $|q|$ ($|q| - 1 + itself$) elements of Q is:*

$$P(e_a \succeq q^+ \subseteq Q) = \sum_{j=|q|}^{|Q|} \left\{ \binom{|Q|-1}{j-1} \cdot \sum_{n \in S} p_n \omega^{j-1}(1-\omega)^{|Q|-j} \right\}$$

The proof is analogous to the proof of Proposition 4.2.4.

Proposition 4.3.5. *In a multiset Q of hDCL elements, the probability that there exists at least one element, which is ancestor of exactly $|q|$ ($|q| - 1 + itself$) elements of Q is:*

$$P(\exists\, e : e \succeq q \subseteq Q) = 1 - (1 - P(e_a \succeq q \subseteq Q))^{|Q|}$$

The proof is analogous to the proof of Proposition 4.2.5.

5 Performance Analysis

This section provides a twofold analysis. State of the art *mOLAP* systems are evaluated with respect to query mapping. Moreover, we show how the analytical model described in Section 4 can facilitate the analysis and confirm the experimental results.

5.1 Simulation Environment

Mobile clients, randomly distributed in a square plane, query a data mart. Queries are propagated periodically using a 802.11 wireless network. When a suitable answer is received, the client poses a new request after an uniformly distributed time span of [0, 5] sec. Each client issues 15 queries.

We used a real but anonymized dataset. It is a data mart consisting of 6 hierarchical dimensions and 819.000 tuples. For each dimension, the number of hierarchical levels and its cardinality are shown in Table 4. We additionally ran experiments with synthetic datasets, but due to the similarity of the results, we present the results for the real dataset only. The *DCL* for this dataset consists of 64 nodes whereas the respective *hDCL* of 2800 nodes.

The query workload was generated as follows. Queries are separated into new queries and drill down or roll up queries. New queries are generated with 40% probability, whereas drill downs and roll ups each with 30% probability. When issuing a new query each dimension has 50% probability to participate in the query. Each level within a dimension has equal probability to be requested with any other level of the same hierarchy.

Table 4. Real data mart metadata

Dimension	Hierarchical levels	Cardinality
A	6	6932
B	4	2212
C	4	3128
D	3	1267
E	1	520
F	1	4

5.2 Exploitation of Derivability

Before presenting our experimental results, we explain how our analytical model can be used to facilitate the evaluation. For reasons of space we restrict our analysis to three discussed probabilities. The rest of them provide similar observations.

First, we use Propositions 4.2.1 and 4.3.1 to compute the subsumption probability of two randomly selected elements residing in the scheduler's queue. The probabilities P_d for each dimension are computed from the server's statistics, as described in Section 4.1. If no statistics are available, the server has to wait, until sufficient statistics about the workload have been measured. For *DCL* mapping the probability is 21,08%, whereas for *hDCL* only 12,77%. *DCL* mapping makes it almost twice as probable that subsumption can be applied.

This result indicates that *DCL* mapping might be optimal, but our analytical model enables a much more detailed analysis. We now consider a waiting queue consisting of $|Q|$ elements. Based on our experimental results we have observed that for a realistic scenario $|Q| \in [20, 40]$. We therefore use $|Q|=30$. We compute the probabilities $P(e_a \succeq q^+ \subseteq Q)$ and $P(\exists e : e \succeq q \subseteq Q)$ as defined in Section 4.

Figure 4 depicts the results of a simple usage of Propositions 4.2.4, 4.2.5, 4.3.4 and 4.3.5. As described before, $P(e_a \succeq q^+ \subseteq Q)$ is an useful probability for *STOBS*, which checks derivations only after the element to be transmitted has been selected. In contrast, $P(\exists e : e \succeq q \subseteq Q)$ is an useful probability for *FCLOS*. The reason for that is that *FCLOS* examines every element for possible derivations. For both probabilities,

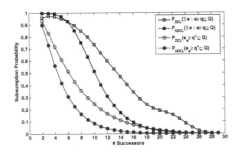

Fig. 4. Subsumption probabilities $P(e_a \succeq q^+ \subseteq Q)$ and $P(\exists e : e \succeq q \subseteq Q)$

we observe the superiority of *DCL* mapping. When $|q|$ becomes relatively high, it was expected that the probability tends to become 0, for both approaches (this could also have been computed from Proposition 4.2.2 and 4.3.2). When $|q|$ is smaller though, *DCL* mapping outperforms its competitor.

5.3   Evaluating mOLAP Dissemination Systems

mOLAP dissemination systems are typically evaluated against the following criteria:

- *Generated Traffic:* Amount of data transmitted by the server. It can be measured as traffic per broadcast or as traffic per issued query (since one broadcast serves multiple queries)
- *Average Query Access Time:* The total period of time a client spends since posing a query until the requested subset is available.
- *Average Energy Consumption:* The energy a client consumes since posing a query until the requested subset is available.

A detailed comparison of *STOBS* and *FCLOS* can be found in [3]. The objective of this paper is not a direct comparison between them, but an evaluation of their performance with respect to query mapping instead. In this context, since both of them employ *DCL* mapping we implemented extensions of them, $FCLOS_{hDCL}$ and $STOBS_{hDCL}$, which map queries to the respective *hDCL*.

From an application perspective, query access time is the most important measure for performance evaluation. Figure 5 demonstrates that the mobile user profits from *DCL* mapping, regardless of the scheduling approach. Note that the times observed include the time needed for local processing of data, whenever this is necessary. The results for energy consumption are quite similar and are omitted due to space restriction.

Figure 6 depicts a metric which translates more directly to the information provided by the analytical framework. It shows the percentage of the waiting queries that are served pro broadcast, a metric directly related to the results of Fig.4. Particularly for *FCLOS*, the difference is immense. With *DCL* mapping, a broadcast serves half of the waiting queries, whereas with *hDCL* mapping only 25% of them.

As far as generated traffic is concerned, it is rational to expect more transmitted bytes per broadcast with *DCL* mapping, since *DCL* nodes always contain fact table data

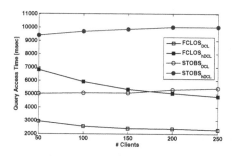

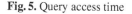

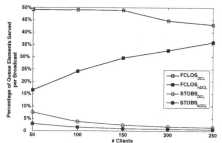

Fig. 5. Query access time

Fig. 6. Percentage of queue elements served per broadcast

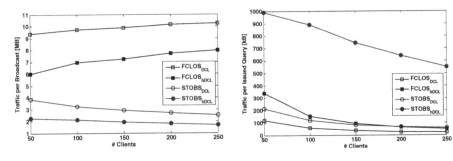

Fig. 7. Traffic per broadcast   **Fig. 8.** Traffic per query

which frequently is a superset of what the client had requested. Figure 7, depicting the amount of transmitted bytes per broadcast, confirms this expectation. *DCL* mapping introduces a huge overhead for both systems, since *DCL* nodes contain extra information. Practically this means that every transmission lasts longer.

Figure 8 on the contrary, depicts the generated traffic per issued query, where a completely different behavior can be observed. *hDCL* mapping appears to cause a huge overhead to the network, especially in the case of *STOBS*. Despite the emerged contradiction, the results are absolutely consistent. *DCL* mapping does indeed invoke a transmission of more data per broadcast cycle, but manages to serve all requests with fewer broadcasts, as indicated by Figure 6. With *DCL* mapping there is always a higher probability that the element to be transmitted is an ancestor of other queue elements. One transmission serves more requests, thus reducing the number of necessary broadcast cycles for a given amount of requests.

6 Related Work

In the broader area of mobile data warehousing, [12] copes with disconnections in hierarchical data warehouses, discussing a variety of architectures for mobile views, from proxy based to non-proxy based systems. In [2], an intelligent cache management method is proposed. *Hand-OLAP*, [13], is a system specifically designed for providing OLAP functionality to users of mobile devices. This proposal focuses mainly on the drawbacks of mobile devices, with emphasis in the small storage space and the frequent disconnections. Advanced OLAP visualization techniques, targeting devices with small screens is the focus in [14]. *SBS*, [15], and its extension *DV-ES*, [16], are *mOLAP* systems very similar to *STOBS*. Finally, [17], deals with ad hoc *mOLAP*.

7 Outlook and Open Issues

This paper deals with the issue of mapping multidimensional queries to aggregation lattices. *mOLAP* systems employ query mapping in order to reduce the number of handled data items and to exploit sub-cube derivability. Previous related work in *mOLAP* systems has not provided any specific answer to the question of whether *DCL* or *hDCL* is more appropriate for query mapping. We identified the exploitation degree of sub-cube derivability as the critical parameter to be considered. Motivated by this fact, we introduced an analytical framework that facilitates the computation of sub-cube derivability probabilities in both lattices. This framework, besides providing the basis for a general evaluation, focuses on the specific domain of *mOLAP* dissemination systems.

The analytical framework revealed, that for a real dataset, sub-cube derivability is optimally exploited in the case of *DCL*. Experimental results for state of the art *mOLAP* systems confirmed, that both server and mobile clients benefit from *DCL* mapping. Although with *DCL* mapping, bigger in size datasets are per broadcast transmitted, in comparison with respective *hDCL* mapping, the optimal exploitation of sub-cube derivability results in more clients being served by each transmission, thus reducing the number of necessary broadcasts. The server profits from the reduction of the generated traffic, while mobile clients experiences reduced access time and energy consumption.

The framework presented considers the dimensionality and number of hierarchical levels of a *MDDB*. Future work should focus on extending the framework by considering additional metadata information, like cardinality of each hierarchical level and size of the database. The intuition behind this is, that in applications other from *mOLAP*, where *MDDBs* are much bigger, it might be beneficial to use the *hDCL* instead.

References

1. Sharaf, M., Chrysanthis, P.: On-demand data broadcasting for mobile decision making. Mobile Networks and Applications 9, 703–714 (2004)
2. Huang, S.M., Lin, B., Deng, Q.S.: Intelligent cache management for mobile data warehouse systems. Journal of Database Management 16(2), 46–65 (2005)
3. Michalarias, I., Lenz, H.J.: Dissemination of multidimensional data using broadcast clusters. In: Chakraborty, G. (ed.) ICDCIT 2005. LNCS, vol. 3816, pp. 573–584. Springer, Heidelberg (2005)
4. Gray, J., Chaudhuri, S., Bosworth, A., Layman, A., Reichart, D., Venkatrao, M., Pellow, F., Pirahesh, H.: Data cube: A relational aggregation operator generalizing group-by, cross-tab, and sub totals. Data Min. Knowl. Discov. 1(1), 29–53 (1997)
5. Gyssens, M., Lakshamanan, L.V.S: Multidimensional data model and query language for infometrics. In: Procceding of the 23rd. VLDB Conference, Athens, pp. 106–115 (1997)
6. Harinarayan, V., Rajaraman, A., Ullman, J.D.: Implementing data cubes efficiently. SIGMOD Rec. 25(2), 205–216 (1996)
7. Kotidis, Y., Roussopoulos, N.: A case for dynamic view management. ACM Transactions on Database Systems 26(4), 388–423 (2001)
8. Baralis, E., Paraboschi, S., Teniente, E.: Materialized views selection in a multidimensional database. In: VLDB '97: Proceedings of the 23rd International Conference on Very Large Data Bases, pp. 156–165. Morgan Kaufmann Publishers Inc., San Francisco (1997)
9. Shukla, A., Deshpande, P., Naughton, J.F., Ramasamy, K.: Storage estimation for multidimensional aggregates in the presence of hierarchies. In: Proceedings of the 22th International Conference on Very Large Data Bases, San Francisco, CA, USA, pp. 522–531. Morgan Kaufmann Publishers Inc., San Francisco (1996)
10. Lenz, H.J., Thalheim, B.: Olap databases and aggregation functions. In: SSDBM '01: Proceedings of the Thirteenth International Conference on Scientific and Statistical Database Management, pp. 91–100. IEEE Computer Society Press, Washington (2001)

11. Vassiliadis, P., Skiadopoulos, S.: Modelling and optimisation issues for multidimensional databases. In: Wangler, B., Bergman, L.D. (eds.) CAiSE 2000. LNCS, vol. 1789, pp. 482–497. Springer, Heidelberg (2000)
12. Stanoi, I., Agrawal, D., Abbadi, A.E., Phatak, S.H., Badrinath, B.R.: Data warehousing alternatives for mobile environments. In: MobiDe '99: Proceedings of the 1st ACM international workshop on Data engineering for wireless and mobile access, pp. 110–115. ACM Press, New York (1999)
13. Cuzzocrea, A., Furfaro, F., Saccam, D.: Hand-olap: a system for delivering olap services on handheld devices. In: Proceedings of ISADS 2003, Pisa, Italy, pp. 213–224 (2003)
14. Maniatis, A., Vassiliadis, P., Skiadopoulos, S., Vassiliou, Y., Mavrogonatos, G., Michalarias, I.: A presentation model and non- traditional visualization for olap. International Journal of Data Warehousing & Mining 1(1), 1–36 (2005)
15. Sharaf, M.A., Chrysanthis, P.K.: Facilitating mobile decision making. In: WMC '02: Proceedings of the 2nd international workshop on Mobile commerce, pp. 45–53. ACM Press, New York (2002)
16. Sharaf, M.A., Sismanis, Y., Labrinidis, A., Chrysanthis, P., Roussopoulos, N.: Efficient dissemination of aggregate data over the wireless web. In: International Workshop on the Web and Databases(WebDB), pp. 93–98
17. Michalarias, I., Becker, C.: Multidimensional querying in wireless ad hoc networks. In: Proceedings of the 2007 ACM symposium on Applied computing, pp. 529–530. ACM Press, New York (2007)

A Statistics Propagation Approach to Enable Cost-Based Optimization of Statement Sequences

Tobias Kraft, Holger Schwarz, and Bernhard Mitschang

Institute of Parallel and Distributed Systems,
University of Stuttgart, Universitätsstraße 38, 70569 Stuttgart, Germany
{kraftts,hrschwar,mitsch}@ipvs.uni-stuttgart.de

Abstract. Query generators producing sequences of SQL statements are embedded in many applications. As the response time of such sequences is often far from optimal, their optimization is an important issue. CGO (Coarse-Grained Optimization) is an appropriate optimization approach that applies rewrite rules to statement sequences. In previous work on CGO, a heuristic, priority-based control strategy was utilized to choose and execute rewrite rules. In this paper, we present an approach to enable cost-based optimization of statement sequences. We show how to exploit histogram propagation and the costing component of the underlying database system for this purpose. Our work extends previous work on histogram propagation. We conclude with experiments demonstrating the effectiveness of our approach.

Keywords: cost-based query optimization, query processing, histograms.

1 Introduction

Many applications such as information retrieval systems, search engines, and business intelligence tools embed query generators. Some of these generators not only produce a single query but a sequence of SQL statements for a complex information request. These sequences compute the final result of a request in a set of subsequent steps where multiple steps can share intermediate results of previous steps as input. These intermediate results are stored in tables that are created as part of the sequence and that exist only temporarily during the execution of the sequence.

As shown in [1], the response time of such sequences is often far from optimal. Therefore, we suggested CGO (Coarse-Grained Optimization), an optimization approach that supports the optimization of statement sequences outside the database system [2][1][3]. CGO is based on a set of rewrite rules considering the fact that statements of a sequence are correlated. These rules transform a statement sequence into an equivalent sequence such that the database system consumes far less resources.

In [1], we introduced a heuristic control strategy to determine the order in which rewrite rules are applied to a statement sequence. Thereto, we assigned

Y. Ioannidis, B. Novikov, and B. Rachev (Eds.): ADBIS 2007, LNCS 4690, pp. 267–282, 2007.

a priority to each rule. Among the rules that can be applied to a sequence the one with the highest priority is being chosen. This is repeated until no more rule applications are possible. This strategy works well but there is still potential for improvements. Therefore, in this paper, we propose a practical approach to compute cost estimates for statement sequences which is the basis for an enhanced cost-based control strategy. Cost estimates would allow to compare several alternative sequences without executing them and to select the presumably most efficient one from the search space that is made up by the set of CGO rewrite rules. Our approach exploits the costing component of the underlying database system which should later on execute the statement sequence. Furthermore, we use histogram propagation to provide statistics for the intermediate-result tables increasing the accuracy and usability of the cost estimates returned by the optimizer of the underlying database system. Thereto, we had to extend previous work on histogram propagation and adapt it to our needs.

So, the main contributions of this paper are:

- An approach to provide cost estimates for statement sequences which exploits histogram propagation and the costing component of the underlying database system.
- Extensions to previous work on histogram propagation. In particular, we unify some comparison-operator implementations, so they can be handled by a generalized operator implementation and we extend the capabilities to treat arithmetic terms, grouping and aggregation.
- Performance evaluations that demonstrate the effectiveness of our approach.

The rest of this paper is organized as follows: In Section 2 we present our approach to cost estimation for statement sequences. Section 3 describes our extensions to histogram propagation. We discuss related work in Section 4. Performance experiments are presented in Section 5. Section 6 concludes this paper.

2   Cost Estimation for Statement Sequences

Figure 1 gives an architectural overview consisting of the cost-based CGO optimizer and the underlying database system. The CGO optimizer comprises the CGO ruleset, a cost-based control strategy and a component that provides cost estimates for statement sequences without executing them. The CGO ruleset can be adopted from the heuristic CGO optimizer but the heuristic control strategy has to be replaced by a cost-based control strategy. The CGO optimizer communicates with the underlying database system via JDBC and StatisticsAPI. StatisticsAPI is a JDBC-based programming interface for DBMS-independent access and management of DBMS statistics. It supports the retrieval and manipulation of histograms as well as the retrieval of cost estimates for arbitrary SQL statements from different DBMSs. So, the CGO optimizer is largely independent from a certain underlying DBMS. More information on StatisticsAPI can be found in [4]. In the following, we focus on the component that provides cost estimates for statement sequences.

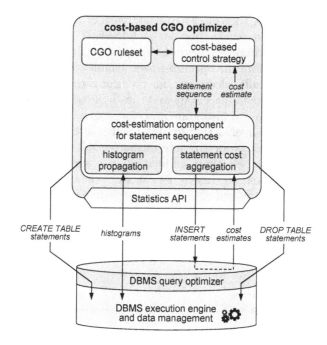

Fig. 1. Architectural overview

CGO addresses statement sequences that are composed of CREATE TABLE, INSERT and DROP TABLE statements. The CREATE TABLE and DROP TABLE statements are used to implement temporary tables. These tables only exist during the execution of the sequence to store intermediate results produced by the INSERT statements. Furthermore, one of the CREATE TABLE statements creates the table that stores the final result of the sequence which is not dropped within the sequence. So, the tables created within the statement sequence do not exist at optimization time and there are also no statistics available for these tables.

Building our own cost model on top of the database system would be no feasible solution. This is due to the fact that the runtime of a statement and therefore an appropriate cost estimate depends on the physical layout of the underlying database as well as on the capabilities and strategies of the query optimizer. That means, we would have to simulate the optimizers of all possibly underlying database systems. We propose a more practical and feasible approach exploiting the capabilities of the optimizer of the underlying database system to estimate costs for single SQL statements. Thus, we sum up the cost estimates for all INSERT statements of a sequence to provide a cost estimate for an entire sequence. To make this work, we have to execute the CREATE TABLE statements that create the intermediate-result tables before we force the optimizer to estimate costs for the INSERT statements; otherwise, the optimizer would raise a failure due to the nonexistence of the intermediate-result tables used in the INSERT statements. When the CREATE TABLE statements have been executed, the optimizer can build an execution plan for each INSERT statement and estimate the costs of this plan. However, no statistics are available for the intermediate-result tables and therefore the optimizer uses default values for the cardinality of these tables and default selectivities for predicates that include attributes of these tables. This results in inaccurate and thus useless cost estimates. To solve this problem, we make use of histogram propagation, i.e., when we propagate the histograms through the queries that form the bodies of the INSERT statements, we can provide histograms for the intermediate-result tables. Histograms are supported and used for selectivity estimation in a similar way by almost all commercial database systems. Hence, we can store the computed histograms in the database catalog of the underlying database system and its optimizer can use them to produce more accurate query plans and cost estimates for the INSERT statements that depend on intermediate-result tables.

The following algorithm shows how the cost-estimation component for statement sequences works:

Input: A statement sequence S.
Output: A cost estimate for S.
(1) $totalcosts = 0$
(2)   **foreach** CREATE TABLE statement c in S
(3)   Execute c on the underlying database system.
(4)   **foreach** INSERT statement i in S (in the order given by S)
(5)   Retrieve a cost estimate for i from the optimizer of the underlying database system by the use of StatisticsAPI. Add this cost estimate to $totalcosts$.
(6)   Translate the body of i into an algebraic tree. Retrieve statistics for the base tables used in this tree from the underlying database system by the use of StatisticsAPI. (Statistics for intermediate-result tables are still available from earlier histogram-propagation steps.) Propagate the histograms through the algebraic tree to retrieve histograms for the target table of i. Store the resulting histograms in the catalog of the underlying database system by the use of StatisticsAPI.
(7)   **foreach** CREATE TABLE statement c in S
(8)   Drop the table that has been created by c.
(9)   **return** $totalcosts$

As the tables created for cost estimation are being dropped after cost estimation, the associated histograms stored in the database catalog also get lost. So, when a statement sequence is being run later on, the histograms derived from histogram propagation no longer exist. However, the algorithm presented above can be modified so that it can be used for executing sequences with these histograms available. Thereto, in line 5 the current INSERT statement i has to be executed. Furthermore, line 7 and 8 have to be modified so that only the intermediate-result tables are being dropped, i.e., the foreach-loop has to iterate over all DROP TABLE statements of the sequence and execute them.

3   Histogram Propagation

In database systems, histograms are constructed by partitioning the data distribution of an attribute into a sequence of mutually disjoint buckets. Each bucket approximates the frequencies and values in its range. So, a histogram is an approximation of an attribute's data distribution. As we want to extract and process histograms from different DBMSs, our bucket and histogram implementation [4] supports unidimensional histograms of various classes and sources, e.g. serial histograms like equi-width histograms and equi-depth histograms as well as non-serial histograms. Thereto, we represent each bucket as a 4-tuple $(low, high, card, dv)$. low and $high$ denote the lower and upper bound of the bucket range. These values are elements of the domain of the approximated data distribution. We take into account that each domain has its own properties and reflect this in the way we treat histograms of attributes of specific domains. dv stands for the number of distinct values in the interval between low and $high$. $card$ is the cardinality of the bucket, i.e., the sum of frequencies of all distinct values that are present in the interval between low and $high$. We have chosen decimal numbers instead of integer numbers as domain for $card$ and dv because integer numbers would increase the inaccuracy of histograms produced by histogram propagation. The $NULL$ value is covered by a special $NULL$ bucket where low and $high$ is set to $NULL$. This eases computations because NULL buckets can be handled similar to 'normal' buckets. Constant values, which are fixed values used in arithmetic terms or for comparison, are also represented by histograms containing a single bucket whose lower and upper bound equals the constant value.

For propagating histograms through the statements of a sequence, we translate the SQL statements into algebraic operator trees. The algebra, that we use, operates on so-called *histogram relations*. A histogram relation is a set of attributes with a histogram assigned to each attribute. So, a histogram relation can be regarded as an approximation of a relation where in this case a relation is a multiset of tuples. Our algebra consists of logical operators because we only use it for the purpose of histogram propagation and not for cost estimation or execution; the algebraic operators are: *projection, selection, cartesian product, union, difference* and *grouping* (including aggregation). This algebra is minimal in the manner that an operator cannot be expressed by a combination of the other operators.

Arithmetic terms that appear in the different clauses of SQL statements are translated into trees of arithmetic operators. These trees are used in *projection* operators. Furthermore, they build the operands of comparison operators. Comparison operators in turn may be the operands of logical operators such as AND, OR or NOT which are used to build complex predicates. Trees that include comparison operators and / or logical operators are used in *selection* operators.

Histogram propagation is done recursively by traversing the algebraic tree in a post-order manner, i.e. the histograms are propagated and adapted from the leaves of the algebraic tree to its root. The same holds for the trees that represent predicates or arithmetic terms and that are used within the algebraic operators., i.e., we step-by-step compute the histograms of all subterms / subpredicates in a post-order manner. The basic histogram-propagation algorithm for the comparison operators and arithmetic operators is very similar due to the fact that all calculations are based on the *Attribute Value Independence Assumption*. This means, we have to enumerate all bucket combinations that can be produced from the histograms affected by an operator and apply the operator to each bucket combination. To calculate the cardinality of a single bucket combination, we have to determine the probability of this specific bucket combination and multiply it with the cardinality of the histogram relation. Thereto, we treat the data distribution as a probability distribution and calculate the proportion of the actual bucket combination at all possible bucket combinations so that the sum of cardinalities of all bucket combinations equals the cardinality of the input histogram-relation. An arithmetic operator produces a single output histogram that represents the result of the arithmetic operator applied to the input histograms. This histogram again might be the input of an arithmetic operator or of a comparison operator. A comparison operator just compares two input histograms and therefore produces no result histogram but computes the selectivity of this comparison. It also modifies the two input histograms so that after the comparison they contain the buckets that fulfill the condition. After the computation of the output histogram or the modification of the input histograms, these histograms are being serialized.

In the following subsections, we present our extensions to previous work on histogram propagation and we present some more details of specific operators and the query translation process.

3.1   Utilization of Interval Arithmetic for Arithmetic Terms

As stated before, the output histogram of an arithmetic operator is calculated by enumerating all bucket combinations from the input histograms and applying the arithmetic operator to each of these bucket combinations. To determine the bounds of a result bucket we make use of interval arithmetic [5]. To give an example, we compute the lower bound $B_O.low$ and the upper bound $B_O.high$ of a bucket B_O resulting from the multiplication of the buckets B_{I1} and B_{I2} as follows (*min* returns the minimum value of the given value set and *max* returns the maximum value):

$$B_O.low = min(B_{I1}.low \cdot B_{I2}.low, B_{I1}.low \cdot B_{I2}.high,$$
$$B_{I1}.high \cdot B_{I2}.low, B_{I1}.high \cdot B_{I2}.high) \tag{1}$$
$$B_O.high = max(B_{I1}.low \cdot B_{I2}.low, B_{I1}.low \cdot B_{I2}.high,$$
$$B_{I1}.high \cdot B_{I2}.low, B_{I1}.high \cdot B_{I2}.high) \tag{2}$$

If one of the input buckets is a *NULL-bucket*, the result bucket is also a *NULL-bucket*.

3.2 Unification of Comparison Operators

Comparison operators compute a selectivity value. Thereto, they enumerate all bucket combinations that can be built up of the two input histograms and apply the predicate to each of these bucket combinations. The selectivity of the comparison operator results from the aggregation of the selectivities of these bucket combinations. This selectivity might be the input of a logical operator. The final selectivity of an operator tree that represents a predicate is used to determine the cardinality of the output histogram-relation of the respective *selection* operator. Presuming the *Attribute Value Independence Assumption*, the histograms of all attributes in the input histogram-relation are adapted to the cardinality of the output histogram-relation keeping the relative data distribution of the histograms in the input histogram-relation. I.e., the cardinality of the buckets is proportionally adapted but the bucket bounds stay the same.

As stated before, a comparison operator also adapts the data distribution of its two input histograms. This is done by eliminating buckets or by altering their bounds, cardinality and number of distinct values. If a predicate consists of a single comparison operator ($=, \neq, \leq, \geq, <,$ or $>$) and one or both of its operands are attributes, the adapted histograms of these attributes can be taken into the output histogram-relation of the corresponding *selection* operator. So, for these attributes not only the cardinality but also the data distribution is adapted within the *selection* operator.

This way of implementing comparison operators offers a flexible usage for different purposes within a *selection* operator. So, we unify different operand-dependent operator implementations in a more generic operator implementation. For example, the same implementation of the comparison operator '$=$' can be used for:

- $\sigma_{A_i=A_j}(R)$: A predicate comparing two attributes.
- $\sigma_{A_i=C}(R)$: A predicate comparing an attribute and a constant value, because constant values are represented by histograms, too.
- $R_1 \bowtie_{A_i=A_j} R_2 \Rightarrow \sigma_{A_i=A_j}(R_1 \times R_2)$: A join-condition. As we can transform a join operation into a cartesian product followed by a selection with the join condition as predicate, we need no separate implementation of a join operator.
- $\sigma_{A_i\ IN\ (C_1,...,C_n)}(R) \Rightarrow R \bowtie_{A_i=A} R_C \Rightarrow \sigma_{A_i=A}(R \times R_C)$: An IN predicate. IN predicates comparing an attribute with a set of values can be realized by a join (cartesian product followed by a selection) with an additional table R_C that contains the values as rows. The additional table is represented by a histogram relation with a single attribute A and a corresponding histogram that contains a single bucket for each value in the value set.

This uniform treatment similarly applies to the other comparison operators. For example, this enables the usage of the comparison operator '$\leq$' for selections comparing two attributes, selections comparing an attribute and a constant value, and non-equi joins. For the latter one, this operator is used in the predicate applied to the result of the cartesian product.

3.3   Heuristics for Grouping and Aggregation

For each bucket combination of the grouping attributes, we first compute the amount of groups that can be built with this bucket combination and the average size of these groups. The aggregation operators get these values (number of groups and average group size) computed for all bucket combinations and the histogram of the attribute whose values should be aggregated as input. The calculation of aggregate values is a sophisticated combinatorial problem. In our case this becomes even more difficult because we store bucket cardinalities as decimal numbers. Furthermore, in real-world data there's always noise and a deviation from the uniform distribution within a bucket. Therefore, there is also some deviation in the group sizes for a single bucket combination. To overcome these problems, we developed some simple heuristics to calculate the aggregates $COUNT$, SUM, AVG, MIN and MAX as long as the attribute that should be aggregated contains none or only negligible few $NULL$ values. E.g., for a given average group size, the aggregate operator SUM sums up the lowest values of the attribute's histogram as the lower bound of the aggregate. I.e., it ascendingly aggregates all buckets starting with the bucket that has the lowest lower bound till the sum of bucket cardinalities reaches the average group size. Thereby, for each of these buckets, it calculates the mean value of the bucket ((lower bound + upper bound)/2) and multiplies the mean value with the bucket cardinality. The sum of these products represents the lower bound. For the upper bound we descendingly sum up the highest values, i.e., we start with the bucket that contains the highest upper bound.

3.4   Normalization of Histograms

The enumeration of bucket combinations in arithmetic operators implicates that in the worst case there are as much buckets in the output histogram of an operator as the product of the number of buckets in the input histograms. Serializing the output histogram may additionally double the number of buckets. So, the deeper an operator tree and the longer a sequence the larger the histograms may get in number of buckets. Furthermore, the number of bucket combinations increases exponentially with the number of grouping attributes in a grouping operator. To cope with all these problems, a normalization step can be added prior to the enumeration phase and / or after an output histogram has been produced. In this normalization step, the number of buckets can be reduced to a given upper limit. Our implementation takes this into account. However, for our experiments, we omitted any normalizations during the propagation process because the number of buckets didn't increase extremely during propagation. The main reason therefor is that selections rather reduce than increase the number of buckets.

3.5   Common Subexpressions

When transforming an SQL statement into an algebraic operator tree, we take into account that the same arithmetic term may appear multiple times in different

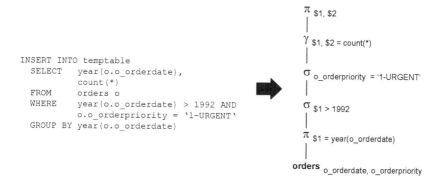

```
INSERT INTO temptable
    SELECT   year(o.o_orderdate),
             count(*)
    FROM     orders o
    WHERE    year(o.o_orderdate) > 1992 AND
             o.o_orderpriority = '1-URGENT'
    GROUP BY year(o.o_orderdate)
```

Fig. 2. A SQL statement that includes a common subexpression and the corresponding algebraic tree

clauses of the SQL statement. Therefore, we add projections to the algebraic operator trees that map common subexpressions to attributes that can be used in subsequent operators. Figure 2 shows a sample SQL statement and the corresponding algebraic operator tree. The arithmetic term `year(o.o_orderdate)` is used in the WHERE clause as well as in the GROUP BY clause and in the SELECT clause. By adding a projection directly above the operator that represents the base table `orders` this term is mapped onto the attribute `$1`. So, the histogram of `year(o.o_orderdate)` which is adapted by the comparison operator in the predicate `year(o.o_orderdate) > 1992` does not get lost but is reused in the subsequent *grouping* operator $\gamma_{\$1,\$2=count(*)}$ and in the *projection* operator $\pi_{\$1,\$2}$. When we do not consider common subexpressions, the term `year(o.o_orderdate)` would be recomputed in these two algebraic operators and the modifications of the corresponding histogram caused by the *selection* operator would get lost.

4   Related Work

Histograms [6] are a widely investigated concept of statistical summaries. They are able to summarize the contents of large tables by approximating the data distribution of the attributes. Past research addressed the efficient computation and usage of histograms. In database systems, histograms can be used for selectivity estimation in the query optimizer [7][8][9][10][11][12][13] or for approximate query answering [14][15]. Below, we discuss the latter one in more detail because we adopt and extend this concept for our approach. Further research addresses the efficient and automatic maintenance of histograms [16] and the usage of feedback from the query execution engine to infer data distributions instead of examining the data itself; thereto, [17] and [7] introduce a new type of multi-dimensional histogram, called STHoles. Commercial database management systems like IBM DB2 [18], Oracle [19] and Microsoft SQL Server [20] store unidimensional histograms in their catalog tables and exploit them for query optimization, but their usage is mostly restricted to selectivity estimation of point queries, range queries and queries containing equi-joins. Furthermore, IBM DB2 and Oracle do not modify and adapt histograms within the execution plans, i.e., they do not make use of histogram propagation [18][19]. [8] and [21] indicate that Microsoft SQL Server uses histogram propagation. Unfortunately, there are no publications that explicitly describe the techniques applied in Microsoft SQL Server or that give an overview of the operators that are supported. Even if some kind of histogram propagation is used in Microsoft SQL Server, histograms for result tables are not available outside the database system and therefore a reuse of these histograms for our purpose is not possible.

Approximate query answering [22] is an approach to provide approximate answers using statistical summaries of the data, such as samples, histograms, and wavelets. Applications where the precise answer is not relevant and early feedback is helpful can benefit from this approach due to the reduction in query response time. We exploit and extend approximate query answering based on histograms for our histogram propagation approach and therefore focus on histogram-based work in the following. [14] and [15] describe how to transform SQL queries working on tables of a database into SQL queries working on the respective histograms. However, this approach is restricted to simple queries with equi-joins and predicates of the form $A\Phi C$ where Φ is a comparison operator, A is an attribute and C is a constant value. Furthermore, only simple aggregate queries are supported where the group by clause must be empty. Otherwise, a special kind of histogram must exist with the combination of grouping attributes as domain and the average aggregate value instead of the frequencies stored in the buckets. Our approach differs from this approach in that we do not map the given SQL queries onto SQL queries. We transform them into operator trees using our own algebra, which is not expressed in SQL. Moreover, we add support for arithmetic terms and support for grouping. The algebra, that we use, is similar to existing relational algebras and multi-set algebras [23][24] but it processes sets of histograms instead of relations. As our implementations of comparison operators enumerate all bucket combinations that can be built from the histograms of the operands, it also supports selections containing predicates with an attribute on both sides of the comparison operator. Thus, we do not need a separate join operator to process joins. Furthermore, this also allows to process non-equi joins. The implementation of the comparison operator '=' is similar to the join operator for histograms presented in [21].

At the moment, all our operator implementations are based on the *Attribute Value Independence Assumption*, i.e., we assume that attributes are not correlated. Furthermore, when comparing two buckets by the comparison operator '=', we presume inclusion in the interval where the two buckets overlap. In comparison to that, a join operation in the approach discussed above recomputes a frequency distribution for the two attributes of the join-condition. These frequency distributions are derived from the corresponding histograms in consideration of the *Uniform Spread Assumption*. In the worst case, the two tables, that represent the frequency distributions and that are computed during the join operation, have the same cardinality as the original tables. Furthermore, strictly

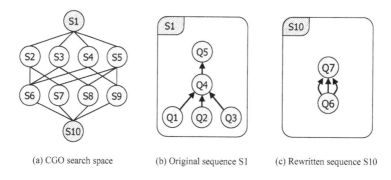

(a) CGO search space   (b) Original sequence S1   (c) Rewritten sequence S10

Fig. 3. Sequences used in the experiments

relying on the *Uniform Spread Assumption* is not always a feasible solution. Assume we have a histogram where all bucket bounds are even integer numbers and another histogram where all bucket bounds are odd integer numbers. For both histograms we further assume that the number of distinct values for a bucket is half the length of the bucket interval. When we follow the approach presented in [14] and [15], which strictly relies on the *Uniform Spread Assumption*, the selectivity of the join of these two histograms will be zero.

To the best of our knowledge there is no publication on histogram-based approximate query answering or histogram propagation that covers all mentioned extensions based on unidimensional histograms under the *Attribute Value Independence Assumption*. Special properties of different domains are also not considered in previous work on histogram propagation, i.e., respective publications are restricted to histograms in the integer domain. We added the mentioned extensions motivated by the fact that data warehouse queries (e.g. TPC-H benchmark [25]) include arithmetic terms and grouping.

5   Experiments

The results of the experiments, that we present in this section, should demonstrate the effectiveness of our costing approach. Furthermore, the experiments should show that histogram propagation can successfully be exploited to provide useful cost estimates for statement sequences.

All experiments were performed on a Windows XP machine with two 1.53 GHz AMD Athlon 1800+ processors and 1 GB main memory. As database system we used DB2 V9.1 where 50% of the main memory was assigned to the buffer pool. The experiments were conducted on a TPC-H database [25] containing TPC-H data created with a scaling factor of 10 ($\approx$ 10 GB of raw data). We employed DB2 to create histograms on all columns of all tables that belong to the TPC-H schema.

A JAVA prototype of the cost-estimation component for statement sequences did the histogram propagation and provided the cost estimates for the sequences that we examined in our experiments. As we mentioned in Section 2, StatisticsAPI

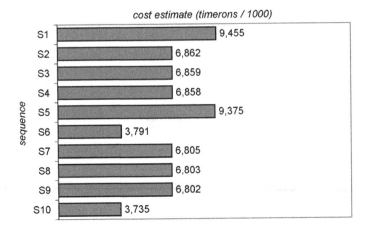

Fig. 4. Cost estimates for parameter value 300 000

was used to retrieve and modify the statistics stored in the underlying database system and to retrieve cost estimates for single statements of a sequence. In DB2 histograms are stored as quantiles, i.e., each bucket stores a cumulative frequency including the frequencies of all previous buckets. When retrieving histograms from DB2, StatisticsAPI transforms these quantile data into our own histogram representation. To store the histograms resulting from histogram propagation, StatisticsAPI has to compact histograms with more than 20 buckets and it has to transform the buckets into quantiles. The number of buckets is reduced by merging adjacent buckets with respect to keeping cardinality errors in the data distribution minimal. Moreover, StatisticsAPI also updates the corresponding logical table statistics and column statistics in the database catalog.

The left side of Figure 3 shows a CGO search space containing 10 statement sequences ($S1$ to $S10$). This search space has been built by applying the CGO rewrite rules to the given statement sequence $S1$ which has been created by the MicroStrategy DSS tool suite. Besides, Figure 3 shows the dependency graph of the original sequence $S1$ and the dependency graph of sequence $S10$ which we obtained by subsequently applying three rewrite rules to the original sequence. Each node Qx in these dependency graphs represents a combination of a CREATE TABLE, an INSERT and a DROP TABLE statement, except the root node which stands for the final result and therefore includes no DROP TABLE statement. In the following, when we talk about Qx we just refer to the INSERT statement of Qx. Statement $Q1$ calculates the turnover ($extendedprice \cdot (1 - discount) \cdot (1 + tax)$) for each line item ordered in 1992 and sums it up for each customer. $Q2$ and $Q3$ provide the same for the years 1993 and 1994. $Q4$ joins these intermediate results and selects customers that have a turnover that is greater than a given constant parameter value in each of the years. $Q5$ joins the result table of $Q4$ with the customer table to look up the customer names. $Q6$ of the rewritten sequence calculates the turnover for each customer and each year between 1992 and 1994 and applies the filter

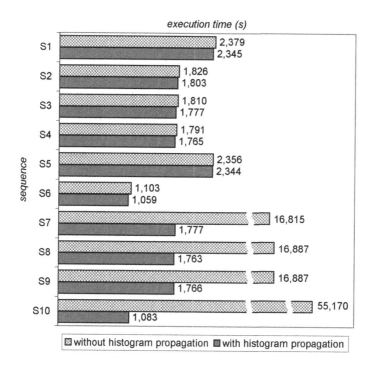

Fig. 5. Execution times for parameter value 300 000 with and without using the results of histogram propagation

on turnover with the given constant parameter value. Hence, the result table of $Q6$ contains the union of $Q1$, $Q2$ and $Q3$ with the predicates of $Q4$ applied. $Q7$ accesses the result table of $Q6$ three times selecting each of the included years once and joins this data with the customer table; so, $Q7$ is not just a merge of $Q4$ and $Q5$. We performed the experiments with three different parameter values for the filter: 100 000, 300 000 and 500 000. A higher parameter value denotes a smaller selectivity.

The different sequences have been executed in isolation with empty buffer pool and empty statement cache. The execution times in Figure 5 and in Figure 7 are the average execution times of 3 runs. In both figures, we distinguish between executions where the propagated histograms for the intermediate-result tables have been stored in the catalog tables of the underlying database and executions without statistics for the intermediate-result tables. Both types of execution show pure execution times, i.e., overhead for the histogram propagation is not included. In our experiments, this overhead is less than 1% for a single sequence. The overhead for propagating histograms and estimating costs for all ten sequences adds up to less than 5% of the execution time of the original sequence $S1$. This overhead is acceptable as we could identify $S10$ as the most efficient sequence resulting in a performance gain of more than 50% in comparison to the original sequence $S1$. The overhead for histogram propagation does

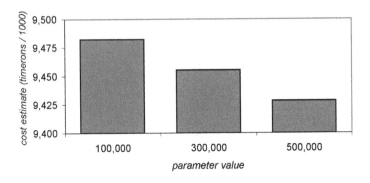

Fig. 6. Cost estimates for sequence $S1$

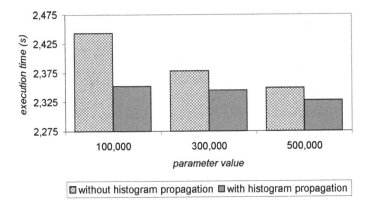

⊠ without histogram propagation ▪ with histogram propagation

Fig. 7. Execution times for sequence $S1$ with and without using the results of histogram propagation

not depend on the execution time of a sequence but on the complexity of the corresponding operator trees, the type of the operators used in these operator trees and the number of buckets in the histograms of the base tables accessed by the sequence. Thus, this approach is profitable for long running sequences or sequences with a high optimization potential. As stated in Section 3.4, the complexity of the calculations during histogram propagation can be reduced by adding a normalization step prior to or after an operator.

Figure 4 shows the cost estimates for all sequences of the search space using 300 000 as parameter value, Figure 5 shows the corresponding execution times. These two figures depict that the retrieved cost estimates are a good indicator for the corresponding execution times when histogram propagation is being used. The extremely long execution times of sequence $S7$, $S8$, $S9$ and $S10$ without histogram propagation are a result of the missing statistics. This effect only appears when 300 000 or 500 000 is used as parameter value in the filter condition; for parameter value 100 000 the execution times of these sequences fit to the corresponding cost estimates, i.e., for this parameter value there is no wide difference between the execution times with and without using the results of histogram propagation. So, this shows how the optimizer of the underlying database system benefits from the statistics available through histogram propagation and that bad plans could be avoided using this information. Figure 6 and Figure 7 show that the cost estimates also reflect the changes in the execution times when different parameter values are used in the filter condition. Execution times for parameter value 100 000 again depict how the statistics that our approach provides for the intermediate-result tables support the optimizer of the underlying database system in finding a good plan. Moreover, without these statistics, the underlying database system provides the same cost estimate for query $Q4$ independent of the parameter value that is used. Thus, histogram propagation is necessary and a feasible solution to retrieve appropriate cost estimates for the INSERT statements that read from intermediate-result tables.

6   Conclusion

In this paper we proposed an approach to retrieve cost estimates for statement sequences by exploiting the optimizer of the underlying database system. We also have shown that histogram propagation is needed to make this approach work. For this purpose, we extended previous work on histogram propagation by the support for arithmetic terms, grouping and aggregation. The experiments showed that this is a feasible approach to support the optimizer of the underlying database system in finding a good execution plan and providing useful cost estimates. Future work addresses a cost-based control strategy within the CGO optimizer.

References

1. Kraft, T., Schwarz, H., Rantzau, R., Mitschang, B.: Coarse-Grained Optimization: Techniques for Rewriting SQL Statement Sequences. In: Proc. VLDB (2003)
2. Schwarz, H., Wagner, R., Mitschang, B.: Improving the Processing of Decision Support Queries: The Case for a DSS Optimizer. In: Proc. IDEAS (2001)
3. Kraft, T., Schwarz, H.: CHICAGO: A Test and Evaluation Environment for Coarse-Grained Optimization. In: Proc. VLDB (2004)
4. Kraft, T., Mitschang, B.: Statistics API: DBMS-independent Access and Management of DBMS Statistics in Heterogeneous Environments. In: Proc. ICEIS (2007)
5. Petkovic, M., Petkovic, L.: Complex Interval Arithmetic and Its Applications. Wiley-VCH, Chichester (1998)
6. Ioannidis, Y.: The History of Histograms (abridged). In: Proc. VLDB (2003)
7. Bruno, N., Chaudhuri, S., Gravano, L.: STHoles: A Multidimensional Workload-Aware Histogram. In: Proc. SIGMOD (2001)
8. Chaudhuri, S.: An Overview of Query Optimization in Relational Systems. In: Proc. PODS (1998)
9. Ioannidis, Y., Christodoulakis, S.: Optimal Histograms for Limiting Worst-Case Error Propagation in the Size of Join Results. ACM Transactions on Database Systems 18(4) (1993)

10. Ioannidis, Y., Poosala, V.: Histogram-Based Solutions to Diverse Database Estimation Problems. Data Engineering Bulletin 18(3) (1995)
11. Ioannidis, Y., Poosala, V.: Balancing Histogram Optimality and Practicality for Query Result Size Eestimation. In: Proc. SIGMOD (1995)
12. Poosala, V., Haas, P., Ioannidis, Y., Shekita, E.: Improved Histograms for Selectivity Estimation of Range Predicates. In: Proc. SIGMOD (1996)
13. Poosala, V., Ioannidis, Y.: Selectivity Estimation Without the Attribute Value Independence Assumption. In: Proc. VLDB (1997)
14. Ioannidis, Y., Poosala, V.: Histogram-Based Approximation of Set-Valued Query-Answers. In: Proc. VLDB (1999)
15. Poosala, V., Ganti, V., Ioannidis, Y.: Approximate Query Answering using Histograms. IEEE Data Engineering Bulletin 22(4) (1999)
16. Gibbons, P., Matias, Y., Poosala, V.: Fast Incremental Maintenance of Approximate Histograms. ACM Transactions on Database Systems 27(3) (2002)
17. Aboulnaga, A., Chaudhuri, S.: Self-tuning Histograms: Building Histograms Without Looking at Data. In: Proc. SIGMOD (1999)
18. IBM Corp.: IBM DB2 Universal Database, Administration Guide: Performance, Version 8.2
19. Oracle Corp.: Oracle Database Performance Tuning Guide, 10g Release 1 (10.1) (2003)
20. Hanson, E., Kollar, L.: Statistics Used by the Query Optimizer in Microsoft SQL Server 2005. Microsoft SQL Server TechCenter (1993)
21. Bruno, N., Chaudhuri, S.: Exploiting Statistics on Query Expressions for Optimization. In: Proc. SIGMOD (2002)
22. Garofalakis, M., Gibbons, P.: Approximate Query Processing: Taming the TeraBytes. In: Proc. VLDB (2001)
23. Grefen, P., de By, R.: A Multi-Set Extended Relational Algebra - A Formal Approach to a Practical Issue. In: Proc. ICDE (1994)
24. Garcia-Molina, H., Ullman, J., Widom, J.: Database Systems: The Complete Book. Prentice Hall PTR, Englewood Cliffs (2001)
25. TPC-H Standard Specification, Revision 2.0.0. (2002), www.tpc.org/tpch

A Fixpoint Approach to State Generation for Stratifiable Disjunctive Deductive Databases

Andreas Behrend

University of Bonn, Institute of Computer Science III,
Römerstr. 164, D-53117 Bonn, Germany
behrend@cs.uni-bonn.de

Abstract. In this paper we present a new fixpoint-based approach to bottom-up state generation for stratifiable disjunctive deductive databases. To this end, a new consequence operator based on hyperresolution is introduced which extends Minker's operator for positive disjunctive Datalog rules. In contrast to already existing model generation methods our approach for efficiently computing perfect models is based on state generation. Additionally, it enhances model state computation based on Minker's operator for positive disjunctive Datalog rules.

Keywords: Deductive Databases, Disjunctive Datalog, View Updating.

1 Introduction

Disjunctive Datalog extends Datalog by disjunctions of literals in facts as well as rule heads. Various applications may benefit from such an extension as it allows for reasoning in the presence of uncertainty [19]. For example, the analysis of view updating requests usually leads to alternative view update realizations which can be represented in disjunctive form [10]. For instance, given an intersection rule $p(\boldsymbol{x}) \leftarrow q(\boldsymbol{x}), r(\boldsymbol{x})$ and the request to delete certain facts from p, the consequence analysis can be performed using the disjunctive rule $p^-(\boldsymbol{x}) \rightarrow q^-(\boldsymbol{x}) \vee r^-(\boldsymbol{x})$ where p^-, q^-, r^- contain tuples to be deleted from the corresponding relation, respectively.

A variety of different semantics for disjunctive databases has been proposed (e.g., Minimal Models [13], Perfect Models [15], Possible Model Semantics [18] or the equivalent Possible World Semantics [3], (partial) Disjunctive Stable Model Semantics [16], and the Well-Founded Semantics [22,5]) which describe the intended meaning of a given rule set as a set of possible conclusions. In this paper we consider the minimal model semantics and its extension to perfect models in the presence of stratifiable recursion. With respect to the minimal model semantics, DNF and CNF representations of conclusions have been proposed, leading to alternative fixpoint approaches for the determination of minimal models: While the model generation approach [8] is based on sets of Herbrand interpretations, the state generation approach [14] uses hyperresolution operating on sets of positive disjunctions. [20] investigates the relationship between these two approaches and shows the equivalence of the obtained conclusions. When computing perfect

Y. Ioannidis, B. Novikov, and B. Rachev (Eds.): ADBIS 2007, LNCS 4690, pp. 283–297, 2007.

models of stratified rules, however, CNF and DNF representations of conclusions appear to be not equally well suited. While an extended model generation approach for perfect model computation based on DNF has been proposed in [20], until now, the only way to obtain the model state of a given database was to employ a model generation procedure first and to transform the resulting conclusions into CNF. In this paper we provide a method which avoids the explicit generation of Herbrand models and is purely based on state generation. Especially when a state representation is preferred with respect to a given application or tends to be more compact than a DNF or model tree representation, our proposed method appears to be more suited than model based ones. In addition to closing an open gap in the theory of disjunctive logic programming, new insights with respect to the application of hyperresolution in this context are obtained allowing to enhance Minker's state generation procedure for positive rules.

This paper is organized as follows: After introducing the syntax and semantics of disjunctive Datalog, we recall a constructive model generation method for stratifiable rules in Section 3. In Section 4 state generation for positive rules is revisited and the problems arising when considering stratifiable rules are discussed. Our new approach to state generation for stratifiable rules is presented in Section 5. The results from this section are employed to enhance model state computation for positive disjunctive rules in Section 6. Finally, the two possible fact representations by means of CNF and DNF are compared in Section 7.

2   Basic Concepts

A disjunctive Datalog *rule* is a function-free clause of the form $A_1 \vee \ldots \vee A_m \leftarrow B_1 \wedge \cdots \wedge B_n$ with $m, n \geq 1$ where the rule's head $A_1 \vee \ldots \vee A_m$ is a disjunction of positive atoms, and the rule's body $B_1, \ldots, B_n$ consists of literals, i.e. positive or negative atoms. We assume all rules to be *safe*, i.e., all variables occurring in the head or in any negated literal of a rule must be also present in a positive literal in its body. If $A \equiv p(t_1, \ldots, t_n)$ with $n \geq 0$ is a literal, we use $\mathtt{pred}(A)$ to refer to the predicate symbol p of A. If $A \equiv A_1 \vee \ldots \vee A_m$ is the head of a given rule R, we use $\mathtt{pred}(R)$ to refer to the set of predicate symbols of A, i.e. $\mathtt{pred}(R) = \{\mathtt{pred}(A_1), \ldots, \mathtt{pred}(A_m)\}$. For a set of rules $\mathcal{R}$, $\mathtt{pred}(\mathcal{R})$ is defined as $\cup_{r \in \mathcal{R}} \mathtt{pred}(r)$. A *disjunctive fact* $f \equiv f_1 \vee \ldots \vee f_k$ is a disjunction of ground atoms f_i with $i \geq 1$. f is called *definite* if $i = 1$. In the following, we identify a disjunctive fact with a set of atoms such that the occurrence of a ground atom A within a fact f can also be written as $A \in f$. The set difference operator can then be used to remove certain atoms from a disjunction while the empty set as result is interpreted as the boolean constant $false$.

Definition 1. *A disjunctive deductive database $\mathcal{D}$ is a pair $\langle \mathcal{F}, \mathcal{R} \rangle$ where $\mathcal{F}$ is a finite set of disjunctive facts and $\mathcal{R}$ a finite set of disjunctive rules such that $\mathtt{pred}(\mathcal{F}) \cap \mathtt{pred}(\mathcal{R}) = \emptyset$. Within a disjunctive database $\mathcal{D} = \langle \mathcal{F}, \mathcal{R} \rangle$, a predicate symbol p is called derived (view predicate), if $p \in \mathtt{pred}(\mathcal{R})$. The predicate p is called extensional (or base predicate), if $p \in \mathtt{pred}(\mathcal{F})$.*

Stratifiable rules are considered only, that is, recursion through negative predicate occurrences is not permitted [1]. Note that in addition to the usual stratification concept for definite rules it is required that all predicates within a rule's head are assigned to the same stratum. A stratification partitions a given rule set such that all positive derivations of relations can be determined before a negative literal with respect to one of those relations is evaluated.

Given a deductive database $\mathcal{D}$, the Herbrand base $\mathcal{H}_\mathcal{D}$ of $\mathcal{D}$ is the set of all ground atoms that can be constructed from the predicate symbols and constants occurring in $\mathcal{D}$ [2,6]. A disjunctive deductive database is syntactically given by a set of facts and rules which we also call the *explicit state* of the database. In contrast to this we define the *implicit state* of a database as the set of all positive and negative conclusions that can be derived from the explicit state.

Definition 2 (Implicit Database State). *Let* $\mathcal{D} = \langle \mathcal{F}, \mathcal{R} \rangle$ *be a stratifiable disjunctive deductive database. The implicit database state* $\mathcal{PM}_\mathcal{D} := \{PM_1, \ldots, PM_n\}$ *of* $\mathcal{D}$ *is defined as the set of perfect models [15] for* $\mathcal{F} \cup \mathcal{R}$:

$$\forall PM_i \in \mathcal{PM}_\mathcal{D} : PM_i = \mathcal{I}_i^+ \uplus \neg \cdot \mathcal{I}_i^-$$

where $\mathcal{I}_i^+, \mathcal{I}_i^- \subseteq \mathcal{H}_\mathcal{D}$ *are sets of ground atoms. The set* $\mathcal{I}_i^+$ *represents the true portion of the perfect model while* $\neg \cdot \mathcal{I}_i^-$ *comprises all true negative conclusions.*

Each perfect model of a database $\mathcal{D}$ partitions the Herbrand base into the set of positive conclusions $\mathcal{I}^+$ and the set of negative conclusions $\mathcal{I}^-$ where $\mathcal{I}^- = \mathcal{H}_\mathcal{D} \setminus \mathcal{I}^+$. Therefore, we allow each perfect model to be represented by the set of true atoms only.

3 Model Generation

We will now recall from [8,9,12,13] a standard model generation approach for determining the semantics of stratifiable disjunctive databases. This approach works on minimal Herbrand interpretations and computes the set of perfect models by iteratively applying the given rule set. We will use the results from this section for a comparison with our new approach to state generation from Section 5. In contrast to a positive Datalog program, there is no unique minimal Herbrand model for a positive disjunctive program. Instead, Minker has shown that there is a set of minimal models which capture the intended semantics of a disjunctive program [13]. In the following, the set of minimal Herbrand models of a given database $\mathcal{D}$ will be denoted by $\mathcal{MM}_\mathcal{D}$. For determining minimal Herbrand models we introduce the operators `min_models` and `min_subset`.

Definition 3. *Let* $\mathcal{I}$ *be an arbitrary multi-set. The operator* `min_subset` *selects all minimal sets contained in* $\mathcal{I}$: `min_subset`$(\mathcal{I}) := \{I \in \mathcal{I} \mid \neg \exists J \in \mathcal{I} : J \subsetneq I\}$.

Definition 4. *Let* $F = \{f_1, \ldots, f_n\}$ *be a set of disjunctive facts.* `min_models` *constructs the set of all minimal models of* F *and is defined as follows:*

`min_models`$(F) :=$ `min_subset`$(\{ \{L_1, \ldots, L_n\} \mid L_i$ *is a literal from fact* $f_i\})$.

In order to formalize the application of disjunctive rules to a set of definite facts, we introduce a consequence operator for determining minimal Herbrand interpretations.

Definition 5. *Let $\mathcal{R}$ be a set of disjunctive rules and $\mathcal{I}$ an arbitrary set of definite facts:*

1. *The single consequence operator $T^m[r]$ defines for each disjunctive rule $r \equiv A_1 \vee \cdots \vee A_l \leftarrow B_1 \wedge \cdots \wedge B_n \wedge \neg C_1 \wedge \cdots \wedge \neg C_k \in \mathcal{R}$ with $l, n \geq 1$ and $k \geq 0$ the set of disjunctive facts which can be derived by a single application of r:*

$$T^m[r](\mathcal{I}) := \{(\,A_1 \vee \cdots \vee A_l)\sigma \mid \sigma \text{ is a ground substitution and}$$
$$\forall i \in \{1, \ldots, n\} : B_i\sigma \in \mathcal{I} \text{ and } \forall j \in \{1, \ldots, k\} : C_j\sigma \notin \mathcal{I}\})$$

2. *The cumulative disjunctive consequence operator $T_{\mathcal{R}}^m$ again models the simultaneous application of all rules in $\mathcal{R}$ to $\mathcal{I}$:*

$$T_{\mathcal{R}}^m(\mathcal{I}) := \bigcup_{r \in \mathcal{R}} T^m[r](\mathcal{I}) \cup \mathcal{I}.$$

Based on these notions we define the model-based consequence operator which computes the minimal Herbrand models of a disjunctive database.

Definition 6. *Let $\mathcal{D} = \langle \mathcal{F}, \mathcal{R} \rangle$ be a disjunctive deductive database. The model-based consequence operator $T_{\mathcal{R}}^{\mathcal{M}}$ is a mapping on sets of ground atoms and is defined for $\mathcal{I} \subseteq 2^{\mathcal{H}_{\mathcal{D}}}$ with $\mathcal{I} = \{I_1, \ldots, I_n\}$ as follows:*

$$T_{\mathcal{R}}^{\mathcal{M}}(\mathcal{I}) := \mathtt{min\_subset}(\{\mathtt{min\_models}(T_{\mathcal{R}}^m(I)) \mid I \in \mathcal{I}\}).$$

The operator $T_{\mathcal{R}}^{\mathcal{M}}$ was originally introduced by Fernández and Minker [9] and is based on the consequence operator $T_{\mathcal{R}}^m$. $T_{\mathcal{R}}^m$ is solely applied to a set of definite facts. New disjunctive (indefinite) facts, however, are generated because of disjunctions within the heads of the rules applied. Therefore, the operators `min_models` and `min_subset` are needed to return to a representation of definite facts only after each cumulative rule application.

In [9] it is shown that $T_{\mathcal{R}}^{\mathcal{M}}$ is monotonic for semi-positive[1] disjunctive databases with respect to a given partial order between sets of Herbrand interpretations. As $\mathcal{H}_{\mathcal{D}}$ is finite, the least fixpoint $\mathtt{lfp}(T_{\mathcal{R}}^{\mathcal{M}}, \mathcal{F})$ of $T_{\mathcal{R}}^{\mathcal{M}}$ exists, where $\mathtt{lfp}(T_{\mathcal{R}}^{\mathcal{M}}, \mathcal{F})$ denotes the least fixpoint $\{\mathcal{M}_1, \ldots, \mathcal{M}_n\}$ of operator $T_{\mathcal{R}}^{\mathcal{M}}$ so that for all $\mathcal{M}_i$ holds: $M_i \models \mathcal{F}$. For correctly evaluating negative literals, the operator $T_{\mathcal{R}}^{\mathcal{M}}$ and thereby the new version of $T_{\mathcal{R}}^m$ is iteratively applied to each stratum of a stratified rule set from bottom to top. This results in the computation of the so-called iterated fixpoint model set.

Definition 7. *Let $\mathcal{D} = \langle \mathcal{F}, \mathcal{R} \rangle$ be a stratifiable disjunctive deductive database and λ a stratification on $\mathcal{D}$. The partition $\mathcal{R}_1 \cup \ldots \cup \mathcal{R}_n$ of $\mathcal{R}$ defined by λ induces a sequence of sets of minimal Herbrand models $\mathcal{MM}_{\mathcal{D}_1}, \ldots, \mathcal{MM}_{\mathcal{D}_n}$ with $\mathcal{MM}_{\mathcal{D}_1} := \mathtt{lfp}(T_{\mathcal{R}_1}^{\mathcal{M}}, \mathcal{F})$, $\mathcal{MM}_{\mathcal{D}_2} := \mathtt{lfp}(T_{\mathcal{R}_2}^{\mathcal{M}}, \mathcal{MM}_{\mathcal{D}_1})$, $\ldots$, $\mathcal{MM}_{\mathcal{D}_n} := \mathtt{lfp}(T_{\mathcal{R}_n}^{\mathcal{M}}, \mathcal{MM}_{\mathcal{D}_{n-1}})$. The iterated fixpoint model set of $\mathcal{D}$ is defined as $\mathcal{MM}_{\mathcal{D}_n}$.*

---

[1] In semi-positive databases negative references in rules are permitted to extensional relations only.

Theorem 1. *Let $\mathcal{D} = \langle \mathcal{F}, \mathcal{R} \rangle$ be a stratifiable disjunctive deductive database and λ a stratification on $\mathcal{D}$, and $\mathcal{R}_1 \cup \ldots \cup \mathcal{R}_n$ the partition of $\mathcal{R}$ defined by λ. Then the set of perfect models $\mathcal{PM}_\mathcal{D}$ of $\mathcal{D}$ is identical with the iterated fixpoint model set $\mathcal{MM}_{\mathcal{D}_n}$ of $\mathcal{D}$.*

Proof. cf. [9]

The definition of the iterated fixpoint model set $\mathcal{MM}_{\mathcal{D}_n} = \{\mathcal{M}_1, \ldots, \mathcal{M}_k\}$ of a given stratified database $\mathcal{D}$ induces a constructive method for determining the intended semantics of $\mathcal{D}$ in form of sets of perfect models. A direct implementation of this method, however, would contain a lot of expensive set operations which are hidden in the otherwise apparently simple definitions above. For illustrating the notations introduced above, consider the following database $\mathcal{D} = \langle \mathcal{F}, \mathcal{R} \rangle$:

$$\mathcal{R}: \quad \begin{aligned} \mathtt{t(X)} \vee \mathtt{s(X)} &\leftarrow \mathtt{p(X)} \wedge \mathtt{r(X)} \\ \mathtt{s(X)} &\leftarrow \mathtt{q(X)} \\ \mathtt{s(X)} &\leftarrow \mathtt{t(X)} \end{aligned} \qquad \mathcal{F}: \quad \begin{aligned} &\mathtt{p(a)} \\ &\mathtt{q(b)} \vee \mathtt{r(a)} \end{aligned}$$

The iterative computation of $\mathtt{lfp}(\mathcal{T}_\mathcal{R}^\mathcal{M}, \mathcal{F})$ then induces the following sets:

$$\begin{aligned} \mathcal{I}_1 &:= \mathtt{min\_models}(\mathcal{F}) = \{ \{p(a),q(b)\}, \{p(a),r(a)\} \} \\ \mathcal{I}_2 &:= \mathcal{T}_\mathcal{R}^\mathcal{M}(\mathcal{I}_1) = \{ \{p(a),q(b),s(b)\}, \{p(a),r(a),t(a)\}, \{p(a),r(a),s(a)\} \} \\ \mathcal{I}_3 &:= \mathcal{T}_\mathcal{R}^\mathcal{M}(\mathcal{I}_2) = \{ \{p(a),q(b),s(b)\}, \{p(a),r(a),s(a)\} \} \\ \mathcal{I}_4 &:= \mathcal{I}_3. \end{aligned}$$

The fixpoint $\mathcal{I}_4$ coincides with the sets of minimal models $\mathcal{MM}_\mathcal{D}$ of $\mathcal{D}$. The computation starts with the minimal set of Herbrand interpretations which respect $\mathcal{D}$. After that the operator $\mathcal{T}_\mathcal{R}^\mathcal{M}$ is iteratively applied to the interpretation obtained in the previous iteration round. $\mathcal{T}_\mathcal{R}^\mathcal{M}$ is not monotonic with respect to the number of minimal Herbrand interpretations due to the application of $\mathtt{min\_subset}$. Nevertheless, [9] shows the monotonicity of $\mathcal{T}_\mathcal{R}^\mathcal{M}$ with respect to a restricted domain of canonical collections of interpretations from the finite $\mathcal{H}_\mathcal{D}$ such that it always reaches its least fixpoint in a finite number of iterations.

4 State Generation

As an alterative to model generation, state generation has been proposed in [11,14] dealing with disjunctive facts only. Given the Herbrand base $\mathcal{H}_\mathcal{D}$ of a database $\mathcal{D}$, the disjunctive Herbrand base $\mathcal{DH}_\mathcal{D}$ is the set of all positive ground disjunctions $A_1 \vee \ldots \vee A_n$ that can be formed by atoms $A_i \in \mathcal{H}_\mathcal{D}$. In order to guarantee finiteness of $\mathcal{DH}_\mathcal{D}$, however, we solely consider disjunctions where each atom occurs only once; that is $i \neq j$ implies $A_i \neq A_j$. Any subset $\mathcal{S}_\mathcal{D}$ of $\mathcal{DH}_\mathcal{D}$ is called a *state* of $\mathcal{D}$. $\mathcal{S}_\mathcal{D}$ is called a *model state* of $\mathcal{D}$ if every minimal model of $\mathcal{D}$ is also a model of $\mathcal{S}_\mathcal{D}$ and every model of $\mathcal{S}_\mathcal{D}$ is a model of $\mathcal{D}$. A model state $\mathcal{MS}_\mathcal{D}$ of $\mathcal{D}$ is called *minimal* if none of its subsets is a model state of $\mathcal{D}$. A non-minimal model state $\mathcal{S}_\mathcal{D}$ must include disjunctive facts which are subsumed by other facts from $\mathcal{S}_\mathcal{D}$ and doesn't provide new information on derivable facts from $\mathcal{D}$. To eliminate those irrelevant facts, we will use the following subsumption operator $\mathtt{red}$.

Definition 8. *The operator* `red` *eliminates all disjunctive facts from a given set F which are subsumed by another fact from F:*

$$\text{red}(F):= \{f \in F \mid \neg \exists f' \in F : f' \subsetneq f\}.$$

Theorem 2. *Let $\mathcal{D}$ be a positive disjunctive database. The minimal model state $\mathcal{MS}_\mathcal{D}$ of $\mathcal{D}$ always exists, is unique and can be obtained from any model state $\mathcal{S}_\mathcal{D}$ of $\mathcal{D}$ by applying* `red`: $\mathcal{MS}_\mathcal{D} = \text{red}(\mathcal{S}_\mathcal{D})$.

Proof. cf. [12].

4.1   State Generation for Positive Rules

For determining the minimal model state of a given positive disjunctive database, Minker and Rajasekar introduced a disjunctive consequence operator [14] based on Robinson's hyperresolution concept [17]. This operator directly works on ground disjunctions and uses factorization to avoid different representations of equivalent disjunctive facts. We will use the operator `norm` for obtaining an equivalent but unique representation of disjunctive facts from an arbitrary set of disjunctions. The operator `norm` maps a disjunctive fact f to a logically equivalent variant f' such that every atom in f' occurs exactly once in f', the boolean constant *false* does not occur in f', and all all atoms in f' are ordered lexicographically. For a set of disjunctive facts F, `norm`(F) is defined as $\cup_{f \in F}$ `norm`(f).

Based on the notions above, we can now define a refined version of the disjunctive consequence operator of Minker and Rajasekar [14] in which each separate rule evaluation step is made explicit. To this end, an auxiliary consequence operator $T^s[r]$ for single rule applications is introduced before defining the cumulative version of the disjunctive consequence operator $T^s_\mathcal{R}$.

Definition 9. *Let $\mathcal{R}$ be a set of positive disjunctive rules and F an arbitrary set of disjunctive facts:*

1. *The single consequence operator $T^s[r]$ defines for each disjunctive rule $r \equiv A_1 \vee \cdots \vee A_l \leftarrow B_1 \wedge \cdots \wedge B_n \in \mathcal{R}$ with $l, n \geq 1$ the set of disjunctive facts which can be derived by a single application of r to F:*

$$T^s[r](F) := \text{red}(\{\text{norm}((A_1 \vee \ldots A_l \vee f_1 \setminus B_1 \vee \cdots \vee f_n \setminus B_n)\sigma) \mid$$
$$\sigma \text{ is a ground substitution and}$$
$$\forall i \in \{1, \ldots, n\} : f_i \in F \text{ and } B_i\sigma \in f_i\})$$

2. *The cumulative disjunctive consequence operator $T^s_\mathcal{R}$ models the simultaneous application of all rules in $\mathcal{R}$ to F. It returns the set of derived facts together with the input set while subsumed facts are eliminated:*

$$T^s_\mathcal{R}(F) := \text{red}(\bigcup_{r \in \mathcal{R}} T^s[r](F) \cup F).$$

Note that it would be sufficient to apply the `red` operator only once within $T^s_\mathcal{R}$ in order to get a subsumption-free set of disjunctive facts. We have additionally included this elimination step into $T^s[r]$, however, in order to keep the set of intermediate results small and to highlight where derivation of subsumed disjunctive facts may occur. In Section 6 we will show how some of these redundant derivations can be avoided due to our new method for state generation of stratifiable rules. Note that even a single rule application may lead to subsumed disjunctive facts as the following example shows:

$$r \equiv a \vee b \leftarrow c \text{ and } \mathcal{S} = \{c \vee b \, , \, c \vee d\}$$

$$T^s[r](\mathcal{S}) = \mathtt{red}(\{\mathtt{norm}(a \vee b \vee b), \mathtt{norm}(a \vee b \vee d)\})$$
$$= \mathtt{red}(\{a \vee b \, , \, a \vee b \vee d\}) = \{a \vee b\}$$

We can now define the semantics of positive disjunctive databases based on $T_\mathcal{R}^s$. $T_\mathcal{R}^s$ is monotonic with respect to an expanded set of resulting disjunctions in which all subsumed facts from $\mathcal{DH}_D$ are additionally included. Therefore, $T_\mathcal{R}^s$ has a unique least fixpoint $\mathtt{lfp}(T_\mathcal{R}^s, \mathcal{F})$ containing $\mathcal{F}$.

Theorem 3. *Let $\mathcal{D} = \langle \mathcal{F}, \mathcal{R} \rangle$ be a positive disjunctive deductive database. The minimal model state $\mathcal{MS}_D$ of $\mathcal{D}$ coincides with $\mathtt{lfp}(T_\mathcal{R}^s, \mathcal{F})$.*

Proof. cf. [11].

The fixpoint characterization of the minimal model state induces a constructive method for determining the semantics of a given disjunctive database by iteratively applying $T_\mathcal{R}^s$, starting from $\mathcal{F}$. As an example, consider the database $\mathcal{D} = \langle \mathcal{F}, \mathcal{R} \rangle$:

$$\mathcal{R}: \quad \mathtt{t(X)} \vee \mathtt{s(X)} \leftarrow \mathtt{p(X)} \wedge \mathtt{r(X)} \qquad \mathcal{F}: \quad \mathtt{p(a)}$$
$$\mathtt{s(X)} \leftarrow \mathtt{q(X)} \qquad\qquad\qquad \mathtt{q(b)} \vee \mathtt{r(a)}$$
$$\mathtt{s(X)} \leftarrow \mathtt{t(X)}$$

The iterative computation of $\mathtt{lfp}(T_\mathcal{R}^s, \mathcal{F})$ then leads to the following sets of minimal states:

$$\mathcal{S}_1 := T_\mathcal{R}^s(\emptyset) \ = \{\mathtt{p(a)} \, , \, \mathtt{q(b)} \vee \mathtt{r(a)}\}$$
$$\mathcal{S}_2 := T_\mathcal{R}^s(\mathcal{S}_1) = \{\mathtt{p(a)} \, , \, \mathtt{q(b)} \vee \mathtt{r(a)}\} \cup \{\mathtt{r(a)} \vee \mathtt{s(b)} \, , \, \mathtt{q(b)} \vee \mathtt{s(a)} \vee \mathtt{t(a)}\}$$
$$\mathcal{S}_3 := T_\mathcal{R}^s(\mathcal{S}_2) = \{\mathtt{p(a)} \, , \, \mathtt{q(b)} \vee \mathtt{r(a)}\} \cup \{\mathtt{r(a)} \vee \mathtt{s(b)} \, , \, \mathtt{q(b)} \vee \mathtt{s(a)}\}$$
$$\mathcal{S}_4 := \mathcal{S}_3.$$

The fixpoint $\mathcal{S}_4$ coincides with the minimal model state $\mathcal{MS}_D$ of $\mathcal{D}$. The computation starts with the base facts $\mathcal{F}$ and then iteratively applies the operator $T_\mathcal{R}^s$ for obtaining new derivable disjunctions. Note that the fact $\{\mathtt{q(b)} \vee \mathtt{s(a)} \vee \mathtt{t(a)}\}$ is eliminated by the **red** operator in iteration round 3 because it is subsumed by the fact $\{\mathtt{q(b)} \vee \mathtt{s(a)}\}$ which has been newly generated in this round. Obviously, model and state generation methods deal with different representations which can be systematically transformed into each other.

Lemma 1. *Let $\mathcal{D}$ be a positive disjunctive database, $\mathcal{MM}_D = \{\mathcal{M}_1, \dots, \mathcal{M}_n\}$ with $n \geq 1$ the set of minimal Herbrand models of $\mathcal{D}$, and $\mathcal{MS}_D$ the minimal model state of $\mathcal{D}$. Then the following equalities hold:*

$$\mathcal{MM}_D = \mathtt{min\_models}(\mathcal{MS}_D) \text{ and } \mathcal{MS}_D = \mathtt{red}(\{\mathtt{norm}(L_1 \vee \dots \vee L_n) \mid L_i \in \mathcal{M}_i\})$$

Proof. cf. [9].

4.2   State Generation for Stratifiable Rules

The semantics of a stratifiable disjunctive database $\mathcal{D}$ is given by its set of perfect models $\mathcal{PM}_\mathcal{D} = \{\mathcal{M}_1, \ldots, \mathcal{M}_k\}$. The corresponding perfect model state $\mathcal{PS}_\mathcal{D}$ again can be derived by simply applying the operators red and norm as shown in Lemma 1, i.e., $\mathcal{PS}_\mathcal{D} = \text{red}(\{\text{norm}(L_1 \vee \cdots \vee L_k) \mid L_i \in \mathcal{M}_i\})$. The question how to determine the corresponding model state without using a model generation approach first has not been answered yet! Thus, we are looking for a state generation method purely based on hyperresolution extending the consequence operator $T_\mathcal{R}^s$ from Section 4.1.

When applying $T_\mathcal{R}^s$ to a positive disjunctive rule, all 'alternatives' of the matching body literals are transferred to the conclusion of the rule. For example, the application of $T^s[r_1](F)$ with $r_1 \equiv p(X) \leftarrow r(X, Y), s(Y)$ to F=$\{r(1, 2), s(2) \vee t(3)\}$ leads to the conclusion $\{p(1) \vee t(3)\}$ where the remaining alternative $t(3)$ of fact $s(2) \vee t(3)$ is included. For negative body literals, hyperresolution would suggest the matching body literals themselves to be propagated to the conclusion instead of their "alternatives". Applying the rule $r_2 \equiv p(X) \leftarrow r(X, Y), \neg s(Y)$ to the state $F = \{r(1, 2), s(2) \vee t(3)\}$ would then yield $\{p(1) \vee s(2)\}$, which together with F corresponds to the perfect model state of the database $\langle F, \{r_2\}\rangle$.

However, this does not always lead to correct derivations with respect to the intended semantics. For instance, applying the rule $r_3 \equiv p(X) \leftarrow a(X), \neg b(X)$ to the state $F' = \{a(1)\}$ would yield $\{p(1) \vee b(1)\}$ while the correct derivation would be simply $\{p(1)\}$. In this case, the negatively referred literal $b(1)$ must not be added to the conclusion in order to avoid the generation of the unsupported Herbrand model $\mathcal{M} = \{b(1)\}$. The reason for this is the non-constructive nature of negation in deductive rules where negative body literals are to be seen as tests of non-existence. Therefore, such literals must never be added to the conclusion of a rule if they occur in no disjunctive fact of the state to which this rule is applied. Otherwise, the generated model states additionally have unsupported Herbrand models which are not present in the set of perfect models.

But even if there were disjunctive facts in the state in which negatively referred atoms occur, the propagation of these atoms to the conclusion sometimes leads to an incorrect model state. Consider the following stratifiable disjunctive database $\mathcal{D} = \langle F'', r_3\rangle = \langle\{a(1) \vee b(1), a(1) \vee c(5)\}, \{p(X) \leftarrow a(X), \neg b(X)\}\rangle$. The set of perfect models of $\mathcal{D}$ is given by $\mathcal{MM}_{\mathcal{D}_n} = \{\{a(1), p(1)\}, \{b(1), c(5)\}\}$ and the corresponding perfect model state $\mathcal{PS}_\mathcal{D}$ is $\{b(1) \vee p(1), c(5) \vee p(1)\} \cup F''$. Suppose now we have modified our single derivation operator T^s to $T^{s'}$ for handling negative body literals in the way proposed above where negatively referred literals are solely added to the conclusion if they do not occur within a disjunction of the given state. However, as the atom $b(1)$ occurs in indefinite facts of F'', the application of $T^{s'}[r_3]$ to F'' would add $b(1)$ to the conclusion of rule r_3 leading to an incomplete set of derived disjunctive facts:

$$\begin{aligned}
T^{s'}[r_3](F'') &= \text{red}(\{\text{norm}(p(1) \vee b(1) \vee b(1)), \text{norm}(p(1) \vee c(5) \vee b(1))\}) \\
&= \text{red}(\{b(1) \vee p(1)\ , \ b(1) \vee c(5) \vee p(1)\} \\
&= \{b(1) \vee p(1)\}
\end{aligned}$$

Instead of the fact $c(5) \vee p(1)$, the non-minimal conclusion $p(1) \vee c(5) \vee b(1)$ is drawn and subsequently eliminated by the subsumption operator red. The correct derivations with respect to $\mathcal{PS_D}$, however, would be obtained again if the negated body literals were not propagated to the conclusion this time.

5   A New State Generation Method for Stratifiable Rules

In Section 4.2 we have shown that for avoiding the derivation of invalid model states using a modified version of the consequence operator T^s from Section 4.1, negative body literals must be added to the conclusion only in specific situations.

5.1   A New Consequence Operator

The basic idea of our approach is to propagate a negative body literal $C_i\sigma$ to the conclusion of a disjunctive rule $r \equiv A_1 \vee \cdots \vee A_l \leftarrow B_1 \wedge \cdots \wedge B_n \wedge \neg C_1 \wedge \cdots \wedge \neg C_m$ iff there exists at least one minimal model $\mathcal{M} \in$ min_models(F) with respect to the given state F which comprises all instantiated positive body literals $B_i\sigma$ together with a $C_j\sigma$, i.e., $\{B_1\sigma, \ldots, B_n\sigma, C_j\sigma\} \subseteq \mathcal{M}$, but none of the instantiated head literals, i.e., $\forall k \in \{1, \ldots, l\} : A_k\sigma \notin \mathcal{M}$. Although there seems to be no need to fire r, there might exist another minimal model $\mathcal{M}' \in$ min_models(F) with $\{B_1\sigma, \ldots, B_n\sigma\} \subseteq \mathcal{M}'$ and $C_j\sigma \notin \mathcal{M}'$ for which the application of r would indeed lead to new correct derivations. Within a state representation it is not possible to directly identify the associated models such that the application of r may lead to logically correct but non-minimal derivations. By adding the instantiated negative body literal $C_j\sigma$ to the conclusion, however, all non-minimal derivations with respect to $\mathcal{M}$ will be eliminated by the subsumption operator within $T^s_{\mathcal{R}}$.

Note that in the last sample database $\langle F'', \{r_3\} \rangle$ from Section 4.2, this condition was not satisfied as the atoms $a(1)$ and $b(1)$ are not contained in any minimal model of F''. Thus, the negated body literal $b(1)$ must not be propagated to the conclusion this time. In the following definition, the single consequence operator $T^s[r]$ for state generation is extended according to the condition mentioned above for correctly handling negative literals.

Definition 10. *Let $\mathcal{R}$ be an arbitrary set of disjunctive rules and F an arbitrary set of disjunctive facts. The single consequence operator $T^s[r]$ defines for each disjunctive rule $r \equiv A_1 \vee \ldots A_l \leftarrow B_1 \wedge \cdots \wedge B_n \wedge \neg C_1 \wedge \cdots \wedge \neg C_m \in \mathcal{R}$ with $l, n \geq 1$ and $m \geq 0$ the set of disjunctive facts which can be derived by a single application of r to F:*

$$T^s[r](F) := \mathrm{red}(\{\mathrm{norm}((A_1 \vee \ldots A_l \vee f_1 \setminus B_1 \vee \cdots \vee f_n \setminus B_n \vee C)\sigma \mid$$
$$\sigma \text{ is a ground substitution and}$$
$$\forall i \in \{1, \ldots, n\} : f_i \in F \wedge B_i\sigma \in f_i \wedge$$
$$\forall j \in \{1, \ldots, m\} : C_j\sigma \notin F \wedge (C_j \in C \Leftrightarrow$$
$$\exists \mathcal{M} \in \mathrm{min\_models}(F) : \{B_1\sigma, \ldots, B_n\sigma, C_j\sigma\} \subseteq \mathcal{M} \wedge$$
$$\forall k \in \{1, \ldots, l\} : A_k\sigma \notin \mathcal{M})\})$$

In addition to the definition of $T^s[r]$, two further conditions are included into the definition of $T^s[r]$ for handling negative literals. For each negatively referred literal C_j the criterion $C_j\sigma \notin F$ is added in order to avoid firing the rule in the presence of corresponding definite facts. Furthermore, the disjunction C is used to add all those negatively referred literals to the conclusion of the given rule which satisfy the condition mentioned above. But the direct reference to minimal models of F seems to be quite problematic as it suggests that the operator works on states and their corresponding models at the same time.

For testing whether there is a model $\mathcal{M} \in \text{min\_models}(F)$ satisfying the conditions above, however, the actual generation of any $\mathcal{M}$ is usually not necessary. For instance, the condition $\{B_1\sigma, \ldots, B_n\sigma, C_j\sigma\} \subseteq \mathcal{M}$ corresponds to the property $T^s_{\{(y \leftarrow B_1 \wedge \cdots \wedge B_n \wedge C_j)\sigma\}}(F) \neq F$ with $y \notin \mathcal{H}_\mathcal{D}$ which can be checked without generating any model of F. In fact, it is not even necessary to apply $T^s_\mathcal{R}$ to all facts in F as for the generation of an unsubsumed fact f_y with $y \in f_y$ solely facts are needed which comprise $B_i\sigma$ or $C_j\sigma$ or an atom A which also appears in a fact $g \in F$ with $B_i\sigma \in g$ or $C_j\sigma \in g$. For eliminating the other (unnecessary) facts from F before testing the property above we employ the following r_σ-restriction.

Definition 11. *Let r_σ be an instantiated disjunctive rule with respect to a ground substitution σ. For an arbitrary state F we define:*

$$F|_{r_\sigma} := \{\, f \in F \mid \text{there is a literal } l \text{ from } r_\sigma \text{ and either } l \in f \text{ or}$$
$$\text{there is a } g \in F \text{ such that } l, l' \in g \text{ and } l' \in f\}$$

Note that $F|_{r_\sigma}$ solely returns facts which directly depend on literals from r_σ. The consideration of transitively connected facts in F, however, is not necessary. Using this notion, the condition above can be refined to $T^s_{\{r_\sigma\}}(F|_{r_\sigma}) \neq F|_{r_\sigma}$ with $r_\sigma \equiv (y \leftarrow B_1 \wedge \cdots \wedge B_n \wedge C_j)\sigma$ meaning that the containment test can be restricted to the usually smaller part $F|_{r_\sigma}$ of F. Even the test whether none of the instantiated head literals $A_1\sigma, \ldots, A_l\sigma$ occurs within a model $\mathcal{M}$ with $\{B_1\sigma, \ldots, B_n\sigma, C_j\sigma\} \subseteq \mathcal{M}$ can be performed over $F|_{r_\sigma}$ with $r_\sigma \equiv (y \leftarrow B_1 \wedge \cdots \wedge B_n \wedge C_j \wedge A_1 \wedge \ldots A_l)\sigma$, only.

5.2 Perfect Model State Generation

The definition of the cumulative consequence operator $T^s_\mathcal{R}$ is analogous to the one already given in definition 9, i.e., $T^s_\mathcal{R}(F) := \text{red}(\bigcup_{r \in \mathcal{R}} T^s[r](F) \cup F)$. This operator is used again for modelling the simultaneous application of all rules in $\mathcal{R}$ to the given state F. Before proving the correctness of $T^s_\mathcal{R}$ which is based on the modified version of $T^s[r]$, we illustrate its application using the following semi-positive database $\mathcal{D} = \langle \mathcal{F}, \{r\}\rangle$:

$$r \equiv t(X) \vee u(X) \leftarrow p(X) \wedge \neg r(X) \qquad \mathcal{F}: \quad p(a) \vee q(1)\,,$$
$$p(b) \vee q(2)\,,\ r(b),$$
$$p(c) \vee q(3)\,,\ r(c) \vee s(3),\ q(3) \vee s(3)$$
$$p(d) \vee q(4)\,,\ r(d) \vee s(4) \vee q(4)$$
$$p(e) \vee q(5)\,,\ r(e) \vee q(5),\ t(e) \vee q(5)$$

The application of $T^s[r](\mathcal{F})$ then leads to the following derivations:

$$T^s[r](\mathcal{F}) = \{t(a) \vee u(a) \vee q(1)\} \ \cup \ \{t(c) \vee u(c) \vee q(3)\} \ \cup$$
$$\{t(d) \vee u(d) \vee q(4) \vee r(d)\} \ \cup \ \{t(e) \vee u(e) \vee q(5)\}$$

The example illustrates five derivation cases which possibly occur when evaluating negative literals. In the first case, the absence of $r(a)$ in any disjunctive fact from $\mathcal{F}$ allows for the derivation of $t(a) \vee u(a) \vee q(1)$ whereas the definite fact $r(b)$ prevents the derivation of $t(b) \vee u(b) \vee q(2)$. In the remaining three cases, the condition $T^s_{\{r_\sigma\}}(\mathcal{F}|_{r_\sigma}) \neq F|_{r_\sigma}$ with $r_\sigma \equiv (y \leftarrow p(X) \wedge r(X))\sigma$ has to be checked for the substitutions $\sigma_3 = \{X \setminus c\}$, $\sigma_4 = \{X \setminus d\}$, and $\sigma_5 = \{X \setminus e\}$, respectively.

The application of $T^s_{r_{\sigma_3}}(\mathcal{F}|_{r_{\sigma_3}})$ with $\mathcal{F}|_{r_{\sigma_3}} = \{\, p(c) \vee q(3), r(c) \vee s(3), q(3) \vee s(3) \,\}$ and $r_{\sigma_3} \equiv y \leftarrow p(c) \wedge r(c)$ derives the disjunction $q(3) \vee s(3) \vee y$ which is subsumed by $q(3) \vee s(3) \in \mathcal{F}|_{r_{\sigma_3}}$. Since $T^s_{r_{\sigma_3}}(\mathcal{F}|_{r_{\sigma_3}}) = \mathcal{F}|_{r_{\sigma_3}}$ it can be concluded that there is no minimal model $\mathcal{M}$ of $\mathcal{F}$ with $\{p(c), r(c)\} \subseteq \mathcal{M}$ and $r(c)$ must not be propagated to the conclusion $t(c) \vee u(c) \vee q(3)$.

The application of $T^s_{r_{\sigma_4}}(\mathcal{F}|_{r_{\sigma_4}})$ with $\mathcal{F}|_{r_{\sigma_4}} = \{p(d) \vee q(4), r(d) \vee s(4) \vee q(4)\}$ and $r_{\sigma_3} \equiv y \leftarrow p(d) \wedge r(d)$ derives the new minimal conclusion $q(4) \vee s(4) \vee y$ such that $T^s_{r_{\sigma_3}}(\mathcal{F}|_{r_{\sigma_3}}) \neq \mathcal{F}|_{r_{\sigma_3}}$. In addition, there cannot be any minimal model of $\mathcal{F}$ which contains one of the instantiated head literals $t(d)$ or $u(d)$ since there is no $f \in \mathcal{F}$ with $t(d) \in f$ or $u(d) \in f$. Consequently, the negatively referred literal $r(d)$ has to be propagated to the conclusion of r_{σ_4} yielding the new fact $t(d) \vee u(d) \vee q(4) \vee r(d)$.

For the last possible substitution σ_5, there is exactly one minimal model $\mathcal{M}$ of $\mathcal{F}$ with $\{p(e), r(e)\} \subseteq \mathcal{M}$ but this time one of the instantiated head literals $t(e)$ is also contained in $\mathcal{M}$. We could employ again $T^s_{r_{\sigma_5}}(\mathcal{F}|_{r_{\sigma_5}})$ with $r_{\sigma_5} \equiv y \leftarrow p(e) \wedge r(e) \wedge t(e)$ for testing whether there is a model which contains $p(e)$, $r(e)$, and $t(e)$. However, this test only shows the existence of such a model. But there might be another minimal model which solely comprises $\{p(e), r(e)\}$ but no head literal. Therefore, all minimal models have to be generated with respect to the subset $\mathcal{F}|_{r'_{\sigma_5}}$ of $\mathcal{F}$ with $r'_{\sigma_5} \equiv y \leftarrow p(e) \wedge r(e) \wedge t(e) \wedge u(e)$ in order to be sure that all models comprising $\{p(e), r(e)\}$ contain at least one instantiated head literal as well. The corresponding models of $\mathcal{F}|_{r'_{\sigma_5}} = \{p(e) \vee q(5), r(e) \vee q(5), t(e) \vee q(5)\}$ are given by $\{\{p(e), r(e), t(e)\}, \{q(5)\}\}$ such that the negatively referred literal $r(e)$ must not be propagated to the conclusion of r_{σ_5} yielding the fact $t(e) \vee u(e) \vee q(5)$.

The set $\mathtt{red}(T^s[r](\mathcal{F}) \cup \mathcal{F})$ represents $\mathtt{lfp}(T^s_{\{r\}}, \mathcal{F})$ and coincides with the perfect model state of $\mathcal{D}$. A model generation approach for this example would have to deal with over 100 minimal models in order to compute the set of perfect models of $\mathcal{D}$. In fact, if CNF is much more compact than DNF as in this example, reasoning with states obviously becomes simpler and thus, more efficient.

Theorem 4. *Let $\mathcal{D} = \langle \mathcal{F}, \mathcal{R} \rangle$ be a disjunctive deductive database where all negative literals in $\mathcal{R}$ refer to base relations, only. The perfect model state $\mathcal{PS}_\mathcal{D}$ of $\mathcal{D}$ coincides with $\mathtt{lfp}(T^s_\mathcal{R}, \mathcal{F})$.*

Proof. (sketch) The correctness of $T^s_\mathcal{R}$ with respect to a given set of rules $\mathcal{R}$ directly follows from the correct evaluation of every single rule r in $\mathcal{R}$. For proving this result, we use the correctness of the model-based consequence operator $T_{\mathcal{R}}^M$ and show that $T_{\mathcal{R}}^s$ and $T_{\mathcal{R}}^M$ return equivalent results if applied to an arbitrary set of disjunctive facts F and its equivalent set of minimal models min_models(F), respectively. For full proofs we refer to [4].

A method for state generation of a stratified database $\mathcal{D}$ ought to make use of a stratification such that each pair $\langle \mathcal{F}, \mathcal{R}_1 \rangle$ and $\langle \mathcal{MS}_{\mathcal{D}_{i-1}}, \mathcal{R}_i \rangle$ $(i = 2, \ldots, n)$ is a semi-positive database. The iterative application of the cumulative consequence operator $T_{\mathcal{R}}^s$ then yields the iterated fixpoint state model of $\mathcal{D}$.

Definition 12. *Let $\mathcal{D} = \langle \mathcal{F}, \mathcal{R} \rangle$ be a stratifiable disjunctive deductive database and λ a stratification on $\mathcal{D}$. The partition $\mathcal{R}_1 \cup \ldots \cup \mathcal{R}_n$ of $\mathcal{R}$ defined by λ induces a sequence of minimal model states $\mathcal{MS}_{\mathcal{D}_1}, \ldots, \mathcal{MS}_{\mathcal{D}_n}$ with $\mathcal{MS}_{\mathcal{D}_1} :=$ lfp($T_{\mathcal{R}_1}^s, \mathcal{F}$), $\mathcal{MS}_{\mathcal{D}_2} :=$ lfp($T_{\mathcal{R}_2}^s, \mathcal{MS}_{\mathcal{D}_1}$), $\ldots$, $\mathcal{MS}_{\mathcal{D}_n} :=$ lfp($T_{\mathcal{R}_n}^s, \mathcal{MS}_{\mathcal{D}_{n-1}}$). The iterated fixpoint state model of $\mathcal{D}$ is defined as $\mathcal{MS}_{\mathcal{D}_n}$.*

Theorem 5. *Let $\mathcal{D} = \langle \mathcal{F}, \mathcal{R} \rangle$ be a stratifiable disjunctive deductive database and λ a stratification on $\mathcal{D}$, $\mathcal{R}_1 \cup \ldots \cup \mathcal{R}_n$ the partition of $\mathcal{R}$ induced by λ, $\mathcal{MM}_{\mathcal{D}_n}$ the iterated fixpoint model of $\mathcal{D}$, and $\mathcal{PS}_{\mathcal{D}}$ the corresponding perfect model state of $\mathcal{D}$. Then the perfect model state $\mathcal{PS}_{\mathcal{D}}$ of $\mathcal{D}$ coincides with the iterated fixpoint state model $\mathcal{MS}_{\mathcal{D}_n}$ of $\mathcal{D}$. In particular, it holds that $\mathcal{MM}_{\mathcal{D}_n} \equiv$ min_models($\mathcal{MS}_{\mathcal{D}_n}$).*

Proof. The proof directly follows from the correctness of the fixpoint computations for each stratum.

The definition of the iterated fixpoint state model $\mathcal{MS}_{\mathcal{D}_n}$ of a given stratified disjunctive database $\mathcal{D}$ induces a constructive method for determining the perfect model state $\mathcal{PS}_{\mathcal{D}}$ of $\mathcal{D}$ which is purely based on model state generation.

6   Enhanced State Generation for Positive Rules

In principle, negative and positive body literals in disjunctive rules are treated in a complementary way; that is, while for positive body literals the resolved disjunction is added to the conclusion, negatively referenced body literals are propagated themselves. As shown in the previous section, however, negatively referenced body literals must not be propagated if this turns a correct and minimal conclusion into a non-minimal one. The question is now whether a similar condition can be defined for the treatment of positive body literals. Consider the following positive database $\mathcal{D} = \langle \mathcal{F}, \{r\} \rangle$ in which the two facts from $\mathcal{F}$ will always lead to non-minimal and thus redundant derivations:

$$r \equiv s(X) \leftarrow p(X) \wedge r(Y) \qquad \mathcal{F}: \ p(a) \vee q(1) \ , \ r(b) \vee p(a) \vee 1(2)$$

The application of $T^s[r](\mathcal{F})$ leads to the following derivations:

$$T^s[r](\mathcal{F}) = \text{red}(\{ \ \text{norm}(s(a) \vee q(1) \vee p(a) \vee 1(2)),$$
$$\text{norm}(s(a) \vee r(b) \vee 1(2) \vee p(a) \vee 1(2))\})$$

Both derivations are redundant as they are subsumed by the two facts in $\mathcal{F}$. Consequently, the operation red eliminates these two derivations during the application of the cumulative consequence operator $T^s_{\{r\}}$. It is easy to see that when considering a ground instantiated rule any disjunctive fact including at least two different body literals will produce subsumed derivations. The following refined version of the single consequence operator $T^s[r]$ for positive rules includes a corresponding criterion for avoiding such redundant derivations.

Definition 13. *Let $\mathcal{R}$ be a set of positive disjunctive rules and $\mathcal{S}$ an arbitrary set of disjunctive facts. The single consequence operator $T^s[r]$ defines for each disjunctive rule $r \equiv A_1 \vee \ldots A_l \leftarrow B_1 \wedge \cdots \wedge B_n \in \mathcal{R}$ with $l, n \geq 1$ the set of disjunctive facts which can be derived by a single application of r to F:*

$$T^s[r](F) := \mathtt{red}(\{\mathtt{norm}((A_1 \vee \cdots \vee A_l \vee f_1 \setminus B_1 \vee \cdots \vee f_n \setminus B_n)\sigma)|$$
$$\sigma \text{ is a ground substitution and } \forall i, j \in \{1, \ldots, n\} :$$
$$f_i \in F \text{ and } B_i\sigma \in f_i \text{ and } (i \neq j \Rightarrow B_j\sigma \notin (f_i \setminus B_i\sigma)\})$$

As the above condition does not affect the evaluation of negative literals, the modification of the consequence operator $T^s[r]$ from Definition 10 is straightforward. The correctness of $T^s[r]$ for the positive case can be seen as follows: Let $r \equiv A \leftarrow B_1 \wedge \cdots \wedge B_n$ be a positive disjunctive rule with $n \geq 2$, f be a disjunctive fact from F with $B_i\sigma \in f$ and $B_j\sigma \in f$ for $j \neq i$, and f' be a further arbitrary disjunctive fact from F with $B_j\sigma \in f'$ possibly identical with f. Applying the two facts to the body of r allows for deriving the two conclusions $(A \vee f \setminus B_i \vee f \setminus B_j \vee \ldots)\sigma$ and $(A \vee f \setminus B_i \vee f' \setminus B_j \vee \ldots)\sigma$. The former is subsumed by f while the latter is subsumed by f' since $B_j\sigma \in f \setminus B_i\sigma$. Thus, during the application of the cumulative consequence operator $T^s_{\{r\}}$ the red-operation will always eliminate these redundant derivations.

7   Discussion

Model states and Herbrand models directly correspond to CNF respectively DNF representations and thus none can be preferred over the other. If for a given example one representation is considerably shorter than the other, [20] favors the corresponding model or state generation method for a more efficient computation based on practical experience with the disjunctive deductive database engine Dislog [21]. What representation type ought to be chosen additionally depends on the application scenario as certain information can be easier retrieved from a CNF than from a DNF representation and vice versa. For instance, the number of models in which a certain fact occurs can be easily determined from a DNF representation while alternative view update realizations are directly given by a disjunctive fact within a CNF representation.

Another aspect is the number of possible derivations when applying disjunctive rules which varies with the chosen fact representation as shown in the following two examples. In the first example, the rule $r \equiv d \leftarrow b$ must be applied n times to the state $F = \{a_1 \vee b, \ldots, a_n \vee b\}$ for deriving the consequences $\{a_1 \vee d, \ldots, a_n \vee d\}$ whereas a single application of r is sufficient for extending the corresponding set of minimal interpretations $\mathcal{I} = \{\{a_1, \ldots, a_n\}, \{b\}\}$. In the second example, a single application of r to the state $F = \{a_1 \vee \cdots \vee a_n, b\}$ is sufficient for deriving the consequence d while it can be applied n times to the corresponding set of minimal interpretations $\mathcal{I} = \{\{a_1, b\}, \ldots, \{a_n, b\}\}$ for extending each interpretation by the new consequence d. Thus, if an application scenario usually deals with substantially more definite than indefinite facts - as is the case for view updating methods - the model state approach leads to a more compact (but not necessarily shorter) representation than the set of minimal Herbrand models inducing again computational advantages with respect to the number of derivations.

Another problem arising from different fact representations is the existence of redundant derivations. In Section 6 we have shown how to avoid certain redundant derivations when applying the consequence operator $T_{\mathcal{R}}^s$ to facts in CNF. However, similar redundancies may also occur during the model generation approach from Section 3 when applying the consequence operator $T_{\mathcal{R}}^M$ to facts in DNF. As an example, consider the following minimal interpretation $I = \{a, b\}$ and the rule $r \equiv b \vee c_1 \vee \cdots \vee c_n \leftarrow a$. As in the (instantiated) conclusion of r the ground literal b occurs which is also contained in I, the resulting derived interpretation $I_1 = \{a, b\} \cup \{b\}$ subsumes all other derivations $I_n = \{a, b\} \cup \{c_n\}$ by $T_{\{r\}}^m(I)$ which are therefore redundantly generated. These derivations are subsequently eliminated by the operation min_subset during the application of $T_{\{r\}}^M$. Similar to the state generation approach, such redundant derivations can be avoided by adding corresponding restrictions into the definition of $T_{\mathcal{R}}^m$.

Thus, the discussion shows that CNF or DNF ought to be chosen depending on the application, but none can be preferred over the other in the general case. Indeed, despite of the enhancements discussed in Section 6, our new model state generation approach does not improve the general complexity for determining perfect models of stratifiable databases $\mathcal{D} = \langle \mathcal{F}, \mathcal{R} \rangle$ [7]. It simply represents an alternative approach for computing perfect models which is based on model states and thus closes a missing gap in the theory of disjunctive logic programs. Additionally, our approach provides insights with respect to the application of hyperresolution in disjunctive databases which allows for avoiding redundant derivations when applying positive disjunctive rules.

8 Conclusion

In this paper, a new method for computing perfect models of stratifiable disjunctive databases has been presented. In contrast to existing model generation procedures, our fixpoint-based approach is the first to be purely based on state generation. It provides new insights into the application of hyperresolution when considering minimal model semantics and allows for enhancing the efficiency of methods for perfect model state generation.

References

1. Apt, K.R., Blair, H.A., Walker, A.: Towards a Theory of Declarative Knowledge. In: Foundations of Deductive Databases and Logic Programming. ch. 2, pp. 89–148. Morgan Kaufmann Publishers, San Francisco (1988)

2. Apt, K.R.: Logic Programming. In: Handbook of Theoretical Computer Science. Formal Models and Semantics, vol. B, pp. 493–574. The MIT Press, Cambridge (1990)
3. Chan, E.P.F.: A Possible World Semantics for Disjunctive Databases. IEEE Transactions on Knowledge and Data Engineering 5(2), 282–292 (1993)
4. Behrend, A.: A Fixpoint Approach to State Generation for Stratifiable Disjunctive Deductive Databases. Technical Report IAI-TR-2006-7, Institut für Informatik III, Universität Bonn (December 2006)
5. Baral, C., Lobo, J., Minker, J.: Generalized Well-founded Semantics for Logic Programs (Extended Abstract). In: Stickel, M.E. (ed.) 10th International Conference on Automated Deduction. LNCS, vol. 449, pp. 102–116. Springer, Heidelberg (1990)
6. Lloyd, J.W.: Foundations of Logic Programming, 2nd edn. Springer, Heidelberg (1987)
7. Eiter, T., Gottlob, G., Mannila, H.: Adding Disjunction to Datalog. In: PODS 1994, pp. 267–278 (1994)
8. Fernández, J.A., Minker, J.: Bottom-Up Evaluation of Hierarchical Disjunctive Deductive Databases. In: ICLP 1991, pp. 660–675 (1991)
9. Fernández, J.A., Minker, J.: Bottom-Up Compuation of Perfect Models for Disjunctive Theories. Journal of Logic Programming 25(1), 33–51 (1995)
10. Sakama, C., Inoue, K.: Updating Extended Logic Programs through Abduction. In: Gelfond, M., Leone, N., Pfeifer, G. (eds.) LPNMR 1999. LNCS (LNAI), vol. 1730, pp. 147–161. Springer, Heidelberg (1999)
11. Lobo, J., Minker, J., Rajasekar, A.: Extending the Semantics of Logic Programs to Disjunctive Logic Programs. In: ICLP 1989, pp. 255–267 (1989)
12. Lobo, J., Minker, J., Rajasekar, A.: Foundations of Disjunctive Logic Programming. MIT Press, Cambridge, Massachusetts (1992)
13. Minker, J.: On Indefinite Databases and the Closed World Assumption. In: Loveland, D.W. (ed.) 6th Conference on Automated Deduction. LNCS, vol. 138, pp. 292–308. Springer, Heidelberg (1982)
14. Minker, J., Rajasekar, A.: A Fixpoint Semantics for Disjunctive Logic Programs. Journal of Logic Programming 9(1), 45–74 (1990)
15. Przymusinski, T.C.: On the Declarative Semantics of Deductive Databases and Logic Programs. In: Foundations of Deductive Databases and Logic Programming, pp. 193–216. Morgan Kaufmann, San Francisco (1988)
16. Przymusinski, T.C.: Stable Semantics for Disjunctive Programs. New Generation Computing 9(3/4), 401–424 (1991)
17. Robinson, J.: Automated Deduction with Hyper-Resolution. International Journal of Computer Mathematics 1, 227–234 (1965)
18. Sakama, C.: Possible Model Semantics for Disjunctive Databases. In: DOOD 1989, pp. 369–383 (1989)
19. Sarma, A.D., Benjelloun, O., Halevy, A., Widom, J.: Working Models for Uncertain Data. In: ICDE 2006 (2006)
20. Seipel, D., Minker, J., Ruiz, C.: Model Generation and State Generation for Disjunctive Logic Programs. Journal of Logic Programming 32(1), 49–69 (1997)
21. Seipel, D., Thöne, H.: DISLOG - A System for Reasoning in Disjunctive Deductive Databases. In: DAISD 1994, pp. 325–343 (1994)
22. Van Gelder, A., Ross, K.A., Schlipf, J.S.: The Well-Founded Semantics for General Logic Programs. Journal of the ACM 38(3), 620–650 (1991)

Graphical Querying of Multidimensional Databases

Franck Ravat, Olivier Teste, Ronan Tournier, and Gilles Zurfluh

IRIT, Institut de Recherche en Informatique de Toulouse
Université Toulouse 3 (Paul Sabatier), 118 route de Narbonne
F-31062 Toulouse CEDEX9, France
{ravat,teste,tournier,zurfluh}@irit.fr

Abstract. This paper provides an answer-oriented multidimensional analysis environment. The approach is based on a conceptual point of view. We define a conceptual model that represents data through a constellation of facts and dimensions and we present a query algebra handling multidimensional data as well as multidimensional tables. Based on these two propositions, in order to ease the specification of multidimensional analysis queries, we define a formal graphical language implemented within a prototype: GraphicOLAPSQL.

1 Introduction

OLAP (On-Line Analytical Processing) [6] systems allow analysts to improve the decision-making process by consulting and analysing aggregated historical business or scientific data. One of the reasons for the absence of standard in Multidimensional DataBases (MDB) modelling is the lack of conceptual level in commercial tools [20]. Indeed, analysts and decision-makers manipulate logical and physical concepts instead of manipulating conceptual elements.

1.1 Context and Motivations

OLAP or multidimensional modelling [13] represents data as points in a multidimensional space with the use of the cube (or hypercube) metaphor. The following figure presents the data of a commercial activity (*imports amounts*) of a company analysed according to three analysis axes (*suppliers, dates, products*).

MDB are modelled through subjects of analysis, named *facts* and analysis axes, named *dimensions* [13]. These structures are grouped into a *star schema* [13]. Facts are groupings of analysis indicators, named *measures*. Dimensions are composed of hierarchically ordered attributes which model the different detail levels of the axis–data granularity. Notice that, in spite of a decade of research in OLAP systems, concepts and systems exist without uniform theoretical basis [17, 20].

OLAP analysis is performed through interactive exploration and analysis of indicators according to several analysis axes. Analysts need a decisional data representation semantically rich, which clearly distinguishes subjects and axes of analysis as well as structure and content. Moreover, this representation must be independent of any implementation choice. In order to facilitate decision-making, analysts need a multidimensional data visualisation interface adapted to the display of the analysis data. A

Y. Ioannidis, B. Novikov, and B. Rachev (Eds.): ADBIS 2007, LNCS 4690, pp. 298–313, 2007.
© Springer-Verlag Berlin Heidelberg 2007

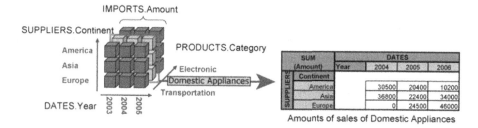

Fig. 1. Cube representation of a MDB, with a slice corresponding to domestic appliances

graphic manipulation language is required, which must be homogeneously positioned between reporting restitution (query) commands and manipulation (analysis) commands.

1.2 Related Works

Without a model based on consensus for multidimensional data, several propositions have been made. Several surveys may be found [24, 1]. According to [24], these models may be classified into two categories. First, works based on the "cube model" [2, 15, 7, 12], have the following issues: 1) weakness in modelling the fact (subject of analysis) or its measures (analysis indicators); 2) little or no conceptual modelling of dimensions (analysis axes) with no explicit capture of their hierarchical structure; and 3) no separation between structure and content. The second category called "multidimensional model" overcomes these drawbacks and is semantically richer. It allows a precise specification of each multidimensional component and notably the different aggregation levels for measures [14, 18, 26, 1]. These models are based on the concepts of fact and dimension possibly with multiple hierarchies. A hierarchy defines a point of view of an analysis axis.

The first works on OLAP manipulation algebras extended relational algebra operators for the cube model [11, 2, 15, 12, 19]. To counter the inadaptability of relational algebra for manipulating multidimensional structures in an OLAP context, numerous works provided operations for specifying and manipulating cubes [4, 5, 18, 1, 8]. These works are not user-oriented for the following reasons: 1) they do not focus on an adapted data structure for displaying decisional data to the user; 2) they are based on partial sets of OLAP operations. Hardly any multidimensional model provides multi-fact and multi-hierarchies as well as a set of associated operations.

Despite of more than a decade of research in the field of multidimensional analysis, very little attention has been drawn on graphical languages. In [5], the authors present a graphical multidimensional manipulation language associated to a conceptual representation of the multidimensional structures. Although the authors define a rather complete manipulation algebra and calculus, the high level graphical language offers very little manipulations in comparison. In [3], the authors offer an intermediate solution, with more manipulations but the system uses complex forms for the multidimensional query specifications. Neither solution provides a restitution interface. [23] and [22] are advanced visualisation tools. The first one offers an impressive pivot table that adapts its display according to the analysis data type, whereas the second offers an arborescent view with multiple scales and very specific manipulations. Here, neither proposition provides a formal description of the manipulation language.

Microsoft Excel Pivot tables, although very expressive restitution interfaces, do not provide many dynamic manipulations. On the other hand, other commercial tools offer extensive manipulations (Business Objects[1], Cognos BI[2], Tableau[3], Targit[4]...). But all these tools display the multidimensional structures of the MDB within an arborescent view, rendering impossible comparative analyses between different subjects sharing analysis axes. Moreover, the representation used mixes completely MDB structures and content. The user completely lacks an adapted conceptual view of the MBD concepts [20].

Nowadays, decision-makers whish to perform their own analyses, but they lack the knowledge to manipulate multidimensional structures with the use of multidimensional query algebras or with adapted procedural query languages. On the other hand, commercial tools provide adapted manipulation languages but lack: 1) rigorous reference to multidimensional operations, 2) a uniform theoretical basis as well as 3) an adapted conceptual view of the multidimensional elements of the underlying MDB. Moreover, these tools sacrifice analysis coherence for analysis flexibility.

1.3 Aims and Contributions

In this context, in order to ensure access to company data, we intend to define a multidimensional OLAP language to ease multidimensional data analysis specification and manipulation. This language must: 1) disregard all logical and implementation constraints; 2) manipulate concepts close to analysts' point of view; 3) be based on a stable theoretical basis to provide consistency and ensure analysis coherence; and 4) provide interaction with the analyst through an incremental and graphical interface.

The paper contributions may be summarized in four points.

1) A conceptual representation of multidimensional data structures. This graphic conceptual view eases users to understand the available analyses supported by the MDB schema. Analysts express queries from the graphic conceptual view, using graphic elements of the MDB schema.

2) A display interface adapted to multidimensional analyses for representing the query results. Contrarily to pivot tables or commercial software output, this interface takes into account the hierarchical nature of multidimensional data in order to ensure analysis consistency. The tabular structure may be directly manipulated by users and analyst may specify complex analyses by incremental definition of complex queries.

3) A user-oriented multidimensional query algebra that uses the model elements as input and the multidimensional table (mTable) as output structure. This algebra is based on a closed minimal core of operators that may be combined together.

4) A graphic language that allows users to express operations using a reduced set of formal primitives. Each primitive handles multidimensional concepts independently of their implantation. This language is complete with regard to the algebraic core and provides incremental manipulations of the mTable, allowing fine analysis tuning.

---

[1] Business Objects XI from http://www.businessobjects.com/
[2] Cognos Business Intelligence 8 from http://www.cognos.com/
[3] Tableau 2 from http://www.tableausoftware.com/
[4] Targit Business Intelligence Suite from http://www.targit.com/

The paper layout is as follows: section 2 defines the concepts and formalisms of the multidimensional model, section 3 presents the algebraic operators and section 4 defines the graphic language.

2   Multidimensional Model: Constellation

In this section we present our multidimensional framework based on a conceptual view displaying MDB structures as a graphical conceptual view as well as a data interface displaying decisional data through a multidimensional table (mTable). Our model allows users to disregard technical and storing constraints, sticks closer to decision-makers' view [10], eases correlation between various subjects of analysis [21, 26] with multiple perspective analysis axes [16]. It also allows a clear distinction structural elements and values and offers a workable visualisation for decision-makers [12]. We invite the reader to consult [25] for a more complete discussion on multidimensional models and definition of the constellation model.

2.1   Concepts

A constellation regroups several subjects of analysis (facts), which are studied according to several analysis axes (dimensions) possibly shared between facts. It extends star schemas [13] commonly used in the multidimensional context.

A *constellation Cs* is composed of a set $F^{Cs}=\{F_1,...,F_m\}$ of facts F_i, a set $D^{Cs}=\{D_1,...,D_n\}$ of dimensions D_i and a function $Star^{Cs}$ linking dimensions and facts together. A fact represents a subject of analysis. It is a conceptual grouping of analysis indicators called measures. It reflects information that has to be analysed according to several dimensions. A *fact F_i* is composed of a set $M^{Fi}=\{M_1,...,M_w\}$ of *measures M_i*. Analysis axes are dimensions seen through a particular perspective, namely a hierarchy. This hierarchy of parameters represents the different graduations of the analysis axis. A *dimension D_i* is composed of a set H^{Di} of hierarchies H_i and a set A^{Di} of attributes (namely parameters and weak attributes). A dimension D_i linked to the fact F_i is noted $D_i \in Star^{Cs}(F_i)$. A *hierarchy H_i* is an ordered list $Param^{Hi}=<p_1,...,p_{np}>$ of *parameters p_i* and $\forall p_i \in Param^{Hi}$, $\forall H_i \in H^{Di} => p_i \in A^{Di}$. Note that p_1 is named *root parameter* while p_{np} is the *extremity parameter*. *Weak attributes* may be associated to a parameter in order to complete its semantic. Note that, within the rest of the paper no distinction will be made between parameters and weak attributes.

Notations. An aggregated measure M_i is noted $f_{AGGi}(M_i)$. $level^H(p_i)$ is the position of p_i in $Param^H$. The notation $p_i \in H$ is a simplified notation for $p_i \in Param^H$. The set of values of a parameter p_i is $dom(p_i)=<v_{min},...,v_{max}>$.

Graphic notations, based on [9], offer a clear global conceptual view. These notations highlight subjects of analysis (facts) and their associated axes of analyses (dimensions and hierarchies). See Figure 2 for an example.

2.2   Multidimensional Table (MTable)

OLAP analysis consists in analysing key performance indicators (measures) according to several analysis axes. As in [12], we offer a tabular visualisation called multidimensional table (mTable), displaying a fact and detailed information of two dimensions.

A *multidimensional table* is defined by $T=(S, C, L, R)$, where $S=(F^S, M^S)$ represents the analysed *subject* through a fact F^S and a set of aggregated measures $M^S=\{f_{AGG1}(M_1),\ldots, f_{AGGv}(M_v)\}$; $C=(DC, HC, PC)$ represents the column analysis axis where $PC=<p^C_{max},\ldots p^C_{min}>$, HC is the *current hierarchy* of column dimension DC; $L=(DL, HL, PL)$ represents the line analysis axis where $PL=<p^L_{max},\ldots p^L_{min}>$, HL is the *current hierarchy* of the line dimension DL; and $R=pred_1 \wedge \ldots \wedge pred_t$ is a normalised conjunction of predicates (restrictions of dimension and fact data). An mTable has the following constraints: $DC\in Star^{Cs}(F^S)$, $DL\in Star^{Cs}(F^S)$, $HC\in H^{DC}$, $HL\in H^{DL}$, $\exists\, p_i\in PC \mid p_i\in HC, \exists\, p_i\in PL \mid p_i\in HL, \exists\, M_i\in M^S \mid M_i\in M^{FS}, F^S\in F^{Cs}$.

The $DISPLAY^{Cs}$ operator constructs an mTable (T_{RES}) from a constellation Cs. The expression is $DISPLAY^{Cs}(F^S, M^S, DL, HL, DC, HC) = T_{RES}$ where: $M^S=\{f_1(M_1),\ldots,f_x(M_x)\}$, $\forall i\in[1..x]$, $M_i\in M^F$ and $T_{RES}=(S_{RES}, L_{RES}, C_{RES}, R_{RES})$ is the output mTable, where: $S_{RES} = (F^S, M^S)$; $L_{RES}=(DL, HL, <p^L_{max}, p^L_{min}>)$, where $p^L_{max}=All$ and $p^L_{min}=p^{DL}_{np}$; $C_{RES}=(DC, HC, <p^C_{max}, p^C_{min}>)$, where $p^C_{max}=All$ and $p^C_{min}=p^{DC}_{np}$; and

$$R_{RES} = \bigwedge_{\forall i, D_i\in Star^C(F^S)} D_i.ALL = 'all'\,.$$

In the following figure is displayed a constellation allowing the analysis of import-ing companies activity as well as their manpower. An mTable shows an analysis ex-ample: total import amounts by years and by continent of origin of the suppliers.

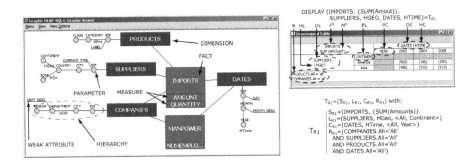

Fig. 2. Example of a constellation and an associated analysis (mTable T_{R1})

3 OLAP Algebra

Multidimensional OLAP analyses consist in exploring interactively constellation data. The following algebra allows manipulation and retrieval of data from a constellation through nested expressions of algebraic operators. The OLAP algebra provides a minimal core of operators, all operators may not be expressed by any combination of other core operators; this ensures the *minimality* of the core. Although there is no consensus on a set of operations for a multidimensional algebra, most papers offer a partial support of seven operation categories see [25] for more details. Based on these categories, our algebraic language allows the presentation of analysis data in an mTable.

- **Drilling:** these operations allow navigating through the hierarchical structure along analysis axes, in order to analyse measures with more or less precision. Drilling upwards (operator: ROLLUP) consists in displaying the data with a coarser level of detail; *e.g.* import amounts analysed by months then analysed by year. The opposite, drilling downwards (DRILLDOWN) consists in displaying the data with a finer level of detail.

- **Selections:** these operations (SELECT) allow the specification of restriction predicates on fact or dimension data. This operation is also known as "slice/dice" [2].

- **Rotations:** these operations allow: changing an analysis axis by a new dimension (ROTATE); changing the subject of analysis by a new fact (FROTATE); and changing an analysis perspective by a new hierarchy (HROTATE).

Some authors have also presented complementary operations:

- **Fact modification:** These operations allow the modification of the set of selected measures. They allow adding (ADDM) or removing (DELM) a measure to the current analysis.

- **Dimension modification:** These operations allow more flexibility in analysis. They allow converting a dimensional element into a subject (PUSH), thus "pushing" a parameter into the subject; or converting a subject into a dimensional element (PULL), thus "pulling" a measure out of the subject. They also allow nesting (NEST). It is a structural reordering operation. It allows changing the order of parameters in a hierarchy but it also allows adding in a hierarchy a parameter from another dimension. The consequence is to be able to display in the bi-dimensional mTable more than two dimensions. This compensates for the 2D limitation of the bi-dimensional table.

- **Ordering:** these operations allow changing the order of the values of dimension parameters or inserting a parameter in another place in a hierarchy. Switching (SWITCH) is an operation that eases analysis allowing regrouping columns or lines together independently of the order of the parameter values in lines or columns. It is a value reordering operation. Notice that switching as well as nesting may "break" hierarchies' visual representation in the mTable. Contrarily to commercial software which allows this operation without warning the user, the use of these specific operators allows a query system to warn the user analysis incoherence risks.

- **Aggregation:** This operation allows the process of totals and subtotals of the displayed data (AGGREGATE). If applied to all displayed parameters, it is equivalent to the Cube operator [11]. The reverse operation is UNAGGREGATE.

The following table describes the symbolic representation (algebraic expression) and the necessary conditions of input and output semantics for each operator. We invite the reader to consult [25] for more detailed specifications.

3.1 Minimal Closed Algebraic Core of OLAP Operators

To ensure *closure* of the core, each operator takes as input a source mTable (T_{SRC}) and produces as output a result mTable (T_{RES}). $T_{SRC}=(S_{SRC}, L_{SRC}, C_{SRC}, R_{SRC})$ where:

$S_{SRC} = (F^S, M^S)$ with $M^S = \{f_1(M_1),...,f_x(M_x)\}$; $L_{SRC} = (DL, HL, PL)$ with
$PL = <All, p_{Lmax},..., p_{Lmin}>$; $C_{SRC} = (DC, HC, PC)$ with $PC = <All, p_{Cmax},..., p_{Cmin}>$; and
$R_{SRC} = (pred_1 \wedge ... \wedge pred_t)$. $T_{RES} = (S_{RES}, L_{RES}, C_{RES}, R_{RES})$.

We have also defined advanced operators as well as set operators but due to lack of space, we invite the reader to consult [25] for complete specification.

Table 1. OLAP Algebraic operators

Operation	Semantics
DRILLDOWN(T_{SRC}, D, p_i) = T_{RES}	
Conditions	$D \in \{DC, DL\}$; (D=DC, $\exists p_i \in HC \mid level(pi) < level(p_{Cmin})$) $\vee$ (D=DL, $\exists p_i \in HL \mid$ level(p_i)<level(p_{Lmin}))
Output	$T_{RES} = (S_{SRC}, L_{RES}, C_{RES}, R_{SRC})$ where: if D=DL then $L_{RES} = (DL, HL, <All, p^L_{lmax},..., p^L_{lmin}, p_i>)$ and $C_{RES} = C_{SRC}$; if D=DC then $L_{RES} = L_{SRC}$ and $C_{RES} = (DC, HC, <All, p^C_{cmax},... p^C_{cmin}, p_i>)$.
ROLLUP(T_{SRC}, D, p_i) = T_{RES}	
Conditions	$D \in \{DC, DL\}$; (D=DC, $\exists p_i \in HC \mid level(p_i) > level(p_{Cmin})$) $\vee$ (D=DL, $\exists p_i \in HL \mid$ level(p_i)>level(p_{Lmin})).
Output	$T_{RES} = (S_{SRC}, L_{RES}, C_{RES}, R_{SRC})$ where: if D=DL then $L_{RES} = (DL, HL, <All, p^L_{lmax},..., p_i>)$ and $C_{RES} = C_{SRC}$; if D=DC then $L_{RES} = L_{SRC}$ and $C_{RES} = (DC, HC, <All, p^C_{cmax},..., p_i>)$.
SELECT(T_{SRC}, pred) = T_{RES} / UNSELECT(T_{SRC})=T_{RES}	
Conditions	pred= $pred_1 \wedge ... \wedge pred_t$, $pred_i$ is a predicate on dom(M_i) or dom(p_i), $M_i \in M^{FS} \wedge p_i \in A^D \wedge$ $D \in Star^{Cs}(F^S)$
Output	$T_{RES} = (S_{SRC}, L_{SRC}, C_{SRC}, R_{RES})$ where R_{RES} = pred. For UNSELECT, $R_{RES} = \varnothing$
ROTATE(T_{SRC}, D_{old}, D_{new}, H_{new}) = T_{RES}	
Conditions	$D_{old} \in \{DC, DL\}$; $D_{new} \in Star^{Cs}(F^s) \wedge H_{new} \in H^{Dnew}$
Output	$T_{RES} = (S_{SRC}, L_{RES}, C_{SRC}, R_{SRC})$ where: if D_{old}=DL then $L_{RES} = (D_{new}, H_{new}, <All, p^{Dnew}_{np}>)$ and $C_{RES} = C_{SRC}$; if D_{old}=DC then $L_{RES} = L_{SRC}$ and $C_{RES} = (D_{new}, H_{new}, <All, p^{Dnew}_{np}>)$. Where p^{Dnew}_{np} is the coarser-granularity parameter
ADDM(T_{SRC}, $f_i(M_i)$) = T_{RES}	
Conditions	$M_i \in M^{Fs} \wedge f_i(M_i) \notin M^S$
Output	$T_{RES} = (S_{RES}, L_{SRC}, C_{SRC}, R_{SRC})$ where: $S_{RES} = (F^S, \{f_1(M_1),..., f_x(M_x), f_i(M_i)\})$.
DELM(T_{SRC}, $f_i(M_i)$) = T_{RES}	
Conditions	$f_i(M_i) \in M^S \wedge \|M^S\| > 1$.
Output	$T_{RES} = (S_{RES}, L_{SRC}, C_{SRC}, R_{SRC})$ where: $S_{RES} = (F^S, \{f_1(M_1),... f_{i-1}(M_{i-1}), f_{i+1}(M_{i+1}),... f_x(M_x)\})$. This operator may not remove the last displayed measure of the subject
PUSH(T_{SRC}, D, p_i) = T_{RES}	
Conditions	$p_i \in H \wedge H \in H^D \wedge D \in Star^{Cs}(F^s)$.
Output	$T_{RES} = (S_{RES}, L_{SRC}, C_{SRC}, R_{SRC})$ where: $S_{RES} = (F^S, \{f_1(M_1),..., f_x(M_x), p_i\})$ and $p_i \notin H$.
PULL(T_{SRC}, $f_i(M_i)$, D) = T_{RES}	
Conditions	$M_i \in M^{Fs} \wedge f_i(M_i) \notin M^S \wedge D \in \{DC, DL\}$.
Output	$T_{RES} = (S_{RES}, L_{RES}, C_{RES}, R_{SRC})$ where: $S_{RES} = (F^S, \{f_1(M_1),..., f_{i-1}(M_{i-1},..., f_x(M_x), p_i\})$ and if D=DL then $L_{RES} = (DL, HL, <All, p_{Lmax},..., p_{Lmin}, f_i(M_i)>)$ and $C_{RES} = C_{SRC}$, if D=DC then $L_{RES} = L_{SRC}$ and $C_{RES} = (DC, HC, <All, p_{Cmax},..., p_{Cmin}, f_i(M_i)>)$
NEST(T_{SRC}, D, p_i, D_{nested}, p_{nested}) = T_{RES}	
Conditions	$D \in \{DC, DL\} \wedge$ (D=DC, $p_i \in PC \vee$ D=DL, $p_i \in PL$) $\wedge D_{nested} \in Star^{Cs}(F^s) \wedge p_{nested} \in A^{Dnested}$
Output	$T_{RES} = (S_{SRC}, L_{RES}, C_{RES}, R_{SRC})$ where: if D=DL then $L_{RES} = (DL, HL, <All, p_{Lmax},..., p_i, p_{nested},..., p_{Lmin}>)$ and $C_{RES} = C_{SRC}$; if D=DC then $L_{RES} = L_{SRC}$ and $C_{RES} = (DC, HC, <All, p_{Cmax},..., p_i, p_{nested},..., p_{Cmin}>)$
SWITCH(T_{SRC}, D, p_i, v_x, v_y) = T_{RES}	
Conditions	$D \in \{DC, DL\} \wedge p_i \in PC \wedge v_x \in dom(p_i) \wedge v_y \in dom(p_i) \wedge dom(p_i) = <v_{min},... v_x,... v_y,..., v_{max}>$.
Output	$T_{RES} = (S_{SRC}, L_{SRC}, C_{SRC}, R_{SRC})$ where: $dom(p_i) = <v_{mi\ n}... v_y... v_x..., v_{max}>$

Table 1. (*continued*)

AGGREGATE(T_{SRC}, D, $f_i(p_i)$) = T_{RES}	
Conditions	D∈{DC, DL}∧ (D=DC, p_i∈PC ∨ D=DL, p_i∈PL) ∧ dom(p_i)=<v_{min},…,v_{max}>, f_i∈{SUM, AVG, MIN, MAX, COUNT...}.
Output	T_{RES}=(S_{SRC}, L_{RES}, C_{RES}, R_{SRC}) where: If D=DL then in L_{RES}, dom(p_i) changes in PL: dom(p_i)=<v_{min},$f_i(v_{min})$,…,v_{max},$f_i(v_{max})$>. Each initial value v_j of p_i is completed by the aggregation value $f_i(v_j)$ and C_{RES}=C_{SRC}; If D=DC then L_{RES}=L_{SRC} and in C_{RES}, dom(p_i) changes in PC: dom(p_i)=<v_{min},$f_i(v_{min})$,…,v_{max},$f_i(v_{max})$>. Each initial value v_j of p_i is completed by the aggregation value $f_i(v_j)$
UNAGGREGATE(T_{SRC}) = T_{RES}	
Conditions	$\exists p_i$∈PL ∨ $\exists p_i$∈PC | dom(p_i)=<v_{min},$f_i(v_{min})$,…,v_{max},$f_i(v_{max})$>.
Output	T_{RES}=(S_{SRC}, L_{RES}, C_{RES}, R_{SRC}) where: $\forall p_i$∈PL and $\forall p_i$∈PC, dom(p_i)=<v_{min},…, v_{max}>, i.e. the previously aggregated values are removed from dom(p_i)

3.2 Example

The decision-maker proceeds with the previous analysis (see T_{R1} in figure 2) by focussing his observations on the average *import amounts* in *2005* for *electronic products*. He also wishes to refine the analysis by visualizing amounts more precisely by *country* of origin of the *suppliers* while modifying the column axis to observe the measures by importing *company*. In order to do so, this complex query is specified by a combination of several basic operators from the algebra core: 1) DRILLDOWN on the SUPPLIER axis; 2) SELECT of 'Electronic' PRODUCTS during year 2005; 3) ADDM for the indicator: AVG(Amount); and 4) ROTATE dimensions: DATES and COMPANY. The complete algebraic expression (1) and the corresponding mTable (T_{R2}) are displayed in the following figure. In the same way, the reverse operation combination is presented (2).

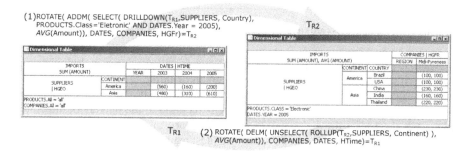

Fig. 3. Example of algebraic manipulations: how to obtain T_{R2} from T_{R1} (1) and vice-versa (2)

4   OLAP Graphic Language

This section presents an incremental graphic language, which is complete with regard to the algebra core and operating directly on the conceptual elements of the constellation model to stay closer to decision-makers view. As core operators may be combined together, algebraic queries may end up being very complex. The graphic language eases such expressions by providing some simple manipulations hiding this complexity to the user.

4.1 Principles

We have developed a tool composed of two interfaces. The first allows the display of a conceptual graph representing the constellation. The second is the visual representation of the mTable. Both are displayed in figure 1. The user specifies the operations using drag and drop actions or contextual menus. Moreover the mTable supports incremental On-Line Analytical Processing; e.g., the mTable components may be removed, replaced and new components may be added. The mTable display adapts itself after each manipulation.

Using drag and drop, the user may select a graphic icon representing a multidimensional element displayed in the graph, drag it onto the mTable and then drop it in one of the mTable zones (see Figure 4-left), thus directly specifying the resulting mTable. The mTable builds itself incrementally as the elements are dropped into position. In order to ensure consistency during multidimensional OLAP analyses, the user is guided throughout the process: incompatible operations with the ongoing analysis are deactivated, i.e. operations that do not meet conditions. This forbids the user to create erroneous analyses. In the same way, the user may also "move" elements by dragging them from the mTable and dropping them into another zone of the same mTable.

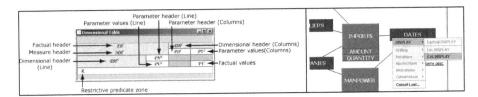

Fig. 4. mTable drop zones (left) and a contextual menu called on the dimension *DATES* (right)

Alternatively, the user may use contextual menus called upon a multidimensional element either displayed in the constellation graph or in the mTable. For example, in Figure 4-right the user designates the DATES dimension to be displayed in columns.

In some cases, in order to ensure the proper execution of the operation, the system might ask the user with dialog boxes complementary information that may not be provided by the graphic operation's context (designated by '?' in formal specifications). Formally, each operation takes as input a component of the multidimensional graph: a fact F, a measure M_i, a dimension D, a hierarchy H or a parameter p_i.

4.2 Graphic Definition of a mTable

An mTable is defined once the three major elements have been specified, i.e. in T_{RES} ($S_{RES} \neq \varnothing \land L_{RES} \neq \varnothing \land C_{RES} \neq \varnothing$). Three of the display-oriented operations allow the specification of the elements of the DISPLAY instruction: DIS_SUBJ to select the subject: F^S and M^S and DIS_COL (respectively DIS_LN) for the definition of the

Table 2. Graphic specification of the DISPLAY operator

Source element of the action (and conditions)	Algebraic equivalent
Subject specification : DIS_SUBJ(E)	
$E=F \mid F \in F^{Cs}$	DISPLAY(N^{Cs}, F^S, M^S, DL, HL, DC, HC) with F^S=F and $M^S=\{f_1?(M_1),\dots f_w?(M_w)\}$)
$E=M_i \mid M_i \in M^F \wedge F \in F^{Cs}$	DISPLAY(N^{Cs}, F^S, M^S, DL, HL, DC, HC) with F^S=F, $M^S=\{f_i?(M_i)\}$
Column specification: DIS_COL(E)	
$E=D \mid \exists\ H \in H^D \wedge D \in D^{Cs}$	DISPLAY(N^{Cs}, F^S, M^S, DL, HL, DC, HC) with DC=D, HC=H?
$E=H \mid H \in H^D \wedge D \in D^{Cs}$	DISPLAY(N^{Cs}, F^S, M^S, DL, HL, DC, HC) with DC=D, HC=H
$E=p_i \mid p_i \in H \wedge H \in H^D \wedge$ $D \in D^{Cs}$	DRILLDOWN(ROLLUP(DISPLAY(N^{Cs}, F^S, M^S, DL, HL, DC, HC), DC, All), DC, p_i) with DC=D, HC=H?
Line specification: DIS_LN(E), same as lines but specifies DL and HL	

The following constraints must be verified: DC$\in$Star^{Cs}(F^S) $\wedge$ DL$\in$Star^{Cs}(F^S).
? = complementary information that might be necessary.

column axis: DC and HC (resp. the line axis: DL and HL). The mTable is built incrementally as the user specifies the different elements. Formal specifications are presented in the following table.

4.3 Graphic Manipulation of a mTable

Once the mTable is defined it switches to alteration mode and all other operations are available. A total of twelve graphic operations split into three operation categories are available: display-oriented operations, mTable modification oprations and drilling operations.

The three previous operations (DIS_SUBJ, DIS_COL and DIS_LN) no longer specify a DISPLAY instruction. Instead they now allow the designation of elements that are to replace previously displayed ones in the mTable. DIS_SUBJ, now allows to replace the previous fact; to add measures (ADDM) and to display a parameter as a measure (PUSH). DIS_COL (respectively DIS_LN) replaces the current elements displayed in columns (resp. lines) with the newly specified one. With these two instructions, the user may also ask to display a measure as a parameter in columns or in lines (PULL). In addition to these instructions, the system also provides three others (INS_LN, INS_COL and DEL). INS_COL (resp. INS_LN) allow to insert an element in columns (resp. in lines) enabling nesting (NEST) and to drilling downwards (DRILLDOWN). Finally the deletion operation (DEL) is a very handy operation allowing to remove a displayed component from the mTable. This operation replaces complex algebraic expressions. These six operations are the contextual menu equivalent of most of the drag and drop actions.

The operations that modify the mTable display structure consist in four operations, identical to their algebraic equivalent (SELECT, SWITCH, AGGREGATE and UNAGGREGATE). Apart from AGGREGATE, the other operations may also be expressed using drag and drop. Table 3 presents the equivalence between graphical operation names and drag and drop actions.

Table 3. Correspondance between graphic operations and drag and drop actions

Graphic operation	Drag and drop action
DIS_SUBJ	drop an element into factual header
DIS_COL	drop an element into column dimensional header
DIS_LN	drop an element into line dimensional header
INS_COL	drop an element into column parameter header[1]
INS_LN	drop an element into line parameter header[1]
DEL	drag an element outside the mTable
SELECT	drag an element into the restriction zone
SWITCH	select a parameter value and drop it on another value of the same parameter
UNAGGREGATE	drag an aggregated parameter value outside the mTable
[1]= if the instruction requires it, the position where the element is dropped within the header is used as complementary information.	

Finally as drilling operations are the heart of multidimensional OLAP analyses, the graphic language has two specific operations (ROLLUP and DRILLDOWN), identical to their algebraic equivalent although part of these operations may be done by the use of the display-oriented operations.

Formal specifications are available in appendix. Unfortunately, due to lack of space, we invite the reader to consult [25] for the complete specification of the graphic language.

4.4 Example

In this section we show the same example as in section 3.2. The analyst wishes to create the mTable T_{R1} then manipulate it in order to obtain T_{R2} (see figure 2). First the analyst executes three operations to create T_{R1}: 1) drag the icon representing the measure *Amount* and dropping it in the factual header and then selecting *SUM* as desired aggregation function; 2) drag the dimension *Dates* and dropping it in the column dimensional header; and 3) drag the dimension *Suppliers* and dropping it in the line dimensional header and select *HGEO* as the hierarchy to be used (see figure 3a). Notice that instead of this last action, the user may drag the parameter *Continent*, thus avoiding specifying the hierarchy. In the figure, one may find the formal specifications and alternative actions. These actions may be expressed through contextual menus or drag and drop actions.

The analyst then modifies T_{R1} with five operations to obtain T_{R2}: 4) drag the parameter *Country* into the right part of the line parameter header (dragging it in the center would replace the previously displayed parameter); 5) (resp. 6) drag the parameter Class (resp. Year) onto the restriction zone and specify ='*Electronic*' (resp. =*2005*); 7) drag the measure *Amount* into the factual header and select the aggregation function *AVG* (Average); and 8) drag the parameter *Region* into the column dimensional header (see figure 3b). After each graphic operation, the mTable updates itself, thus the decision-maker may stop or decide to change his analysis after each new operation.

In the following figure blue zones (light grey) underneath each arrow are displayed by the mTable according to the context of the action. This indicates to the user where

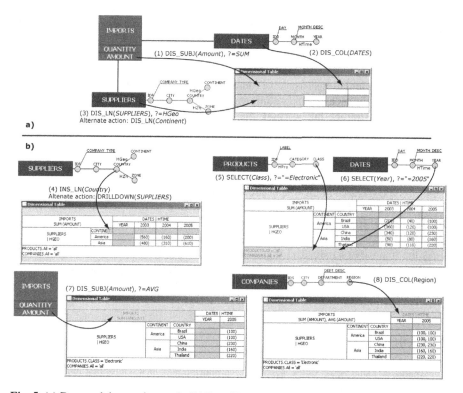

Fig. 5. (a) Drag and drop actions to build T_{R1}; (b) successive actions to generate T_{R2} from T_{R1}

the dropped element will go, thus, for example, if it will be inserted within already displayed elements (action 4 in figure 3b) or if it will replace one or all a set of displayed elements (action 8 in figure 3b).

Manipulations with drag and drop actions and contextual menus are complete with regard to the algebraic core. Due to lack of space, we invite the reader to consult [25] for more details.

5 Conclusion

The aim of this paper is to define a user-oriented multidimensional OLAP analysis environment. In order to do so we have defined a conceptual model modelling multidimensional elements with the concepts of fact, dimension and hierarchy and representing them through a constellation. Based on this model, we defined an OLAP query algebra, formally specifying operations executed by decision-makers. The algebra is composed of a core of operators that may be combined to express complex analysis queries. As such a language is inappropriate for users, we defined a graphical query language based on this algebra. This language allows decision makers to express multidimensional OLAP analysis queries and it is complete with regard to the algebraic core.

Compared to actual solutions, our proposition has the advantages of providing: 1) a global view of the analysis data (with the constellation graph); 2) manipulation of high-level concepts (direct manipulation of the conceptual elements); 3) analysis correlations (analysing different facts with common dimensions); 4) analysis coherence during manipulations (usage of hierarchies within the mTable); and if necessary 5) as much flexibility as actual solutions, i.e. disregarding analysis coherence (usage of the NEST operator), but warning the decision-maker of incoherence risks. As in [20], we argue that the use of conceptual representations eases query specifications for decision-makers. Moreover, the graphical language must be as expressive and as robust as complex multidimensional algebraic query languages.

The algebraic and graphic languages have been validated by their implementation in a prototype (figures in this paper are screen captures). The prototype is based on an implementation of the multidimensional concepts in a ROLAP environment with the DBMS Oracle *10g*. The client is a java application composed of over a hundred classes and other components such as JGraph and Javacc.

As the majority of industrial implementations are relational, for a future work, we intend to complete this proposition by the specification of a transformation process, mapping multidimensional operations into an optimal combination of relational operators.

References

1. Abelló, A., Samos, J., Saltor, F.: YAM2: a multidimensional conceptual model extending UML. Information Systems (IS) 31(6), 541–567 (2006)
2. Agrawal, R., Gupta, A., Sarawagi, S.: Modeling Multidimensional Databases. In: 13th Int. Conf. Data Engineering (ICDE), pp. 232–243. IEEE Computer Society, Los Alamitos (1997)
3. Böhnlein, M., Plaha, M., Ulbrich-vom Ende, A.: Visual Specification of Multidimensional Queries based on a Semantic Data Model. In: Vom Data Warehouse zum Corporate Knowledge Center (DW), pp. 379–397. Physica-Verlag, Heidelberg (2002)
4. Cabibbo, L., Torlone, R.: Querying Multidimensional Databases. In: Cluet, S., Hull, R. (eds.) Database Programming Languages. LNCS, vol. 1369, pp. 319–335. Springer, Heidelberg (1998)
5. Cabibbo, L., Torlone, R.: From a Procedural to a Visual Query Language for OLAP. In: 10th Int. Conf. on Scientific and Statistical Database Management (SSDBM), pp. 74–83. IEEE Computer Society, Los Alamitos (1998)
6. Codd, E.F.: Providing OLAP (On Line Analytical Processing) to user analysts: an IT mandate, Technical report, E.F. Codd and Associates (1993)
7. Datta, A., Thomas, H.: The cube data model: a conceptual model and algebra for on-line analytical processing in data warehouses. Decision Support Systems (DSS) 27(3), 289–301 (1999)
8. Franconni, E., Kamble, A.: The GMD Data Model and Algebra for Multidimensional Information. In: Persson, A., Stirna, J. (eds.) CAiSE 2004. LNCS, vol. 3084, pp. 446–462. Springer, Heidelberg (2004)
9. Golfarelli, M., Maio, D., Rizzi, S.: The Dimensional Fact Model: A Conceptual Model for Data Warehouses. International Journal of Cooperative Information Systems (IJCIS) 7(2-3), 215–247 (1998)

10. Golfarelli, M., Rizzi, S., Saltarelli, E.: WAND: A Case Tool for Workload-Based Design of a Data Mart. In: 10th National Convention on Systems Evolution for Data Bases, pp. 422–426 (2002)
11. Gray, J., Bosworth, A., Layman, A., Pirahesh, H.: Data Cube: A Relational Aggregation Operator Generalizing Group-By, Cross-Tab, and Sub-Total. In: 12th Int. Conf. on Data Engineering (ICDE), pp. 152–159. IEEE Computer Society, Los Alamitos (1996)
12. Gyssen, M., Lakshmanan, L.V.S.: A Foundation for Multi-Dimensional Databases. In: 23rd Int. Conf. on Very Large Data Bases (VLDB), pp. 106–115. Morgan Kaufmann, San Francisco (1997)
13. Kimball, R.: The Data Warehouse Toolkit: Practical Techniques for Building Dimensional Data Warehouses, 2nd edn. John Wiley & Sons, Inc., Chichester (2003)
14. Lehner, W.: Modeling Large Scale OLAP Scenarios. In: Schek, H.-J., Saltor, F., Ramos, I., Alonso, G. (eds.) EDBT 1998. LNCS, vol. 1377, pp. 153–167. Springer, Heidelberg (1998)
15. Li, C., Wang, X.S.: A Data Model for Supporting On-Line Analytical Processing. In: 5th Int. Conf. on Information and Knowledge Management (CIKM), pp. 81–88. ACM, New York (1996)
16. Malinowski, E., Zimányi, E.: Hierarchies in a multidimensional model: From conceptual modeling to logical representation. J. of Data & Knowledge Engineering (DataK) 59(2), 348–377 (2006)
17. Niemi, T., Hirvonen, L., Jarvelin, K.: Multidimensional Data Model and Query Language for Informetrics. Wiley Periodicals 54(10), 939–951 (2003)
18. Pedersen, T., Jensen, C., Dyreson, C.: A foundation for capturing and querying complex multidimensional data. Information Systems (IS) 26(5), 383–423 (2001)
19. Rafanelli, M.: Operators for Multidimensional Aggregate Data. In: Multidimensional Databases: Problems and Solutions. ch. 5, pp. 116–165. Idea Group Inc. (2003)
20. Rizzi, S., Abelló, A., Lechtenbörger, J., Trujillo, J.: Research in data warehouse modeling and design: dead or alive? In: 9th Int. Workshop on Data Warehousing and OLAP (DOLAP), pp. 3–10. ACM, New York (2006)
21. Sapia, C., Blaschka, M., Höfling, G.: Dinter: Extending the E/R Model for the Multidimensional Paradigm. In: Kambayashi, Y., Lee, D.-L., Lim, E.-p., Mohania, M.K., Masunaga, Y. (eds.) Advances in Database Technologies. LNCS, vol. 1552, pp. 105–116. Springer, Heidelberg (1999)
22. Sifer, M.: A Visual Interface Technique for Exploring OLAP Data with Coordinated Dimension Hierarchies. In: Int. Conf. on Information and Knowledge Management (CIKM), pp. 532–535. ACM, New York (2003)
23. Stolte, C., Tang, D., Hanrahan, P.: Polaris: A System for Query, Analysis, and Visualization of Multidimensional Relational Databases. IEEE Trans. Vis. Comput. Graphics (TVCG) 8(1), 52–65 (2002)
24. Torlone, R.: Conceptual Multidimensional Models. In: Rafanelli, M. (ed.) Multidimensional Databases: Problems and Solutions. ch. 3, pp. 69–90. Idea Group Inc. (2003)
25. Tournier, R.: OLAP model, algebra and graphic language for multidimensional databases. Scientific Report n° IRIT/RR—2007-6–FR, IRIT, Université Paul Sabatier (Toulouse 3), France
26. Trujillo, J.C., Luján-Mora, S., Song, I.: Applying UML for designing multidimensional databases and OLAP applications. In: Siau, K. (ed.) Advanced Topics in Database Research, vol. 2, pp. 13–36. Idea Group Publishing (2003)

A Appendix

This section provides tables with the formal specifications of the graphic operations. Each operation takes as input a component of the multidimensional graph: E (F, M_i, D, H, p_i) or a specific element of the mTable: X. They may also require complementary information notified by '?'. Only column operation specifications are specified (line specification are identical.

Table 4. Formal specification of the display-oriented graphic operations

Source element of the action (and conditions)	Algebraic equivalent
Operation for displaying as subject: DIS_SUBJ(E)	
$E=F^S$	History$^{(1)}$ (T_{SRC}, DL, History(T_{SRC}, DC, DISPLAY(N^{CS}, F^S, $\{f_1(M_1),\dots f_w(M_w)\}$, DL, HL, DC, HC))), $\forall i \in [1..w]$, $M_i \in M^{FS}$
$E=F_{new} \| F_{new} \neq F^S \wedge DC \in Star^{Cs}(F^{new})$ $\wedge DL \in Star^{Cs}(F^{new})$	History$^{(1)}$(T_{SRC}, DL, History(T_{SRC}, DC, DISPLAY(N^{CS}, F_{new}, $\{f_1(M_1),\dots f_w(M_w)\}$, DL, HL, DC, HC))), $\forall i \in [1..w]$, $M_i \in M^{Fnew}$
$E=M_i \| M_i \in M^S$	ADDM(T_{SRC}, f_i?(M_i)) with f_i?(M_i)$\notin M^S$
$E=M_i \| M_i \in M^S \wedge M_i \in M^{FS}$	ADDM(T_{SRC}, f_i?(M_i))
$E=M_i \| M_i \in M^{FS} \wedge M_i \in M^{Fnew} \wedge$ $DC \in Star^{Cs}(F^{new}) \wedge DL \in Star^{Cs}(F^{new})$	History$^{(1)}$(T_{SRC}, DL, History(T_{SRC}, DC, DISPLAY(N^{CS}, F_{new}, $\{f_i$?(M_i)$\}$, DL, HL, DC, HC)))
$E=p_i \| \forall p_i \in A^{Dnew} \wedge D_{new} \in Star^{Cs}(F^S)$	PUSH(T_{SRC}, D_{new}, p_i)
Operation for displaying in columns: DIS_COL(E) (DIS_LN for lines)	
$E=M_i \| M_i \in M^{FS}$	PULL(T_{SRC}, f_i?(M_i), C)
$E=DC$	ROLLUP(T_{SRC}, DC, p^{DC}_{np}), the display of HC is reset on the most general parameter: $p_{Cmin}= p_{Cmax}=p^{DC}_{np}$
$E=D_{new} \| D_{new} \neq DC \wedge D_{new} \in Star^{Cs}(F^S)$	ROTATE(T_{SRC}, DC, D_{new}, H?), with $H \in H^{Dnew}$
$E=HC$	ROLLUP(T_{SRC}, DC, p^{DC}_{np}), the display of HC is reset on the most general parameter: $p_{Cmin}= p_{Cmax}=p^{DC}_{np}$
$E=H_{new} \| H_{new} \neq HC \wedge H_{new} \in H^{DC}$	ROTATE(T_{SRC}, DC, DC, H_{new}) $\|H^{DC}\|>1$, i.e. $\exists H_{new} \in H^{DC} \| H_{new} \neq HC$
$E=H_{new} \| H_{new} \notin H^{DC} \wedge H_{new} \in H^{Dnew} \wedge$ $D_{new} \in Star^{Cs}(F^S)$	ROTATE(T_{SRC}, DC, D_{new}, H_{new})
$E=p_i \| p_i \in PC \vee (p_i \notin PC \wedge p_i \in HC)$	DRILLDOWN(ROLLUP(T_{SRC},DC,All),DC, p_i)
$E=p_i \| p_i \notin HC \wedge p_i \in H_{new} \wedge H \in H^{DC}$	DRILLDOWN(ROLLUP(ROTATE(T_{SRC}, DC, DC, H_{new}), DC,All),DC, p_i) $\|H^{DC}\|>1$, i.e. $\exists H_{new} \in H^{DC} \| H_{new} \neq HC$
$E=p_i \| p_i \in A^{DC} \wedge p_i \in A^{Dnew} \wedge$ $D_{new} \in Star^{Cs}(F^S)$	DRILLDOWN(ROLLUP(ROTATE(T_{SRC}, DC, D_{new} ,H_{new}), D_{new},All), D_{new}, p_i), with $H_{new} \in H^{Dnew}$
Operation for inserting in columns: INS_COL(E) (INS_LN for lines)	
$E=M_i \| M_i \in M^{FS}$	PULL(T_{SRC}, f_i?(M_i), DC)
$E=DC$	DRILLDOWN(…(DRILLDOWN(ROLLUP(T_{SRC}, DC, p^{DC}_{np}), DC, p^{DC}_{np-1})…), DC, p^{DC}_1) all parameters of HC are displayed: PC = Param^{HC}
$E=D_{new} \| D_{new} \neq DC \wedge D_{new} \in Star^{Cs}(F^S)$	ROTATE(T_{SRC}, DC, D_{new}, H_{new}?), with $H_{new} \in H^{Dnew}$
$E=HC$	DRILLDOWN(…(DRILLDOWN(ROLLUP(T_{SRC}, DC, p^{DC}_{np}), DC, p^{DC}_{np-1})…), DC, p^{DC}_1) all parameters of HC are displayed: PC = Param^{HC}
$E=p_i \| p_i \in PC \wedge PC \neq \{All, p_i\}$	NEST(DRILLDOWN(…(DRILLDOWN(ROLLUP(T_{SRC}, DC, p_{i+1}), DC, p_{i-1})…), DC, p_{Cmin}), DC, p_j?, DC, p_i), with $\{p_{i+1}, p_i, p_{i-1},\dots,p_{Cmin}\} \in PC$

Table 4. (*continued*)

$E=p_i \mid p_i \notin PC \wedge p_i \in HC$	DRILLDOWN(T_{SRC}, DC, p_i), if $\text{level}^{HC}(p_i)<\text{level}^{HC}(p_{Cmin})$ or DRILLDOWN(...(DRILLDOWN(ROLLUP(T_{SRC}, DC, p_i), DC, p_{i+1})...), DC, p_{Cmin}), if $\text{level}^{HC}(p_i)>\text{level}^{HC}(p_{Cmin})$ [2]
$E=p_i \mid p_i \notin HC \wedge p_i \in H \wedge H \in H^{DC}$	NEST(T_{SRC}, DC, p_j?, DC, p_i), with $p_j \in PC$
$E=p_i \mid p_i \notin A^{DC} \wedge p_i \in A^{Dnew} \wedge D_{new} \in Star^{Cs}(F^S)$	NEST(T_{SRC}, DC, p_j?, D_{new}, p_i), with $p_j \in PC$

Delete operation: DEL(E) ou DEL(X) [3]	
$E=F^S$	History[1](T_{SRC}, DL, History(T_{SRC}, DC, DISPLAY(N^{CS}, F_{new}?, $\{f_1(M_1),...,f_w(M_w)\}$, DL, HL, DC, HC))), $\forall i \in [1..w]$ $M_i \in M^{Fnew}$ and $DC \in Star^{Cs}(F_{new})$ and $DL \in Star^{Cs}(F_{new})$
$E=f_i(M_i) \mid f_i(M_i) \in M^S$	DELM(T_{SRC}, $f_i(M_i)$)
$E=DC$	ROTATE(T_{SRC}, DC, D_{new}?, H?), $D_{new} \in Star^{Cs}(F^S)$, $H \in H^{Dnew}$
$E=HC$	ditto above: DEL(DC)
$E=p_i \mid p_i \in PC \wedge PC \neq <All, p_i>$	DRILLDOWN(...(DRILLDOWN(ROLLUP(T_{SRC}, DC, p_{i+1}), DC, p_{i-1})...), DC, p_{Cmin}), $\{p_{i+1}, p_i, p_{i-1},...,p_{Cmin}\} \in PC$
$X=R$ (zone de restriction) [3]	SELECT(T_{SRC}, F^S.All='all' $\wedge$ ($\bigwedge_{D_i \in Star^{Cs}(F^S)} D_i.All = 'all'$))

[1] = History(T_{old}, obj, T_{new})=T_R is producing T_R by applying on T_{new} the history of the operations that were applied in T_{old} on obj (dimension or fact).

[2] = p_i not displayed ($p_i \notin PC \wedge p_i \in HC$), p_{Ci+1} (respectively p_{Ci-1}) is the attribute immediately inferior (resp. superior) to p_i in PC: $\text{level}^{HC}(p_{Ci-1})=\text{level}^{HC}(p_i)-1$ (resp. $\text{level}^{HC}(p_{Ci+1})=\text{level}^{HC}(p_i)+1$);

[3] = This operation is done on the restriction zone (R) of the mTable;

Table 5. Formal specification of the algebra-oriented graphic operations

Source element of the action (and conditions)	Algebraic equivalent
Data restriction: SELECT(E)	
$E=M \mid M \in M^{FS}$	SELECT(pred?), pred is a predicate on dom(M_i)
$E=p_i \mid p_i \in A^D$, $D \in Star^{Cs}(F^S)$	SELECT(pred?), pred is a predicate on dom(p_i)
Line/Column value inversion: SWITCH(X) [1]	
$X=val_x \mid val_x \in dom(p_i)$, $p_i \in PC$	SWITCH(T_{SRC}, DC, p_i, val_x, val_y?), with $val_y \in dom(p_i)$
Data aggregation: AGGREGATE(E) and UNAGGREGATE(E)	
$E=p_i \mid p_i \in PC$, $dom(p_i)=<v_1,...,v_x>$	AGGREGATE(T_{SRC}, DC, f_i?(p_i))
$E=p_i \mid p_i \in PC$, $dom(p_i)=<v_1,f_i(v_1),...,v_x,f_i(v_x)>$	UNAGGREGATE(T_{SRC})

Table 6. Formal specification of the graphic drilling operations

Source element of the action (and conditions)	Algebraic equivalent
Drilling downwards: DRILLDOWN(E)	
$E=DC \mid \exists p_i \in HC \mid \text{level}^{HC}(p_i)=\text{level}^{HC}(p_{Cmin})-1$	DRILLDOWN(T_{SRC}, DC, p_{Cmin-1})[1]
$E=HC \mid \exists p_i \in HC \mid \text{level}^{HC}(p_i)=\text{level}^{HC}(p_{Cmin})-1$	DRILLDOWN(T_{SRC}, DC, p_{Cmin-1})[1]
$E=p_i \mid \text{level}^{HC}(p_i)>\text{level}^{HC}(p_{Cmin})$	DRILLDOWN(T_{SRC}, DC, p_i)
Drilling upwards: ROLLUP(E)	
$E=DC \mid \exists p_i \in HC \mid \text{level}^{HC}(p_i)=\text{level}^{HC}(p_{Cmin})+1$	ROLLUP(T_{SRC}, DC, p_{Cmin+1})[1]
$E=HC \mid \exists p_i \in HC \mid \text{level}^{HC}(p_i)=\text{level}^{HC}(p_{Cmin})+1$	ROLLUP(T_{SRC}, DC, p_{Cmin+1})[1]
$E=p_i \mid \text{level}^{HC}(p_i)<\text{level}^{HC}(p_{Cmin})$	ROLLUP(T_{SRC}, DC, p_i)

[1] p_{Cmin-1} (resp. p_{Cmin+1}) is the parameter immediately inferior (resp. superior) to p_{Cmin} in the list Param^{HC} (i.e. $\text{level}^{HC}(p_{Cmin-1})+1=\text{level}^{HC}(p_{Cmin})=\text{level}^{HC}(p_{Cmin+1})-1$).

Incremental Validation of String-Based XML Data in Databases, File Systems, and Streams

Beda Christoph Hammerschmidt[1], Christian Werner[3], Ylva Brandt[3],
Volker Linnemann[2], Sven Groppe[2], and Stefan Fischer[3]

[1] Oracle Corporation, 400 Oracle Parkway, 4OP408, Redwood Shores, CA 94065, USA
beda.hammerschmidt@oracle.co
[2] Institute of Information Systems, University of Luebeck, Germany
{groppe,linnemann}@ifis.uni-luebeck.de
[3] Institute of Telematics, University of Luebeck, Germany
{werner,brandt,fischer}@itm.uni-luebeck.de

Abstract. Although the native (tree-like) storage of XML data becomes more and more important there will be an enduring demand to manage XML data in its textual representation, for instance in relational structures or file systems. XML data has to be wellformed by definition and additionally, in many cases, it has to be valid according to a given XML schema. Because the XML column types are often derived from text types (e.g. CLOBs) guaranteeing well-formedness as well as validity is not trivial. And even worse, for frequently modified data it is usually too expensive to re-validate the whole XML data after each update – but waiving re-validation may lead to inconsistencies and malfunctions of applications. In this paper we present a schema-aware pushdown automaton (i.e. a stack machine) that validates an XML string/stream. Using an element/state-index, the pushdown automaton is able to re-validate local modifications of the data while guaranteeing overall validity. Update operations (e.g. SQLXML, XQuery updates) are validated before executing them.

1 Introduction and Motivation

The storage of XML documents and data in database management systems is a daily task, although no single storage technology has prevailed. Depending on the application constraints one may choose a native storage system (e.g. [10,27]) or (object-) relational systems that were enabled for XML. A common approach is to shred the XML data and map it to corresponding tables. A not less popular approach is to store the XML data in a textual or binary representation in a special column type or in the file system.

Here, the original structure of the XML data is preserved. Therefore, XML query languages like XPath and XQuery and XML processors like XSLT engines can be applied in a more natural way compared to the shredding approach that needs to reconstruct the XML data or to rewrite the instructions to SQL. In addition, this representation as a sequence of tags is the natural choice when transmitting XML over networks. Commercial database management systems like the Oracle Database 10g, the Microsoft SQL Server, and IBM DB2 offer these two storing approaches although native storage of XML data in relational databases has been propagated recently in [4] and in [22]. The differences

Y. Ioannidis, B. Novikov, and B. Rachev (Eds.): ADBIS 2007, LNCS 4690, pp. 314–329, 2007.
© Springer-Verlag Berlin Heidelberg 2007

and advantages of different storage techniques have been discussed extensively. They are not in the focus of this paper. We concentrate on the problem of efficient and partial validation of XML strings stored in a specialized string column type, in the file systems, or as a stream in network communication.

In many applications the XML data refers to an XML Schema [34] that defines the structure and the used data types. Whenever a new row with XML content is added into a table it is validated by a parser that processes the whole XML data. Modifications of the data may violate the XML Schema; this has to be avoided by re-validating the data. In spite of the fact that the data is most often modified locally (e.g. adding an author child to a book element) the validating parser processes the whole data. This raises high computation costs, especially when storing large pieces of XML data. In consequence, the validation is not performed after each modification leading to possibly inconsistent data that cannot be corrected.

In this paper we present a novel approach for validating XML data in its textual representation. Our approach

- is time efficient, since the validating automaton is schema-aware and created only once. The validation time is linear in the size of the XML data. Schema violations are detected immediately because no transition is found in our parser automaton for the next (invalid) tag.
- is memory efficient: unlike DOM-based approaches we do not load the full XML data into main memory. The only memory is a stack that may grow linearly in the depth of the XML data.
- supports recursive schemes (see Section 2).
- allows the *local* and *incremental* validation of the XML string. Using a combined element/state-index we enable direct access to the states in a Push-Down Automaton (PDA) and to elements in the XML data. This is particularly advantageous for XML data that is frequently updated, since only this part has to be revalidated while guaranteeing global validity. Update operations are validated before executing them.
- can be used for various purposes where XML data is represented as a string or a stream and needs to be wellformed and valid. Examples include text-based database columns, file systems, and network communications (e.g. SOAP messages).

The remainder of this paper is organized as follows: we start with an overview on automata used for validation of XML data in Section 2. Section 3 introduces and formalizes the validation problem for streaming XML data. Our pushdown automaton is extended with an element/state-index in Section 4. We present experimental results in Section 5 and finish this paper with a conclusion in Section 6.

2 Related Work

Related work in the context of XML validation and automata theory is manifold. [28] deals with different classes of complexity for recursive and non-recursive DTDs. As real world's XML Schemes may be recursive (e.g. to model dynamic hierarchies), we ignore non-recursive (less expressive) approaches in this paper. The work [29] introduces the difference of validation and *strong* validation for streaming XML data. Strong validation includes the test of well-formedness whereas non-strong validation assumes that the data is well-formed and fails when this is not the case. Because XML updates are usually given in a textual representation (e.g. SQLXML, XQuery updates) checking well-formedness is a must and implies that the automaton uses a memory (e.g. a stack) to match opening tags and closing tags. Therefore, automata without any memory cannot check well-formedness of XML data represented as a string. Additionally, even if the XML data is well-formed a stack may be required to check the validity of some schemes (see Example 1 in Section 3).

The idea of a pushdown automaton for strong validation is presented in [29], but in contrast to our approach, the whole XML data has to be validated leading to significant performance degradation. Additionally, [29] does not present an implementation or algorithm.

[8] presents a one-counter automaton for strong validation of XML documents. As a single counter variable is used instead of a stack, only a subset of DTDs can be validated, namely the so called *(restricted) one-counter languages*. An example for a language (schema) that cannot be validated is $a^n b^m \overline{b}^m \overline{a}^n$ where a, b are opening tags and $\overline{a}, \overline{b}$ are the corresponding closing tags. Using a stack it is possible to count and match the two different tags, with a single variable it is impossible. Additionally, [8] supports only global validation.

A multitude of approaches for partial validation of XML data after updates have been presented (e.g. [2,3,5,6,20,21,25]). All these approaches assume that the XML data accords to the DOM model, i.e. it is represented as a tree of nodes. This representation is inherently well-formed and enables the direct and efficient navigation within the nodes. For instance, it is possible to access all children of a given node. The DOM model works well for native XML database management systems where the tree-like representation is preserved. In contrast, we focus on the string representation of XML data as it is used in XML column types, message systems (e.g. SOAP) or SQLXML update commands. In those systems the XML data is represented as a sequence of tags and values. The sequence can be seen as the result of a preorder traversal of the corresponding XML tree.

[12] presents a schema validation technique without any automata. In this approach the XML data is represented as a set of rows in one dedicated XPath-aware table. Therefore, it is closer to the XML shredding approach and the results are hardly transferable to the XML column type.

Our previous works [13,14,15,16,17] deal with (autonomous) indexing and update issues in *native* XML storage systems representing XML data using the DOM-model.

The pushdown automaton presented in this paper is also used to convert an XML stream into a binary representation in order to compress it. This is achieved by adding binary symbols to all transitions in the automaton. Details on how we compress XML data, in particular SOAP messages, by using a pushdown automaton can be found in [32].

3   The XML Validation Problem

We assume that the reader is familiar with basic terms of formal languages and automata theory like regular grammars and finite state machines as well as context-free grammars

```
<xsd:schema xmlns:xsd="http://www.w3.org/2001/XMLSchema">
 <xsd:element name="a" type="A"/>
 <xsd:complexType name="A">
  <xsd:choice>
   <xsd:element name="b" minOccurs="0">
    <xsd:complexType>
     <xsd:choice>
      <xsd:element ref="a" minOccurs="2" maxOccurs="2"/>
      <xsd:element name="c" type="xsd:int"/>
     </xsd:choice>
    </xsd:complexType>
   </xsd:element>
  </xsd:choice>
 </xsd:complexType>
</xsd:schema>
```

Fig. 1. Recursive XML Schema description

and pushdown automata (finite state machines with a stack). We refer the reader to [18] for fundamentals in automata theory and formal languages. We motivate the complexity of the XML Schema validation problem by a simple informal example:

The XML Schema document presented in Figure 1 consists of three different elements a, b and c. As a maps to b and b maps to a, this is a recursive schema allowing an arbitrary depth of the XML data. We present a more compact representation of this grammar in Figure 2. As one can see the a element under b appears always twice ($minoccurs = maxoccurs = 2$). Therefore, we have a *left* a-element and a *right* a-element that belong together. When reading the XML string an automaton has to remember whether an a-element is the left one (right a is mandatory) or the right one (no more a-elements allowed). As the XML data has an arbitrary depth, the automaton has to remember arbitrary many of these pairs — this is only possible if we use an additional memory like a stack. Even if we assume well-formedness of the XML data to be validated, we cannot support this XML Schema (or the corresponding DTD) without a stack. The approaches for XML stream/string validation that rely only on a finite automaton without further memory call this kind of schema a *not recognizable* schema [8,29]. DOM-tree based validators relying on tree-automata do not need a stack because they can check directly whether a b-element has two a children or not.

3.1   Formalization

We represent an XML data instance (also known as XML document) as a sequence of opening and closing tags with their content. We formalize XML and XML Schema as follows:

Definition 1 (XML data instance). *An XML data instance $x \in \Sigma^*$ is a sequence of arbitrary many symbols $\sigma \in \Sigma$ with $\Sigma = \Sigma_t \cup \Sigma_c$. Σ_t is the tag-alphabet whereas $\Sigma_c = \{a..z, A..Z, 1..9\}$ represents the alphabet for the textual content* [1]. *Σ_c^* denotes an arbitrary sequence of symbols of Σ_c including the empty sequence. We identify closing tags with a line: $\overline{a}$ is the closing tag of a. Opening tags are summarized in Σ_t^+ whereas*

---

[1] One could use the Unicode symbols for Σ_c, as well.

closing tags are in Σ_t^-. A tag corresponds to an element name; given a tag t one may get its name with the function $name(t)$.

In our example it holds that $\Sigma_t^+ = \{a, b, c\}$, $\Sigma_t^- = \{\bar{a}, \bar{b}, \bar{c}\}$, and $\Sigma_t = \Sigma_t^+ \cup \Sigma_t^-$.

Please be aware that this generic definition allows the formulation of non-well-formed and invalid XML data.

We formalize XML Schema as a Regular Tree Grammar (RTG) G. [24] shows that and how an XML Schema can be mapped to a RTG. A regular tree grammar $G = (N, T, P, S)$ is a 4-tuple consisting of a set of non-terminal symbols N, a set of terminal symbols T, a set of production rules P and a set of start symbols $S \subseteq N$. The language defined by G is denoted with $\mathcal{L}(G)$.

All XML data types result in non-terminal symbols (written in capital letters or prefixed with xsd:). All possible element names result in terminal symbols (small letters). The set of start symbols $S \subseteq N$ contains all non-terminals belonging to elements that are declared on the top-level of the XML Schema document.

The production rules P reflect the structure of valid documents. The lefthand side of a rule $p \in P$ is a non-terminal symbol defining an element type. It can be determined by the function $type(p) : P \to N$.

The righthand side consists of an element name (a terminal symbol $\in T$) that corresponds to the type. It can be accessed using the function $name(p) : P \to T$

The element name is followed by a regular expression r which describes the content model of this element, i.e. types of child elements. Each regular expression element maps to types in N, simple XML Schema types (e.g. xsd:int), or to nothing (ϵ).

For our example, we depict the conversion result in Figure 2. See [24] for details about the conversion algorithm.

$$G = (\{A, B, C, \text{xsd:int}\}, \{a, b, c\}, P, \{A\}) \text{ with}$$
$$P = \{A \to a\,(B|\varepsilon),$$
$$B \to b\,(AA|C),$$
$$C \to c\,(\text{xsd:int}) \}$$

Fig. 2. Regular tree grammar of the schema in Figure 1

The rules in our example have the following meaning: The A element type is represented by the element name a and contains one B-type or ($|$) or nothing (ϵ), whereas B contains either two A types or one C type containing an integer value. Hence, G is equivalent to the Schema in Figure 1.

The content models of G can be transferred to automata theory:

Lemma 1. *For each regular expression in a production rule $p \in P$ there exists a finite state machine accepting the content model of $type(p)$. The set of all finite state machines is denoted by $\mathcal{FSM}$. Given a $p \in P$ one gets its corresponding FSM with the function $fsm(p) : P \to \mathcal{FSM}$.*

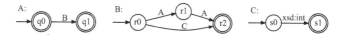

Fig. 3. Set of finite state machines generated from the regular tree grammar of Figure 2

We give no proof for this lemma as it is standard automata theory. Please see [18] describing how to transform regular expressions into FSMs. The corresponding state machines for our example are presented in Figure 3.

3.2 Construction of the XML Schema-Aware Pushdown Automaton

XML is a derivative of the Dyck-languages (parenthesis languages) that can be described by context-free grammars [7]. Therefore, the pushdown automaton is the appropriate model for string processing of XML data. We present the construction of the pushdown automaton (PDA) in the Appendix (Section A). The pushdown automaton as presented here is a slight modification of the automaton as presented in [32] in the context of compression techniques for SOAP messages.

Our PDA starts with the special symbol Z as the only stack content. The PDA accepts a document if the remaining input sequence and the stack are empty.

The main idea of the PDA is to check the content models of the element types of the regular tree grammar by simulating the nested execution of the FSMs. For this reason, each opening and closing tag is represented by one dedicated state.

The stack's content is used to memorize the path to the current tag: States representing the current position in the content model are pushed on the stack while reading the XML data. The transitions capture valid tag sequences, i.e. the restrictions of the content models. The PDA constructed for our sample grammar is presented in Figure 4.

The first symbol of a transition's label represents a tag $t \in \Sigma_t$ to be read from the XML input x, the second symbol represents the state of the content model to be popped from the stack (please see Section A for details), and the last symbol(s) are the states to be pushed on the stack when using this transition. For instance, if we read an opening tag a the symbol q_o is pushed on the stack. Because a may have an empty content model we are able to read the closing tag $\bar{a}$ afterwards. The transition checks if the previous tag was a by popping q_0 from the stack.

Please note that no special treatment for terminating the execution of the automaton is needed because the automaton accepts a document if both the remaining input sequence and the stack content are empty. For a - documents this is the case when the corresponding $CloseA$ state is reached.

4   Efficiently Validating Updates

When processing an XML input string the computational complexity is linear ($O(n)$) with respect to the string's length n. Usually $O(n)$ complexity is not appropriate if the size of the XML data exceeds a certain degree. In this section we describe how we adapt our approach to incrementally validate XML strings *before* performing an update

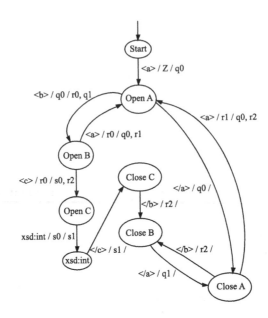

Fig. 4. PDA constructed from the RTG of Figure 2

operation. Incremental validation means that only the modified part is validated in its local context. In this way the complexity becomes linear in the size of the modification.

We assume that an update operation is either an insertion of a new element (with its children) or the deletion of an element (including all its children). Formally, an update operation o is represented as a triple $o = (path, type, content)$, where *path* is an XPath expression selecting one to many context elements in the XML data, *type* indicates what kind of update it is, and *content* is the textual representation of a new element to be inserted. An example for such an update operation might be the following insert statement – formulated in XQuery Update [33,35]:

```
do insert
<category>Harddisks</category>
into /site/categories
```

Here, it holds that *path=/site/categories, type=append* and
content=`<category>Harddisks</category>`.

In order to incrementally validate the XML data we do not read it from the beginning but need direct access to an offset position. For this reason, we use an index whose entries reference open-tags in the XML string. This is technically done by the use of character offset values. The keys of the index's entries are simple path expressions. Simple path expressions contain the child-axis and element names. The Strong DataGuide [11] is an example for such an index structure with the characteristic that *all* possible simple path expressions are covered. In our work we relax this characteristic and assume that at least *some* path expressions are indexed without regarding how to select them.

(We have discussed the problem of finding suitable XML indexes for a given application in [14,15]).

The main idea of our approach is to extend the index entries in such a way that they reference not only into the XML string but also to the corresponding states in the PDA while providing the top element of the stack at the same time. According to the construction of the PDA, the top element denotes the initial state of the corresponding finite state machine. The idea is to keep 'snapshots' of the PDA execution in the index. For this reason we define the index as follows:

Definition 2 (Element/State-Index). *The element/state-index consists of a set of entries* $\mathcal{E}$. *Each entry* $e \in \mathcal{E}$ *is a 2-tuple* $e = (key, ref)$ *where key is a simple path expression and ref* = $(offsetlist, state, stackelement)$ *a reference: offsetlist is a list of character distances each of which is a number indicating the character distance from the root; state* $\in Q$ *is the corresponding state in the PDA, and stackelement* $\in \Gamma$ *is the top element of the corresponding stack content. Q denotes the set of all states and* Γ *is the set of all stack symbols. We describe details in the Appendix.*

We need only the top stack element denoting the initial state of the corresponding finite state machine because regular tree grammars are context free, i.e. we can analyze the content of an element independently of the position of the element within the document. For example, the behaviour of the PDA basically is the same if we analyze the first a-element or the second a-element within a b-element. The stack content differs only in parts that are irrelevant for analyzing the a-element.

An overview of an element/state-index is given in Figure 5. For example, if a b-element with the path /a/b is updated, only the new b-element has to be checked as follows: The PDA skips the root b-element and starts in state $OpenB$ with $r0$ as the only stack element. After successfully checking the new b-element, the PDA stops in state $CloseB$ with an empty stack.

The index is populated when inserting data into the XML column: The string representation of the data is processed once by the PDA in order to check validity. While processing the tag sequence we check if the current tag belongs to a simple path expression to be indexed or not. If this is the case we insert the current offset, state and topmost stack symbol as a new entry into the element/state-index. An efficient retrieval of entries is enabled when using a search structure like a B-tree or hashtable.

As the XML data is a hierarchy of elements, we illustrate it as a tree (triangle). The trees inside the XML data represent subtrees (elements in elements). Please note that the XML data is stored as a sequence of tags in the column and not as a tree of element nodes like in native XML DBMS.

When performing an update operation $o = (path, type, content)$ it is likely that its path expression *path* is not covered by an entry in the index. However, if all elements selected by *path* are descendants of an index element we can still use that index to accelerate the validation process. Whether an indexed path expression can be used or not is determined using the XPath containment algorithm of Miklau and Suciu [23].

It detects if elements selected by a path expression p are a subset of those selected by p'. In our case $p = path$ and p' denotes a path expression of the index entries

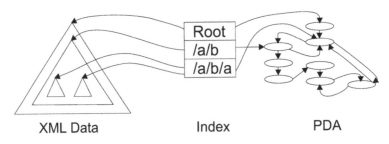

The table shows the index for the automaton given in Figure 4 and for the document `<a> <b> <a> </a> <a> </a> </b> </a>`:

key	ref		
	offsetlist	*state*	*stackelement*
Root	1	Start	Z
/a/b	4	Open B	r0
/a/b/a	7, 14	Open A	q0

Fig. 5. The element/state-index referencing XML data and PDA states

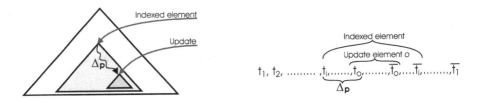

Fig. 6. Finding the update position in the XML data using the index

(the keys). If $p \subseteq p'$ we may use that entry. Of course, in case of $p \subset p'$, i.e. $p \neq p'$ we have to check for each element of the offset list whether the corresponding data up to this position corresponds to p. If there is no index entry with $p \subseteq p'$, we must process the XML data from the root resulting in a conventional validation.

In the following we assume that a suitable index entry is found. Figure 6 illustrates the situation: The index refers to an open-tag element t_i in the XML data x. The corresponding closing tag element $\overline{t_i}$ can be found easily because x is wellformed.

$\triangle_p$ is the relative simple path expression starting from the indexed element t_i and leads to the element t_o that is selected by the update operation. $\triangle_p$ can be calculated by applying XPath difference functions (for details see [13]) and consists of a sequence of navigational steps to child elements.

The idea of our approach is to use the PDA for validation and to pretend that the modifying operation o has already been executed. The validation takes place from t_i to $\overline{t_i}$. The offset in x to t_i is fetched from the index; the PDA is initialized with the settings belonging to that entry. When processing x using the PDA we follow the reference and initialize the PDA and its stack. We now start the processing of x until we reach t_o. We have reached the element selected by the update operation o if the next tag in x corresponds to the last location step in $\triangle_p$. Before reading that tag t_o we memorize the current state of the PDA with its stack content. The next steps of our algorithm depend on the type of the modification operation o:

- Delete: If o is a delete operation all tags in the sequence $t_o, ...\overline{t_o}$ are to be removed. Again, the closing tag $\overline{t_o}$ can be identified because x is well-formed. We skip all tags in that sequence and continue to read x using the PDA until we reach $\overline{t_i}$. If the PDA could process this shortened input, then x would be still valid after performing the deletion. The offsets in the index' entries have to be updated because the update may have changed the position of some tags. This is done by subtracting the length of the deleted sequence from all offsets that are greater than the position of t_o.
- Append: The PDA reads the full sequence $t_o, ...\overline{t_o}$ and stops. The next tags are the content of the update operation o. After reading them the PDA resumes to read from x until reaching $\overline{t_i}$. If the PDA accepts that extended input, then x is still valid after performing the append operation. Again, we have to update the offsets by adding the length of the content to those offsets that are larger than the position of $\overline{t_o}$.
- Insert: This case is very similar to the previous one with the difference that the new element is not a sibling but a child. Therefore, the insertion takes place after reading t_o, if the new element is the first child, or just before reading $\overline{t_o}$, if it is the last child. Analogously, the index offsets have to be adapted.

As this algorithm only reads the tags from t_i to $\overline{t_i}$, we avoid a complexity of O(length of the whole document). Instead, we achieve a complexity of O(length of the modified part of the document). Our solution is also able to handle the <xsd:all> construct efficiently: The content model A & B & C is split up into all possible sequences: ABC | ACB | BAC | BCA | CAB | CBA. On the one hand this solution obviously leads to larger automata, on the other hand it always preserves a runtime complexity of O(length of the changed part of the document). See [32] for the details.

Querying XML Data Using the Index. The element/state-index can also be used to accelerate the path expressions when executing querying database operations. Although it is not in the focus of this paper, we present the basic idea that is similar to the incremental validation: the best index entry is determined referencing to positions in the XML data; Δ_p is calculated and executed on these positions leading to the tags that are selected by the query. We refer to [14,13] for the details.

5   Experiments

In order to demonstrate the benefits of our schema-aware pushdown automaton for validation we compare it with other approaches. We use the XMark [26] benchmark tool to generate XML data of different sizes. The XMark data models an auction scenario. Its schema consists of roughly 80 element types and allows the recursive nesting of some elements without restricting the documents depth. Because the XMark schema is provided by a DTD, we convert it to an equivalent XML Schema using Sun's Trang [30].

The experiments were performed on an Intel P4, 2.67 GHz with 1 GB RAM. The construction of the PDA took 270 ms. It consists of 237 states and 627 transitions.

Table 1. Time to validate the XMark sample data

	PDA generation	100 KB	1 MB	10 MB	100 MB
XML Spy	n/a	1000 ms	4200 ms	28000 ms	n/a
JDOM Parser	n/a	1010 ms	1322 ms	3890 ms	24690 ms
Sax Parser	n/a	622 ms	1035 ms	2122 ms	14210 ms
Xenia global	270 ms	87 ms	370 ms	1061 ms	12648 ms
Xenia local	270 ms	22 ms	25 ms	25 ms	26 ms

We have created XML documents of different sizes and have compared the times for validation with the two well-known parsers JDOM [19] and SAX [9], the XML Editor XMLSpy [1] and our approach called Xenia. We summarize the results of our experiments in Table 1.

The measured values are averaged over ten runs. The 100 MB document could not be processed by the XMLSpy due to memory constraints. As one can see in the columns for 100 KB and 10 MB, Xenia is significantly faster because the validating automaton is already built whereas the other parsers have to read and process the schema information. For larger XML data the measured times are comparable to the SAX parser. Both Xenia and SAX do not need to keep the whole document in the memory because they operate stream-oriented. The JDOM parser and XML Spy keep the whole XML data in memory.

The experiments for incremental validations after local updates are presented in the last row of the table. The updated element has a size of 20 kB. As one can see, the time to validate the update is almost constant and not dependent on the total size of the XML data. This is because only the modified part (which is the same in all cases) has to be checked. It does not make any difference whether we deal with an insert or a delete operation.

The time for constructing the PDA is given separately in the first column of Table 1 because in our approach the PDA has to be generated only once for each schema. The same PDA is used for all updates of documents having the same schema.

6   Conclusion

We have seen that conventional parsers consume linear time when validating an XML string/stream. Especially for larger data sets that are periodically updated the costs for re-validation become prohibitively high. Waiving the validation may be inevitable but is unwanted in most cases. In this paper, we introduce a schema-aware pushdown automaton used for validation. Using an index structure captures snapshots of this automaton while providing direct access into the XML data. Using this index, the costs for re-validation are in the size of the update operation because only a small part of the XML data is processed by the automaton. Additionally, the validation is performed before updating the data, so that invalid data cannot be created. Experimental results show the efficiency of our approach.

To the best of our knowledge our approach is the first one for partially validating XML strings after local updates. This is a highly relevant topic for implementing efficient non-native database management systems which use special XML column types or for storing XML data in the file system.

References

1. Altova. XMLSpy. URL: http://www.altova.com
2. Balmin, A., Papakonstantinou, Y., Vianu, V.: Incremental validation of XML documents. ACM Trans. Database Syst. 29(4), 710–751 (2004)
3. Barbosa, D., Mendelzon, A.O., Libkin, L., Mignet, L., Arenas, M.: Efficient Incremental Validation of XML Documents. In: ICDE '04: Proceedings of the 20th International Conference on Data Engineering, Washington, DC, USA, pp. 671–682. IEEE Computer Society Press, Los Alamitos (2004)
4. Beyer, K., Cochrane, R., Josifovski, V., Kleewein, J., Lapis, G., Lohman, G., Lyle, B., Özcan, F., Pirahesh, H., Seemann, N., Truong, T.: System RX: One Part Relational, One Part XML. In: Proceedings of the 2005 ACM SIGMOD International Conference on Management of Data, Baltimore, Maryland, USA, June 14-16 2005, pp. 347–358. ACM Press, New York (2005)
5. Bouchou, B., Alves, M.H.F.: Updates and Incremental Validation of XML Documents. In: DBPL, pp. 216–232 (2003)
6. Bouchou, B., Alves, M.H.F., Laurent, D., Duarte, D.: Extending Tree Automata to Model XML Validation Under Element and Attribute Constraints. In: ICEIS (1), pp. 184–190 (2003)
7. Brüggemann-Klein, A., Wood, D.: Balanced context-free grammars, hedge grammars and pushdown caterpillar automata. In: Extreme Markup Languages (2004)
8. Chitic, C., Rosu, D.: On validation of XML streams using finite state machines. In: WebDB '04: Proceedings of the 7th International Workshop on the Web and Databases, pp. 85–90. ACM Press, New York, NY, USA (2004)
9. Megginson, D.: Simple API for XML. URL: http://www.saxproject.org/
10. Fiebig, T., Helmer, S., Kanne, C.-C., Moerkotte, G., Neumann, J., Schiele, R., Westmann, T.: Anatomy of a native XML base management system. VLDB Journal 11(4), 292–314 (2002)
11. Goldman, R., Widom, J.: DataGuides: Enabling Query Formulation and Optimization in Semistructured Databases. In: VLDB'97, Proceedings of 23rd International Conference on Very Large Data Bases, pp. 436–445 (1997)
12. Grust, T., Klinger, S.: Schema validation and type annotation for encoded trees. In: Proceedings of the First International Workshop on XQuery Implementation (XIME-P), Paris, France, June 2004, pp. 55–60 (2004)
13. Hammerschmidt, B.C.: KeyX: Selective Key-Oriented Indexing in Native XML-Databases. Dissertation zum Dr.-Ing., Institut für Informationssysteme, Technisch-Naturwissenschaftliche Fakultät, Universität zu Lübeck, October, DISDBIS 93, Akademische Verlagsgesellschaft Aka GmbH, Berlin 2006, ISBN 3-89838-493-4 (2005)
14. Hammerschmidt, B.C., Kempa, M., Linnemann, V.: A selective key-oriented XML Index for the Index Selection Problem in XDBMS. In: Galindo, F., Takizawa, M., Traunmüller, R. (eds.) DEXA 2004. LNCS, vol. 3180, Springer, Heidelberg (2004)
15. Hammerschmidt, B.C., Kempa, M., Linnemann, V.: Autonomous Index Optimization in XML Databases. In: Proceedings of the International Workshop on Self-Managing Database Systems (SMDB 2005), Tokyo, Japan, April 8-9 2005, pp. 56–65 (2005)
16. Hammerschmidt, B.C., Kempa, M., Linnemann, V.: On the Intersection of XPath Expressions. In: Proceedings of the 9th International Database Engineering & Application Symposium (IDEAS 2005), Montreal, Canada, July 25-27, 2005 (2005)
17. Hammerschmidt, B.C., Linnemann, V.: The Index Update Problem for XML Data in XDBMS. In: Proceedings of the 7th International Conference on Enterprise Information Systems (ICEIS 2005), Miami, USA, pp. 27–34 (2005)
18. Hopcroft, J.E., Motwani, R., Ullman, J.D.: Introduction to Automata Theory, Languages, and Computation. Addison Wesley Publishing Company, Reading (2001)

19. Hunter, J., McLaughlin.: JDOM 1.0. URL: http://www.jdom.org/
20. Sang-Kyun, K., Myungcheol, L., Kyu-Chul, L.: Immediate and Partial Validation Mechanism for the Conflict Resolution of Update Operations in XML Databases. In: Meng, X., Su, J., Wang, Y. (eds.) WAIM 2002. LNCS, vol. 2419, pp. 387–396. Springer, Heidelberg (2002)
21. Sang-Kyun, K., Myungcheol, L., Kyu-Chul, L.: Validation of XML Document Updates Based on XML Schema in XML Databases. In: Mařík, V., Štěpánková, O., Retschitzegger, W. (eds.) DEXA 2003. LNCS, vol. 2736, pp. 98–108. Springer, Heidelberg (2003)
22. Liu, Z.H., Krishnaprasad, M., Arora, V.: Native Xquery processing in Oracle XMLDB. In: Proceedings of the 2005 ACM SIGMOD International Conference on Management of Data, Baltimore, Maryland, USA, June 14-16 2005, pp. 828–833. ACM Press, New York (2005)
23. Miklau, G., Suciu, D.: Containment and equivalence for a fragment of XPath. Journal of the ACM 51(1), 2–45 (2004)
24. Murata, M., Lee, D., Mani, M., Kawaguchi, K.: Taxonomy of XML schema languages using formal language theory. ACM Trans. Inter. Tech. 5(4) (2005)
25. Papakonstantinou, Y., Vianu, V.: Incremental Validation of XML Documents. In: Calvanese, D., Lenzerini, M., Motwani, R. (eds.) ICDT 2003. LNCS, vol. 2572, pp. 47–63. Springer, Heidelberg (2002)
26. Schmidt, A., Waas, F., Kersten, M.L., Carey, M.J., Manolescu, I., Busse, R.: XMark: A Benchmark for XML Data Management. In: Proceedings of the International Conference on Very Large Data Bases (VLDB), Hong Kong, China, pp. 974–985 (2002)
27. Schöning, H.: Tamino - A DBMS designed for XML. In: Proceedings of the 17th International Conference on Data Engineering, Heidelberg, Germany, April 2-6, 2001, pp. 149–154. IEEE Computer Society, Los Alamitos (2001)
28. Segoufin, L.: Typing and querying XML documents: some complexity bounds. In: PODS '03: Proceedings of the twenty-second ACM SIGMOD-SIGACT-SIGART symposium on Principles of database systems, pp. 167–178. ACM Press, New York (2003)
29. Segoufin, L., Vianu, V.: Validating streaming XML documents. In: PODS '02: Proceedings of the twenty-first ACM SIGMOD-SIGACT-SIGART symposium on Principles of database systems, pp. 53–64. ACM Press, New York (2002)
30. Sun Microsystems, Inc. Trang: Multi-format schema converter based on RELAX NG (May 2006), URL: http://www.thaiopensource.com/relaxng/trang.html
31. Thompson, H.S., Beech, D., Maloney, M., Mendelsohn, N.: XML Schema part 1: Structures 2 edn. W3C Recommendation (October 2004), URL: http://www.w3.org/TR/xmlschema-1
32. Werner, C., Buschmann, C., Brandt, Y., Fischer, S.: Compressing SOAP Messages by using Pushdown Automata. In: Proceedings of the IEEE International Conference on Web Services, Chicago, USA, September 2006, IEEE Computer Society Press, Los Alamitos (2006)
33. World Wide Web Consortium (W3C). XQuery Update Facility Requirements (2005), URL: http://www.w3.org/TR/xquery-update-requirements/
34. World Wide Web Consortium (W3C). XML Schema (2006) URL: http://www.w3.org/XML/Schema
35. World Wide Web Consortium (W3C). XQuery Update Facility (2006) URL: http://www.w3.org/TR/2006/WD-xqupdate-20060711/

A Appendix

In this section, we describe how the pushdown automaton is constructed by using a regular tree grammar $G = (N, T, P, S)$. As an initial step, we construct the finite state machines representing the content models of G. First, we define some required functions:

Definition 3 (Stack machine functions). *The states of an FSM $f \in \mathcal{FSM}$ can be determined by the function $states(f)$. Accepting states can be retrieved with the function $states_{accept}(f)$, the initial state with the function $state_{init}(f)$. Given a state $q \in states(f)$ the function $out(q)$ returns all outgoing transitions of q and the function $in(q)$ returns all incoming transitions. Each transition t has a label corresponding to an element type $\in N$ that can be determined by the function $type(t)$. The state to which t maps can be retrieved by $dest(t)$.*

The crucial point now is that the FSMs focus on element-nodes (tree-model) whereas our input is an XML string (sequence of symbols). We use a pushdown automaton (PDA) that simulates the execution of the FSMs while processing XML data as a string. This is done by pushing the reached opening states of the FSM onto the stack: The stack represents the path to the current tag. The proof that for each DTD d there exists a PDA accepting $\mathcal{L}(d)$ has been provided by Segoufin and Vianu ([29]). We now give a detailed construction algorithm for XML Schema. However, we first take a closer look at the tag sequence:

Lemma 2. *Any XML string x consists of a sequence of tag-symbols $\in \Sigma_t$ and their values in Σ_c^*. For any arbitrary two consecutive symbols $t_1, t_2 \in \Sigma_t$ there are five different cases for a well-formed sequence x:*

1. $t_1 \in \Sigma_t^+$ *is the first symbol in x* $\qquad$ (e.g. $<root>$) *[root case]*
2. $t_1 \in \Sigma_t^+ \wedge t_2 \in \Sigma_t^- \wedge name(t_1) = name(t_2)$

$\qquad\qquad\qquad\qquad\qquad\qquad\qquad$ (e.g. $<a></a>$) *[leaf case]*
3. $t_1 \in \Sigma_t^+ \wedge t_2 \in \Sigma_t^+$ $\qquad\qquad$ (e.g. $<a><b>$) *[child case]*
4. $t_1 \in \Sigma_t^- \wedge t_2 \in \Sigma_t^-$ $\qquad\qquad$ (e.g. $</a>$) *[parent case]*
5. $t_1 \in \Sigma_t^- \wedge t_2 \in \Sigma_t^+$ $\qquad\qquad$ (e.g. $<a>$) *[sibling case]*

For the sake of simplicity, we ignore all symbols of the textual content alphabet Σ_c. As XML Schema types can be defined by a regular expression[2] it is obvious that any sequence $\sigma_1, ... \sigma_n \in \Sigma_c^*$ can be checked by an automaton whether or not it corresponds to a given XML Schema type.

We generate the pushdown automaton (PDA) accepting $\mathcal{L}(G)$ as follows:

Formally, a pushdown automaton K is a 6-tuple; $K = (Q, \Sigma, \Gamma, \Psi, q_{start}, Z)$ where Q is a set of states, Σ is the input alphabet, Γ is the stack alphabet, Ψ is a set of transitions of the form $(Q \times \Sigma \times \Gamma) \rightarrow (Q \times \Gamma^*)$, $q_{start} \in Q$ is the initial state, and $Z \in \Gamma$ is the initial symbol of the stack.

The transition rules Ψ define what symbols in Σ and Γ have to be read in a state in Q in order to move to the next state while writing onto the stack. The PDA accepts when the input sequence has been completely read and the stack is empty.

Definition 4 (PDA accepting $\mathcal{L}(G)$). *The PDA has one dedicated initial state named s_{start}. For each tag $t \in \Sigma_t$ there is one corresponding state $q_t \in Q$. The set of all tag-states is denoted by Q_{Σ_t}. For each XML Schema type (e.g. xsd:int) there is one dedicated state[3]. These type states are denoted by Q_{type}. Now we define $Q = \{q_{start}\} \cup$*

---

[2] E.g. the regular expression $(-|\varepsilon)(0|1|...|9)^*$ defines the decimal type. See also [31].
[3] The latter states represent sub-automata for checking the textual content.

Algorithm 1. PDA Construction Algorithm

1: **for all** $p \in P$ **do**
2:     $tag \leftarrow element(p)$;
3:     $fsm \leftarrow fsm(p)$;
4:     $s_i \leftarrow state_{init}(fsm)$
5:     $S_a \leftarrow states_{accept}(fsm)$

6:     **if** $type(p) \in S$ **then**   ▷ **Root-case**
7:         AddTransition: $(s_{start}, tag^+, z) \rightarrow (state(tag^+), [s_i, z])$;
8:     **end if**

9:     **for all** $q \in states(fsm)$ **do**

10:        **if** $q = s_i \wedge q \in S_a$ **then**   ▷ **Leaf-case**
11:            AddTransition: $(state(tag^+), tag^-, q) \rightarrow (state(tag^-), [\,])$;
12:        **end if**

13:        **if** $q = s_i$ **then**   ▷ **Child-case**
14:            **for all** $t \in out(q)$ **do**
15:                $type_t \leftarrow type(t)$
16:                $tag_t \leftarrow tag(type_t)$
17:                $q_{dest} \leftarrow dest(t)$
18:                AddTransition: $(state(tag^+), tag_t^+, q) \rightarrow$
                       $(state(tag_t^+), [\,state_{init}(fsm(type_t)), q_{dest}])$;
19:            **end for**
20:        **end if**

21:        **if** $q \in S_a$ **then**   ▷ **Parent-case**
22:            **for all** $t \in in(q)$ **do**
23:                $type_t \leftarrow type(t)$
24:                $tag_t \leftarrow tag(type_t)$
25:                AddTransition: $(state(tag_t^-), tag^-, q) \rightarrow (state(tag^-), [\,])$;
26:            **end for**
27:        **end if**

28:        **for all** $t_{in} \in in(q)$ **do**
29:            **for all** $t_{out} \in out(q)$ **do**   ▷ **Sibling-case**
30:                $type_{in} \leftarrow type(t_{in})$
31:                $type_{out} \leftarrow type(t_{out})$
32:                $tag_{out} \leftarrow tag(type_{out})$
33:                $tag_{in} \leftarrow tag(type_{in})$
34:                AddTransition: $(state(tag_{in}^-), tag_{out}^+, q) \rightarrow$
                       $(state(tag_{out}^+), [\,state_{init}(fsm(type_{out})), dest(t_{out})])$;
35:            **end for**
36:        **end for**
37:    **end for**
38: **end for**

$Q_{\Sigma_t} \cup Q_{type}$. *The input alphabet $\Sigma = \Sigma_t \cup \Sigma_c$ contains tags and content. The function state(t), where $t \in \Sigma_t$, returns the corresponding state $\in Q$. We do not impose any restrictions on the stack alphabet Γ. The transitions Ψ are generated by Algorithm 1.*

Algorithm 1 is split into five parts – corresponding to the five tag-sequence cases. While iterating over the production rules of the regular tree grammar G (line 1) it is checked which one of the five cases holds. The PDA simulates the run of the set of FSMs by pushing and popping corresponding FSM-states onto/from the stack.

If an element is defined as top-level element in the XML Schema it appears as a start-symbol in G. Transitions from the initial state q_{start} to the corresponding open-tag states are created (line 7).

We iterate over all states in the FSM which are generated in line 3. If the initial state is also a final state, then it has an empty content model, i.e. it is a leaf – a closing tag is read directly after the opening tag (line 11). Nothing has to be pushed onto the stack because the subsequence for the tag has been processed completely.

Outgoing transitions of the initial states of the FSMs map to the first element of their content model (child). Therefore, there is a transition in the PDA to the open-tag states of each of such children (line 18). The transitions of the parent-case are generated inversely (line 25) based on the final states of the FSMs. Here, nothing has to be pushed onto the stack because the sub-sequence belonging to the closing tag has been processed completely.

For all states in the FSMs having an incoming and outgoing transition we create a transition in the PDA that corresponds to the sibling-case. Here, the PDA reads the closing-tag of an element a and afterwards the opening-tag of an element b (line 34).

When some XML input is given, the automaton reads it tag by tag and switches into corresponding states. When the input is invalid there will be no transition or the top-most state on the stack does not match the requirements. In this case, the input cannot be fully read and the input is rejected.

The complexity of this construction algorithm is $O(n^3)$ with n being the size of states in the FSMs. However, the construction of the PDA is performed only once when defining the column type.

Combining Efficient XML Compression with Query Processing

Przemysław Skibiński[1] and Jakub Swacha[2]

[1] University of Wrocław, Institute of Computer Science,
Joliot-Curie 15, 50-383 Wrocław, Poland
inikep@ii.uni.wroc.pl
[2] The Szczecin University, Institute of Information Technology in Management,
Mickiewicza 64, 71-101 Szczecin, Poland
jakubs@uoo.univ.szczecin.pl

Abstract. This paper describes a new XML compression scheme that offers both high compression ratios and short query response time. Its core is a fully reversible transform featuring substitution of every word in an XML document using a semi-dynamic dictionary, effective encoding of dictionary indices, as well as numbers, dates and times found in the document, and grouping data within the same structural context in individual containers. The results of conducted tests show that the proposed scheme attains compression ratios rivaling the best available algorithms, and fast compression, decompression, and query processing.

Keywords: XML compression, XML searching, XML transform, semi-structural data compression, semi-structural data searching.

1 Introduction

Although Extensible Markup Language (XML) did not make obsolete all the good old data formats, as one could naively expect, it has become a popular standard, with many useful applications in information systems.

XML has many advantages, but for many applications they are all overshadowed by just one disadvantage, which is XML verbosity. But verbosity can be coped with by applying data compression. Its results are much better if a compression algorithm is specialized for dealing with XML documents.

XML compression algorithms can be divided into two groups: those which do not allow queries to be made on the compressed content, and those which do. The first group focuses on attaining as high compression ratio as possible, employing the state-of-the-art general-purpose compression algorithms: Burrows-Wheeler Transform – BWT [2], and Prediction by Partial Match – PPM [18]. The problem with the most effective algorithms is that they require the XML document to be fully decompressed prior to processing a query on it.

The second group sacrifices compression ratio for the sake of allowing search without a need for full document decompression. This can be accomplished by compressed pattern matching or by partial decompression. The former consists in

Y. Ioannidis, B. Novikov, and B. Rachev (Eds.): ADBIS 2007, LNCS 4690, pp. 330–342, 2007.

compressing the pattern and searching for it directly in the compressed data. As both the compressed pattern and data are shorter than their non-compressed equivalents, the search time is significantly reduced. However, the compressed pattern matching works best with a context-free compression scheme, such as Huffman coding [9], which hurts compression ratio significantly.

Partial decompression means some parts of the document have to be decompressed, but not the whole of it. Context-aware algorithm can be used, as long as it makes use only of the limited-range data correlation. LZ77-family algorithms [21] are ideal candidates here. Although they do not attain as high compression ratios as BWT or PPM derivatives, they are much more effective than context-free compression schemes. And although they do not allow queries to be processed as fast as using compressed matches, the processing time is much shorter than in case of schemes requiring full decompression.

In this paper we describe a new algorithm designed with compression effectiveness as primary concern, and search speed as secondary one. Thus it is particularly suited for huge XML datasets queried with moderate frequency.

We begin with discussion of existing XML compression algorithms' implementations. Then we describe the proposed algorithm, its main ideas and its most significant details. Finally, we present results of tests of an experimental implementation of our algorithm, and draw some conclusions.

2 XML Compression Schemes

2.1 Non-query-supporting Schemes

The first XML-specialized compressor to noticeably surpass in efficiency general-purpose compression schemes was XMill [11]. Its success was due to three features. The first of them is splitting XML document content into three distinct parts containing respectively: element and attribute symbol names, plain text and the document tree structure. Every part has different statistical properties, therefore it helps compression to process them with separate models.

The second feature of XMill is to group contents of same XML elements into so-called containers. Thus similar data are stored together, helping compression algorithms with limited history buffer, such as LZ77 derivatives.

The third one is to encode each container using a dedicated method, exploiting the type of data stored within it (such as numbers or dates). What makes this feature not so useful is that XMill requires the user themself to choose methods to encode specific containers. Such human-aided compression can hardly be regarded as practical.

XMill originally used LZ77-derived gzip to compress the transform output. Although newer version added support for BWT-based bzip2 and PPM implementations, yet in these modes XMill succumbs to other programs employing such algorithms.

The first published XML compression scheme to use PPM was XMLPPM [3]. XMLPPM replaces element and attribute names with their dictionary indices, removes closing tags as they can be reconstructed in a well-formed XML document only provided their positions are marked. The most important XMLPPM feature is 'multiplexed hierarchical modeling' which consists in encoding data with four distinct PPM models: one for element and attribute names, one for element structure, one for attribute values, and one for element contents. In order to exploit some correlation between data going to different models, the previous symbol, regardless the model it belongs to, is used as a context for the next symbol.

XMLPPM was extended into SCMPPM [1], in which a separate PPM model is maintained for every XML element. This helps only in case of large XML documents, as every PPM model requires a due number of processed symbols to become effective. The main flaw of SCMPPM is its very high memory usage (every new model is initialized with 1 MB of allocated memory).

In 2004 Toman presented Exalt [20], one of the first algorithms compressing XML by inferring a context-free grammar describing its structure. His work led to AXECHOP by Leighton et al. [10]. However, even the latter fails to overcome XMLPPM both in terms of compression ratio and time.

In 2005 Hariharan and Shankar developed XAUST [8], employing finite-state automata (FSA) to encode XML document structure. Element contents are put into containers and encoded incrementally with arithmetic coding based on a single statistical model of order 4 (i.e., treating at most 4 preceding symbols as the context for the next one). The published results show XAUST to beat XMLPPM on some test files, yet XAUST has this great drawback that it requires the compressed XML document to be valid and its document type definition (DTD) to be available, as the FSA are constructed based on it.

2.2 Query-Supporting Schemes

As it has been stated in the introduction, adding support for queries decreases compression effectiveness. Therefore, most of the algorithms described below attain compression ratios worse than the original XMill.

XGrind [19] was probably the first XML compressor designed with fast query processing on mind. XGrind forms a dictionary of element and attribute names based on DTD. It uses a first pass over a document to gather statistics so that Huffman trees could be constructed. During the second pass, the names are replaced with respective codewords, and the remaining data are encoded using several Huffman trees, distinct for attribute values and PCDATA elements. An important property of XGrind, mainly thanks to the use of Huffman encoding, is that its transformation is homomorphic, which means that the same operations (such as parsing, searching or validating) can be performed on the compressed document as on its non-compressed form.

XPress [14] extends XGrind with binary encoding of decimal numbers and improves speed of path-based queries using a technique called Reverse Arithmetic Encoding.

XQzip [5] was the first query-supporting XML compressor to beat XMill, although slightly. It separates structure and data of XML documents. The structure is stored in a form of Structural Indexing Tree (SIT) in order to speed up queries, whereas data are grouped in containers, which are partitioned into small blocks (for the sake of partial decompression) and those are finally compressed with gzip. Recently decompressed blocks are kept in cache so that multiple queries are sped up. According to its authors, XQzip processes queries 12 times faster than XGrind.

XQueC [15] was clearly focused on search speed rather than compression efficiency. Like other schemes, it splits XML documents into structure and data parts, groups element contents in containers (based on each element's path from the document root), and forms a dictionary of element and attribute names. Additionally, for the sake of faster query processing, XQueC stores a tree describing document structure and an index – a structural summary representing all possible paths in the document.

XSeq [12] uses Sequitur, a grammar-based compression algorithm to compress both document structure and data, and allows compressed pattern matching. According to its authors, XSeq processes queries even faster than XQzip, but it does not attain compression ratios of XMill.

XBzip [7] employs XBW transform (based on BWT), which transposes XML document structure into its linear equivalent using path-sorting and grouping. Data are sorted accordingly with the structure. The two resulting tables (structure and data) are then compressed using PPM algorithm. XBzip can work in two modes. In the default, non-query-supporting mode, it can attain compression ratios even higher than XMLPPM. In the second, query-supporting mode (required by the XBzipIndex utility) it splits data into containers, and creates an FM-index (a compressed representation of a string that supports efficient substring searches) for each of them. The query processing times are very short, but storing the indices inflates the compressed document size by as much as 25-100%.

3 The QXT Transform

3.1 Overview

Our recent work on highly effective but non-query-supporting XML compression scheme dubbed XWRT [17] led us to awareness of the importance of fast query processing in XML documents. In practice, it is often desirable both to store data compactly, and retain swift query response time.

QXT stands for Query-supporting XML Transform. QXT has been designed combining the best solutions of XWRT with query-friendly concepts in order to make it possible to process queries with partial decompression, while avoiding to hurt compression effectiveness significantly.

For QXT, the input XML document is considered to be an ordered sequence of n tokens:

$$Input=(t_1 \cdot t_2 \cdot \ldots \cdot t_n).$$

QXT parses the input classifying every encountered token to one of generic token classes:

$$TokenClass(t) \in \{ Word, EndTag, Number, Special, Blank, Char \}.$$

The *Word* class contains sequences of characters meeting the requirements for inclusion in the dictionary, *EndTag* contains all the closing tags, *Number* – sequences of digits, *Special* – sequences of digits and other characters adhering to predefined patterns, *Blank* – single spaces between *Word* tokens, and the *Char* class contains all the remaining input symbols.

The *Word* class has two token subclasses: *StartTag* contains all the element opening tags, whereas *PlainWord* all the remaining *Word* tokens.

The *StartTag* and *EndTag* tokens define the XML structure. The *StartTag* tokens differ from *PlainWord* tokens in that they redirect the transform output to a container identified by the opened element's path from the document root. *EndTag* tokens are replaced with a one-byte flag and bring the output back to the parent element's container.

3.2 Handling the Words

A sequence of characters can only be identified as a *Word* token if it is one of the following:

- *StartTag* token – a sequence of characters starting with '<', containing letters, digits, underscores, colons, dashes, or dots. If a *StartTag* token is preceded by a run of spaces, they are combined and treated as a single token (useful for documents with regular indentation);
- a sequence of lowercase and uppercase letters ('a'–'z', 'A'–'Z') and characters with ASCII codes from range 128–255; this includes all words from natural languages using 8-bit letter encoding;
- URL prefix – a sequence of the form 'http://domain/', where domain is any combination of letters, digits, dots, and dashes;
- e-mail – a sequence of the form 'login@domain', where 'login' and 'domain' are combinations of letters, digits, dots, and dashes;
- XML entity – a sequence of the form '&data;', where data is any combination of letters (so, e.g., character references are not included);
- attribute value delimiter – sequences '="' and '">';
- run of spaces – a sequence of spaces not followed by a *StartTag* token (again, useful for documents with regular indentation).

The list of *Word* tokens sorted by descending frequency composes the dictionary. QXT uses a semi-dynamic dictionary, that is it constructs a separate dictionary for every processed document, but, once constructed, the dictionary is not changed during XML transformation. It would be problematic to use a universal static dictionary with predefined list of words as it is hard to find a word set relevant across a wide range of real-world XML documents.

Every *Word* token is replaced with its dictionary index. The dictionary indices are encoded using symbols which are not existent in the input XML document. There are two modes of encoding, chosen depending on the attached back-end compression algorithm. In both cases, a byte-oriented prefix code is used; although it produces slightly longer output than, e.g., bit-oriented Huffman coding, the resulting data can be easily compressed further, which is not the case with the latter.

In the Deflate-friendly mode, the set of available symbols is divided into three disjoint subsets: *OneByte*, *TwoByte*, *ThreeByte*. The *OneByte* symbols are used to encode the most frequent *Word* tokens; one symbol can represent one token, so only

|*OneByte*| tokens can be encoded this way. The *TwoByte* symbols are used as a prefix for another byte, allowing to encode |*TwoByte*|·256 tokens in this way. Finally, the *ThreeByte* symbols are used as a prefix for another two bytes, allowing to encode |*ThreeByte*|·65536 tokens in this way.

In the LZMA-friendly mode, the set of available symbols is divided only into two disjoint subsets: *Prefix, Suffix*. The *Prefix* symbols signalize the beginning of a codeword. The codeword can be but a *Prefix* symbol, or a *Prefix* symbol followed by one or two *Suffix* symbols. This way there are |*Prefix*| one-byte codewords available for the most frequent *Word* tokens, |*Prefix*|·|*Suffix*| two-byte codewords for typical *Word* tokens, and |*Prefix*|·|*Suffix*|^2 three-byte codewords for rare *Word* tokens.

As the single *Blank* tokens can appear only between two *Word* tokens, they are simply removed, as they can be reconstructed on decompression provided the exceptional positions where they should not be inserted are marked.

The *Char* tokens are left intact.

3.3 Handling the Numbers and Special Data

Every *Number* token (decimal integer number) n is replaced with a single byte whose value is $\lceil \log_{256}(n+1) \rceil + 48$. The actual value of n is encoded as a base-256 number. A special case is made for sequences of zeroes preceding another *Number* token – these are left intact.

Special token represent specific types of data made up of combination of digits and other characters. Currently, QXT recognizes following *Special* tokens:

- dates between 1977-01-01 and 2153-02-26 in YYYY-MM-DD (e.g. "2007-03-31", Y for year, M for month, D for day) and DD-MMM-YYYY (e.g. "31-MAR-2007") formats;
- times in 24-hour (e.g., "22:15") and 12-hour (e.g., "10:15pm") formats;
- value ranges (e.g., "115-132");
- decimal fractional numbers with one (e.g., "1.2") or two (e.g., "1.22") digits after decimal point.

Dates are replaced with a flag and encoded as a two bytes long integer whose value is the difference in days from 1977-01-01. To simplify the calculations we assume each month to have 31 days. If the difference with the previous date is smaller than 256, another flag is used and the date is encoded as a single byte whose value is the difference in days from the previous date.

Times are replaced with a sequence of three bytes representing respectively: the time flag, hour, and minutes.

Value ranges in the format "x–y" where $x < 65536$ and $0 < y - x < 256$ are encoded in four bytes: one for the range flag, two for the value of x, and one for the difference $y - x$.

Decimal fractional numbers with one digit after decimal point and value from 0.0 to 24.9 are replaced by two bytes: a flag and their value stored as fixed point integer. In case of those with two digits after decimal point, only their suffix, starting from the decimal point, is considered to be *Special* token, and replaced with two bytes: a flag and the number's fractional part stored as an integer.

3.4 Implementation

The architecture of QXT implementation is presented on Fig. 1.

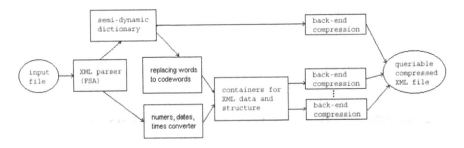

Fig. 1. QXT processing scheme

The QXT implementation contains a fast and simple XML parser written exclusively for this application. The parser does not build any trees, but it treats input XML document as one-dimensional data. It has small memory requirements, as it only uses a stack to trace opening and closing tags.

QXT works in two passes over input data. In the first pass, a dictionary is formed and the frequency of each of its items is computed. The complete dictionary is stored within the compressed file, so this pass is unnecessary during decompression, making the reverse operation faster.

In the second pass, the actual transform takes place, data are parsed into tokens, respectively encoded, and placed into separate containers, depending on their path from the document root.

The containers are memory-buffered until they exceed a threshold of 8 MB (same as in XMill) when they are compressed with general-purpose compression algorithm and written to disk. The containers are compressed in 32 KB blocks, thus allowing partial decompression of blocks of such length.

QXT could be combined with any general-purpose compression algorithm, but the requirement of fast decompression suggests using a LZ77-derivative. We chose two algorithms of this kind: Deflate [6] (well known from zip, gzip, and plenty of other applications) and LZMA (known from the 7-zip utility [22]), whose optimal match parsing significantly improves compression ratio at the cost of much slower compression (decompression speed is not much affected).

Query execution starts with reading the dictionary from the compressed file; the dictionary can be cached in memory in case of multiple queries. Next, the query processor resolves which containers might contain data matching the query. The required containers are decompressed with the general-purpose algorithm, and the transformed data are then searched using the transformed pattern. Only the matching elements are decoded to the original XML form; of course, counting queries do not require the reverse transform at all.

4 Experimental Results

4.1 Test Preparation

The primary objective of the tests was to measure the performance of an experimental implementation of the QXT algorithm written in C++ by the first author and compiled with Microsoft Visual C++ 6.0. This implementation allows to use Deflate or LZMA as the back-end compression algorithm.

For comparison purposes, we included in the tests publicly available XML compressors: XMill (version 0.7, which was found to be the fastest; switches: -w -f), XMLPPM (0.98.2), XBzip (1.0), and SCMPPM (0.93.3); the old XGrind was omitted as, despite multiple efforts, we were not able to port it to the test system. We have extended the list with general-purpose compression tools: gzip (1.2.4; uses Deflate) and LZMA (4.42; -a0), employing the same algorithms as the final stage of QXT, to demonstrate the improvement from applying the XML transform.

So far there is no publicly available and widely respected XML test file corpus, therefore, we have based our test suite on those files from the corpus proposed in [15] that were publicly available, improving it with several files from the University of Washington *XML Data Repository* [13]. The resulting corpus represents a wide range of real-world XML applications; it consists of the following varied XML documents:

- *DBLP*, bibliographic information on major computer science journals and proceedings,
- *Lineitem*, business order line items from the 10 MB version of the TPC-H benchmark,
- *Mondial*, basic statistical data on countries of the world,
- *NASA*, astronomical data,
- *Shakespeare*, a corpus of marked-up Shakespeare plays,
- *SwissProt*, a curated protein sequence database,
- *UWM*, university courses.

Detailed information for each of the documents is presented in Table 1 (see next page); it includes: file size (in bytes), number of elements, number of attributes, number of distinct element types, and the maximum structure depth.

The tests were conducted on an Intel Core 2 Duo E6600 2.40 GHz system with 1024 MB memory and two Seagate 250 GB SATA drives in RAID mode 1 under Windows XP 64-bit edition.

Table 1. Basic properties of the XML documents used in tests

Name	File size	Elements	Attributes	Max. depth
DBLP	133 862 735	3 332 130	404 276	6
Lineitem	32 295 475	1 022 976	1	3
Mondial	1 784 825	22 423	47 423	5
NASA	25 050 288	476 646	56 317	8
Shakespeare	7 894 983	179 690	0	7
SwissProt	114 820 211	2 977 031	2 189 859	5
UWM	2 337 522	66 729	6	5

4.2 Compression Ratio and Time

Table 2 lists the bitrates (in output bits per input character, hence the smaller the better) attained by the tested programs on each of the test files. Some table cells are empty, as XMill and SCMPPM declined to compress some of the test files, whereas XBzip failed to finish the compression of two files due to insufficient memory.

Apart from the general-purpose compression tools, the experimental QXT implementation was the only program to decode all the compressed files accurately. In case of the other programs, some output files were shortened, some had misplaced white characters (space-preserving modes were turned on where possible). In some applications, even the latter can be a grave flaw, as this makes it impossible to verify the original file integrity with a cyclic redundancy check or hash functions.

Table 2. Compression results (in bpc)

File	Gzip	LZMA	XMill	XML PPM	SCM PPM	XBzip	XBzip Index	QXT Deflate	QXT LZMA
DBLP	1.463	1.049	1.250[a]	0.857[a]	**0.741**	–[d]	–[d]	0.925	0.753
Lineitem	0.721	0.461	0.380	0.273[a]	**0.244**	0.248[a]	0.332	0.285	0.245
Mondial	0.767	0.586	–[c]	0.467[b]	–[c]	**0.404**[a]	0.681	0.608	0.407
NASA	1.208	0.818	1.011	0.729[a]	–[c]	0.698[b]	1.085	0.769	**0.607**
Shakespeare	2.182	1.786	2.044	1.367[a]	1.354[a]	**1.350**[b]	1.688	1.505	1.354
SwissProt	0.985	0.540	0.619	0.465[a]	0.426	–[d]	–[d]	0.500	**0.384**
UWM	0.553	0.389	0.382	**0.259**[a]	0.274	0.282	0.446	0.323	0.281
Average	1.126	0.804	–	0.631	–	–	–	0.702	**0.576**

Remarks: (a) Decoded file was not accurate, (b) Decoded file was shorter than the original, (c) Input file was not accepted, (d) Compression failed due to insufficient memory.

The obtained compression results are very favorable for QXT. The QXT+LZMA attained the best ratio in case of two files and was only slightly worse than the best scheme in case of four other files. Notice that this was achieved even though QXT supports queries, whereas the other programs do not (with the sole exception of XBzipIndex), and it uses LZ77-derived compression algorithm, whereas the best of the remaining programs employ more sophisticated PPM-based algorithms.

Compared to the general-purpose compression tools, the proposed transform improves XML compression on average by 28% in case of LZMA and 37% in case of Deflate.

Fig. 2 shows the ratio of the compressed file size to the original size for the four biggest files in the suite, attained by the tested XML-specialized compressors. Smaller columns represent better results.

Table 3 contains the compression and decompression times measured on *Lineitem*, the longest file compressed by all the tested programs.

The results show that QXT is slower than XMill, but when compared to the most effective compressors, QXT+Deflate was found to be almost about three times faster than XBzip, and seven times faster than SCMPPM.

In case of QXT+LZMA, the XML transform is faster than the back-end compression algorithm, and reduces the file size so much, that a large improvement in compression speed can be observed.

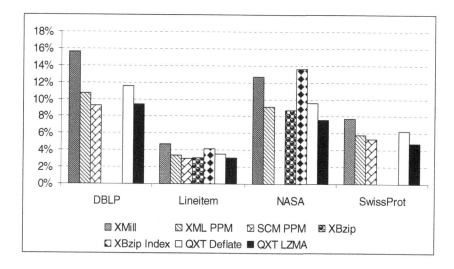

Fig. 2. Compression results for selected files

Table 3. Compression and full decompression times for the *Lineitem* file

Time of	Gzip	LZMA	XMill	XML PPM	SCM PPM	XBzip	XBzip Index	QXT Deflate	QXT LZMA
Compression	0.98	23.50	1.31	2.66	14.24	6.33	11.59	3.06	5.95
Decompression	0.20	0.83	0.22	3.19	13.16	4.51	6.26	1.12	1.34

Remarks: The times listed were measured on the test platform, and are total user times including program initialization and disk operations.

4.3 Query Processing Time

For query processing evaluation, we used the *Lineitem* and *Shakespeare* files (we could not obtain results for XBzipIndex on *DBLP*). Queries *S1*, *S2*, and *L1* are from [5], but as XBzipIndex did not support another query, we contrived the remaining queries on our own. Table 4 lists all the queries used in our tests.

Table 4. Test queries

Id	*Shakespeare* queries
S1	/PLAY/ACT/SCENE/SPEECH/SPEAKER
S2	/PLAY/ACT/SCENE/SPEECH[SPEAKER = "PHILO"]
S3	/PLAY/ACT/SCENE/SPEECH[SPEAKER = "AMIGA"]
Id	*Lineitem* queries
L1	/table/T/L_TAX
L2	/table/T/L_COMMENT
L3	/table/T/[L_COMMENT = "slowly"]

Table 5 contains measured times of processing the test queries. Its first column identifies the query. The next two columns contain query processing time on a non-compressed XML without an external index (column 2), and using one (column 3). The sgrep structural search utility (version 1.92a) was used to obtain these values. Column 4 contains XBzipIndex's query time, and the remaining columns present detailed results for QXT: the part of the file (in percents) that had to be decompressed in the course of query processing (column 5), the time it took to accomplish decompression (column 6 for Deflate, and 7 for LZMA), the time taken by the actual search on the decompressed data (column 8), and the sum of the two (columns 9 and 10), constituting the total query processing time.

Table 5. Query processing times (in seconds)

| Id | sgrep | | XBzip Index | Decom-pressed part | QXT | | | | |
| | Raw XML | Inde-xed XML | | | Decompression | | Search time | Total query time | |
					Deflate	LZMA		Deflate	LZMA
S1	0.313	0.047	0.047	15.7%	0.051	0.025	0.023	0.074	0.048
S2	0.282	0.032	0.061	15.7%	0.051	0.025	0.023	0.074	0.048
S3	0.266	0.031	0.063	10.9%	0.036	0.017	0.000	0.036	0.017
L1	1.031	0.063	0.047	5.5%	0.062	0.083	0.036	0.098	0.119
L2	1.063	0.078	0.047	8.4%	0.095	0.127	0.055	0.150	0.182
L3	1.047	0.047	0.063	8.4%	0.095	0.127	0.055	0.150	0.182
Avg	0.667	0.050	0.055	10.8%	0.065	0.067	0.032	0.097	0,099

Remarks: The times listed were measured on the test platform, and are total user times including program initialization and disk operations.

Average query processing time of QXT was about 0.1 second. This seems to be an acceptable delay for an average user. This is over six times faster than the raw XML search time. The transformed data are searched much faster than non-compressed XML because they are shorter. The decompression does not waste all of this gain because an average query requires only 10% of the file to be decompressed. Moreover, the word dictionary makes it possible to immediately return negative results for queries looking for words non-occurring in the document (see the instant search results for query S3).

The primary purpose of QXT is effective compression, so it does not maintain any indices to the document content. Therefore, we did not expect QXT to beat index-based search times. Indeed, both sgrep's indexed search and XBzipIndex are over 50% faster. However, a look at the compression ratio of XBzipIndex, signifying a 25% bigger storage requirements than QXT+LZMA, helps to realize that these two schemes pursue two different goals.

To demonstrate the scale of improvement over a non-query-supporting schemes that perform full decompression followed by search on the decompressed document, QXT's average query time on *Shakespeare* (0.037 s) is over ten times shorter than XMill's (0.381 s) and over hundred times shorter compared to the time required by SCMPPM (3.834 s).

5   Conclusions

Due to the huge size of XML documents used nowadays, their verbosity is considered a troublesome feature. Until now, there have been but two options: effective compression for the price of a very long data access time, or quickly accessible data for the price of mediocre compression.

The proposed QXT scheme ranks among the best available algorithms in terms of compression efficiency, far surpassing its rivals in decompression time.

The most important advantage of QXT is the feasibility of processing queries on the document without a need to have it fully decompressed. Thanks to the nature of the transform, the measured query processing times on QXT-transformed documents were several times shorter than on their original format. We did not include any indices in QXT as they require significant storage space, so using them would greatly diminish the compression gain which had the top priority in the design of QXT. Still, we may reconsider this idea in a future work.

QXT has many nice practical properties. The transform is completely reversible, the decoded document is an accurate copy of the input document. The transform requires no metadata (such as XML Schema or DTD) nor human assistance. Whereas SCMPPM and XBzip may require even hundreds of megabytes of memory, the default mode of QXT uses only 16 MB, irrespectively of the input file size (using LZMA requires additionally a fixed buffer of 84 MB for compression and 10 MB for decompression). The compression is done on small blocks, so in case of data damage, the data loss is usually limited to just one such block. Moreover, QXT is implemented as a stand-alone program, requiring no external compression utility, XML parser, nor query processor, thus avoiding any compatibility issues.

Acknowledgements

Szymon Grabowski is the co-author of XWRT, the non-query-supporting XML compression scheme on which QXT has been based.

References

1. Adiego, J., de la Fuente, P., Navarro, G.: Merging Prediction by Partial Matching with Structural Contexts Model. In: Proceedings of the IEEE Data Compression Conference, Snowbird, UT, USA, p. 522 (2004)
2. Burrows, M., Wheeler, D.J.: A block-sorting data compression algorithm. SRC Research Report 124. Digital Equipment Corporation, Palo Alto, CA, USA (1994)
3. Cheney, J.: Compressing XML with multiplexed hierarchical PPM models. In: Proceedings of the IEEE Data Compression Conference, Snowbird, UT, USA, pp. 163–172 (2001)
4. Cheney, J.: Tradeoffs in XML Database Compression. In: Proceedings of the IEEE Data Compression Conference, Snowbird, UT, USA, pp. 392–401 (2006)
5. Cheng, J., Ng, W.: XQzip: querying compressed XML using structural indexing. In: Proceedings of the Ninth International Conference on Extending Database Technology, Heraklion, Greece, pp. 219–236 (2004)

6. Deutsch, P.: DEFLATE Compressed Data Format Specification version 1.3. RFC1951(1996), http://www.ietf.org/rfc/rfc1951.txt

7. Ferragina, P., Luccio, F., Manzini, G., Muthukrishnan, S.: Compressing and Searching XML Data Via Two Zips. In: Proceedings of the International World Wide Web Conference (WWW), Edinburgh, Scotland, pp. 751–760 (2006)

8. Hariharan, S., Shankar, P.: Compressing XML documents with finite state automata. In: Farré, J., Litovsky, I., Schmitz, S. (eds.) CIAA 2005. LNCS, vol. 3845, pp. 285–296. Springer, Heidelberg (2006)

9. Huffman, D.A.: A Method for the Construction of Minimum-Redundancy Codes. Proc. IRE 40, 9, 1098–1101 (1952)

10. Leighton, G., Diamond, J., Muldner, T.: AXECHOP: A Grammar-based Compressor for XML. In: Proceedings of the IEEE Data Compression Conference, Snowbird, UT, USA, pp. 467–467 (2005)

11. Liefke, H., Suciu, D.: XMill: an efficient compressor for XML data. In: Proceedings of the 19th ACM SIGMOD International Conference on Management of Data, Dallas, TX, USA, pp. 153–164 (2000)

12. Lin, Y., Zhang, Y., Li, Q., Yang, J.: Supporting efficient query processing on compressed XML files. In: Proceedings of the ACM Symposium on Applied Computing, Santa Fe, NM, USA, pp. 660–665 (2005)

13. Miklau, G.: XML Data Repository, University of Washington (2004), http://www.cs.washington.edu/research/xmldatasets/www/repository.html

14. Min, J.-K., Park, M., Chung, C.: A Compressor for Effective Archiving, Retrieval, and Updating of XML Documents. In: Proceedings of the 2003 ACM SIGMOD International Conference on Management of Data, San Diego, CA, USA, pp. 122–133 (2003)

15. Ng, W., Lam, W.-Y., Cheng, J.: Comparative Analysis of XML Compression Technologies. World Wide Web 9(1), 5–33 (2006)

16. Skibiński, P., Grabowski, S., Deorowicz, S.: Revisiting dictionary-based compression. Software – Practice and Experience 35(15), 1455–1476 (2005)

17. Skibiński, P., Grabowski, S., Swacha, J.: Fast transform for effective XML compression. In: Proceedings of the IXth International Conference CADSM 2007, pp. 323–326. Publishing House of Lviv Politechnic National University, Lviv, Ukraine (2007)

18. Shkarin, D.: PPM: One Step to Practicality. In: Proceedings of the IEEE Data Compression Conference, Snowbird, UT, USA, pp. 202–211 (2002)

19. Tolani, P., Haritsa, J.: XGRIND: a query-friendly XML compressor. In: Proceedings of the 2002 International Conference on Database Engineering, San Jose, CA, USA, pp. 225–234 (2002)

20. Toman, V.: Syntactical compression of XML data. In: Presented at the doctoral consortium of the 16th International Conference on Advanced Information Systems Engineering, Riga, Latvia (2004), http://caise04dc.idi.ntnu.no/CRC_CaiseDC/ toman.pdf

21. Ziv, J., Lempel, A.: A Universal Algorithm for Sequential Data Compression. IEEE Trans. Inform. Theory 23, 3, 337–343 (1977)

22. 7-zip compression utility, http://www.7-zip.org

Fast User Notification in Large-Scale Digital Libraries: Experiments and Results

H. Belhaj Frej[1], P. Rigaux[2], and N. Spyratos[1]

[1] LRI, Univ. Paris-Sud, 91400 Orsay
{hanen,spyratos}@lri.fr
[2] LAMSADE, Univ. Paris-Dauphine,75016 Paris
philippe.rigaux@dauphine.fr

Abstract. We are interested in evaluating the performance of new matching algorithms for user notification in digital libraries (DL). We consider a subscription system which continuously evaluates queries over a large repository containing document descriptions. The subscriptions and the document descriptions rely on a taxonomy that is a hierarchically organized set of terms. The digital library supports insertion, update and removal of a document. Each of these operations is seen as an *event* that must be notified only to those users whose subscriptions match the document's description. The paper proposes a notification algorithm dedicated to taxonomy-based DLs, addresses computational issues and report a full set of experiments illustrating the advantages of the approach.

Keywords: Digital libraries, personalization, publish/subscribe systems, notifications, matching algorithms.

1 Introduction

User notification constitutes one of the key elements to the development of large scale data retrieval and dissemination systems. The notification services are increasingly widespread in various types of applications. They allow the users to register their favorite queries in the form of subscriptions and inform them whenever an event that affects the content of the DL matches their subscriptions.

Several powerful algorithms have been proposed in the recent literature to support the matching process. See for instance [8,5,20], and the section devoted to related work. The common feature of these algorithms is to consider a data model where the objects of interest are described with single-valued attributes. Moreover the domain of these attributes is an unstructured set of values. In practice, applications that rely on a tree-structured domain for objects descriptions are very common. This is the case, among many other examples, of e-commerce applications such as Ebay (*http://www.ebay.com*) which proposes a classification of objects based on a `category` attribute. Category `Baby` is divided into several sub-categories (`clothing, toys, equipments,` · · ·). Each sub-category is itself divided into other sub-categories, and so on. A user may either subscribe to a general concept (e.g., `Baby`) or to a specialized one (e.g., `Baby >> bed`).

Y. Ioannidis, B. Novikov, and B. Rachev (Eds.): ADBIS 2007, LNCS 4690, pp. 343–358, 2007.
© Springer-Verlag Berlin Heidelberg 2007

This introduces naturally a dependency on the set of subscriptions: if an event must be notified to a user that subscribed to `Baby >> bed`, then any user that subscribed to `Baby` must be notified as well. A *matching algorithm* for such applications should take into account the attributes defined over tree-structured domains to benefit from this kind of property.

In the present paper, we introduce and evaluate matching algorithms that explicitly support the definition of subscriptions over tree-valued attributes. Our approach can be combined with several of the existing solutions proposed in the literature, and applies to a wide range of applications, including those previously mentioned. For concreteness we focus in the rest of this paper on digital libraries (DL) and adopt the digital library model discussed in [9]. We assume that the digital library maintains a users' repository which contains information about the users and their subscriptions. Moreover, we assume that the digital library relies on a taxonomy of terms to which both the authors and the users adhere. The authors use the taxonomy in order to describe their documents and the users in order to define their subscriptions. A user must be informed, or *notified* whenever an event matching his subscription occurs at the DL.

The problem addressed in this paper can be stated as follows: assuming a large number of subscriptions over tree-structured domains and a high rate of events, how to design fast algorithms for finding those subscribers that should be notified when an event occurs at the DL? We address computational issues, and conduct an extensive performance evaluation. Our analytical and experimental results show that our approach enjoys many important features. It presents a reasonable storage cost (at most twice the number of subscriptions), and achieves a nice trade-off between the performances of subscriptions (i.e., adding new users) and notifications (i.e., alerting users of new incoming events of interest).

In what follows, we first give in Section 2 some preliminary definitions. Section 3 describes our data structures and algorithms and Section 4 presents our experimental setting and results. Related work is given in Section 5 while Section 6 concludes the paper.

2   Preliminary Definitions

As we mentioned in the introduction, we assume a single taxonomy over which the following basic concepts of a digital library are defined: document description, event and user subscription. This section summarizes the model presented in [9].

2.1   Basic Concepts

The library taxonomy, denoted $(T, \preceq)$ is a set of keywords or *terms* T, together with a *subsumption relation* over them. This subsumption relation, denoted $\preceq$, is a reflexive and transitive binary relation. Given two terms, s and t, if $s \preceq t$ then we say that s is *subsumed* by t, or that t *subsumes* s. We represent the taxonomy as a tree, where the nodes are the terms and there is an arrow from term t to term s iff t subsumes s. Figure 1 shows an example of a taxonomy, in

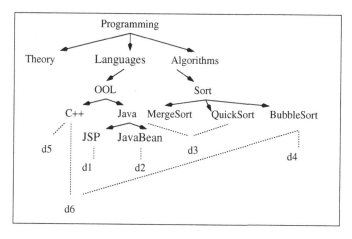

Fig. 1. A repository

which the term Languages subsumes the term OOL, the term Java subsumes the term JavaBeans, and so on.

Documents are added to the library by authors who submit, for each document, a description of the content together with an identifier (say, the document's URI) allowing to access the document's content. The document description is a reduced set of terms from the taxonomy. A set of terms is reduced if none of the terms it contains is subsumed by another term. For example, if the document contains the quick sort algorithm written in java then the terms QuickSort and Java can be used to describe its content. In this case the set of terms {QuickSort, Java} is a reduced set of terms and constitutes the description of the document. During registration of a document d, the library stores in the DL repository, a pair (t, d) for each term t appearing in the description of d.

Figure 1 shows an example of repository. The dotted lines indicate the pairs (t, d) of the repository, relating terms from a taxonomy with documents.

A user is represented by an identifier together with a subscription. A subscription is also a reduced set of terms from the taxonomy. When a user registers a subscription S, he is firs notified of all the documents present in the repository that match S, denoted $ans(S)$. Let $tail(t)$ stands for the set of all terms in the taxonomy T strictly subsumed by t, that is $tail(t) = \{s \mid s \preceq t\}$, and R stands for the repository of the digital library. We compute $ans(S)$ as follows:

Case 1: S is a single term t from T, i.e., $S = t$
    if $tail(t) = \emptyset$ **then** $ans(S) = \{d \mid (t, d) \in R\}$
    else $ans(S) = \bigcup \{ans(s) \mid s \in tail(t)\}$
Case 2: S is a conjunction of terms, i.e., $S = t_1 \wedge t_2 \cdots \wedge t_n$
    $ans(S) = ans(t_1) \cap ans(t_2) \cdots \cap ans(t_n)$

As an example, consider the subscription $s = \{\texttt{C++}, \texttt{Sort}\}$. Referring to Figure 1 and applying the above definition, we find $ans(S) = \{d_5, d_6\} \cap \{d_3, d_4, d_6\} = \{d_6\}$.

The answer to a susbscription changes over time, as new documents are inserted in the library, or existing documents are modified or deleted. These changes that occur at the library over time are precisely what we call *events*. An event is represented by the description of the document being inserted, modified, or removed. When an event occurs the system must inform, or *notify* each user whose subscription is matched by the event.

Definition 1 (Matching). *Let e be an event and S a subscription. We say that e matches S if the following holds: (i) A document d is removed and $d \in ans(S)$ before the event occurs. (ii) A document d is inserted and $d \in ans(S)$ after the event has occurred.*

The case of document modification is treated as a deletion followed by an insertion. Clearly, when an event e occurs, the system must decide which users are to be notified. A naive approach is to examine *every* subscription S and test whether e matches S. However, if the set of subscriptions is large, and/or the rate of events is high, the system might quickly become overwhelmed. In what follows, we shall refer to this simplistic approach as NAIVE. We introduce now a more sophisticated solution which relies on a refinement relation among queries.

2.2   Subscription Refinement

Testing whether an event matches a subscription is basically a set membership test (i.e. testing whether a document belongs to a given set of documents). The idea that we exploit here is the following: if we have to perform test membership for every set in a collection of sets, we can save computations by starting with maximal sets first (maximality with respect to set inclusion). Indeed, if a document does not belong to a maximal set then we don't need to test membership for any of its subsets.

In order to implement this idea, we need to define first a notion of refinement between subscriptions. In fact, we need a definition that translates the following intuition: if subscription S_1 refines subscription S_2 then every event that matches S_1 also matches S_2.

Definition 2 (Refinement Relation). *Let S_1 and S_2 be two subscriptions. We say that S_1 is finer than S_2, denoted $S_1 \sqsubseteq S_2$, iff $\forall t_2 \in S_2, \exists t_1 \in S_1 | t_1 \preceq t_2$.*

In other words, S_1 is finer than S_2 if every term of S_2 subsumes some term of S_1. For example, the subscription $S_1 = \{$QuickSort, Java, BubbleSort$\}$ is finer than $S_2 = \{$Sort, OOL$\}$, whereas S_2 is not finer than S_1.

Intuitively, refining a subscription can be seen as imposing tighter matching constraints. There exist two possible ways of doing so: by simply adding some terms to a subscription, or by replacing a term in the subscription by one of its descendants in the taxonomy. Figure 2 shows an example of the refinement relation.

If S is a subscription and d is a document with description D, then $d \in ans(S)$ iff $D \sqsubseteq S$. Referring to the graph of Fig. 2, we can now explain how we exploit the

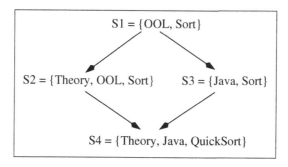

Fig. 2. The refinement relation

refinement relation in our approach: if an event e does not match the subscription S_1, then it cannot match any of the subscriptions S_2, S_3 or S_4. Therefore the idea is to first evaluate e with respect to S_1 (which is the most general subscription): if the matching is successful, then the evaluation continues with respect to both S_2 and S_3 (otherwise evaluation stops); and if the evaluation is successful for either S_2 or S_3, then evaluation continues with respect to S_4 (otherwise evaluation stops). Assuming that, in general, a large fraction of all events will fail to match with any particular subscription, this strategy is expected to save a significant number of computations.

If the sets of terms S_1 and S_2 are reduced sets then the refinement relation just defined becomes a partial order. In fact, it can be shown that the set $\mathcal{S}$ of all subscriptions becomes an upper semi-lattice, as stated in the following proposition.

Proposition 1 ([18]). $(\mathcal{S}, \sqsubseteq)$ *is an upper semi-lattice.*

2.3   The Subscription Tree

The set of user-submitted subscriptions can be organized as a directed acyclic graph that we call the *subscription graph*. Its acyclicity follows from the fact that the refinement relation is a partial order (up to subscription equivalence). However, although the subscription graph is acyclic, it may have several roots (i.e. more than one maximal element). Henceforth, we shall assume that the subscription graph has a single root, by adding to it (if necessary) the lub of all maximal subscriptions.

Now, it turns out that in order to test refinement it is sufficient to maintain one and only one path from the root to every node (the root being the most general subscription). This follows from the observation that if a path to a node S is successful (i.e., an event e matches all the predecessors of S along the path), then *every* path leading to S will be successful as well. As a consequence, it is sufficient to test matching along just one path. As this holds for every node in the subscription graph, it is sufficient to construct a spanning tree [10,17] of the subscription graph in order to be able to test matching for *every* node in the graph.

As there are in general many spanning trees, the question is how to choose the "best" one. To this end, we introduce the notion of filtering rate. The *filtering rate* of a subscription S, denoted $\sigma(S)$, is the probability of an event matching S. It is computed based on the *selectivity* $\sigma(t)$ of a term t, which represents the probability for the description of a document to contain either t or a term subsumed by t. So a "best" spanning tree will be one that optimizes the amount of filtering during a matching test. We call such a tree a minimal spanning tree. From the point of view of our application, selecting the minimal spanning tree is tantamount to selecting, for each subscription S, the most filtering path leading to S. See [9] for more details.

2.4   The Matching Process

Whenever a new event e arrives, the algorithm scans the tree top-down, starting from the root of the tree. The main procedure, MATCH(N), is called recursively and proceeds as follows:

1. if e does not match N, the scan stops; there is no need to access the children of N;
2. else the users of the bucket associated to N can be notified while MATCH is called recursively for each child of N.

The cost of the algorithm is strongly influenced by the average number of children of a node (fanout). If this number is very large, many of the children will not refine an event e, and this results in useless evaluations of the refinement relation. When the fanout of the subscriptions tree decreases the global amount of filtering out increases, and our algorithm is expected to greatly reduce the number of "dead-ends" during the tree traversal.

3   Data Structures and Algorithms

We now detail the construction of the subscription tree and the optimization of the matching process.

3.1   Evaluation of $\sqsubseteq$ and LUB

The evaluation of the refinement relation and the computation of the lub of a set of subscriptions are the basic operations involved in the maintenance of the subscription graph. Given two subscriptions S_1 and S_2, a naive implementation compares each term in S_1 with each term in S_2 and runs in $O(|S_1| \times |S_2|)$, both for $\sqsubseteq$ and LUB. We use an optimized computation which relies on an appropriate encoding of the terms of the taxonomy and avoids the Cartesian product of the naive solution. Its cost is linear in the size of the subscriptions.

Our encoding extends the labelling scheme presented in [2] and further investigated in [7]. In our labelling scheme the successors of each node are assumed to be linearly ordered (by their position, say from left to right), and a node in the taxonomy tree is identified by its position with respect to the parent node. The label $label(t)$ of a term t is obtained by concatenating the label of the parent followed by the the position of the term (with respect to the parent). For example, referring to Figure 1, if 1 is the label of the node `Programming`, then 1.1, 1.2 and 1.3 are respectively the labels of its son nodes, `Theory`, `Languages` and `Algorithms` (see Figure 3). This encoding defines a total lexicographic order $<_l$ on the set of terms.

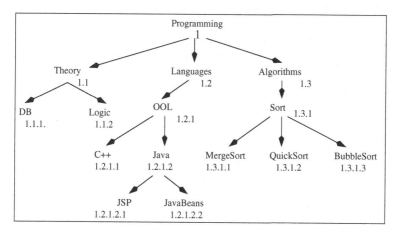

Fig. 3. The taxonomy labelling

Evaluating the subsumption relation $t_1 \preceq t_2$, using this encoding, reduces to checking whether $label(t_2)$ is a prefix of $label(t_1)$. The label of the least upper bound of two terms, $lub_\preceq(t_1, t_2)$, is the longest common prefix of $label(t_1)$ and $label(t_2)$. Recall that since $(T, \preceq)$ is a tree, this least upper bound always exists.

A subscription S is encoded as the list of the labels of its terms, sorted in lexicographic order. For example the subscription {`Theory, Algorithms`} is encoded [1.1,1.3]. The order of the term labels in the labelling of a subscription helps to reduce the number of computations required to evaluate the refinement relation $S_1 \sqsubseteq S_2$, and also to compute the lub $\text{LUB}(S_1, S_2)$, since merge-like algorithms can be applied. Thanks to the technique, $\text{REFINE}(S_1, S_2)$ runs in $O(max(|S_1|, |S_2|))$, and $\text{LUB}(S_1, S_2)$ in $O(|S_1| + |S_2|)$.

3.2 Insertion Algorithm

The insertion algorithm constructs incrementally a tree $\mathcal{T}_S$, where each node consists of a subscription S, associated with a *bucket* containing the set of user identifiers that have subscribed S. Initially the tree consists of the subscription {$root_T$}, containing the root term of the taxonomy, with an empty bucket. When a user u subscribes to S, one first searches the location of S in the tree. Our first goal is to find a location such that the path leading from the root to S enjoys a high filtering rate. A secondary goal is to limit the fanout of the subscriptions tree by clustering the subscriptions that are close to one another. An insertion is performed in two steps:

- **Candidate parent selection**
  A node N in $\mathcal{T}_S$ is a *candidate parent* for S if the following conditions hold: (i) $S \sqsubseteq N$, and (ii) for each child N' of N, $S \not\sqsubseteq N'$, i.e., S strictly refines N but does not refine any child of N.

  The algorithm performs a top-down search, looking for a candidate parent. Starting from the root, it chooses at each level the most selective child which refines N. When such a child no longer exists the candidate parent is found. Note that this is an heuristic which avoids to follow an unbounded number of paths in the tree, but does not guarantee that the "best" candidate parent (i.e., the most selective one) is found.

- **Lub selection**
  Once the candidate parent N is found, the second step inserts S as a child or grandchild of N as follows. First, for each child N' of N, one computes $\text{LUB}(S, N')$ and keeps only those lubs which refine N. Now:

  1. if at least one such lub $l = \text{LUB}(S, N')$ has been found, the most selective one is chosen, and a new subtree $l(S, N')$ is inserted under the parent of N';
  2. else S is inserted as a child of N'.

Note that when the lub of S and a node N' is computed, we know for sure that this lub does not refine any sibling N'' of N', otherwise $\text{LUB}(N', N'')$ would have been inserted in the tree in the first place.

3.3   Removal Algorithm

A leaf in the subscription tree whose bucket becomes empty can be removed (note that an empty *internal* node can still play the role of a filter, and must be kept in the structure). The removal of an empty leaf S, with parent node P, is outlined below:

1. first compute the lub L of the siblings of S (if S has no sibling, then $L = \emptyset$), and remove S from the children of P; then:
2. (a) if S has at least one sibling, the second step depends on the bucket of P: if it is empty, P is replaced by L, else L becomes the child of P.
   (b) else S has no sibling, and P may become an empty leaf in turn if its bucket is empty. The procedure must then be called recursively and bottom-up.

In the worst case, the removal of S may affect all the nodes along the path from S up to the root. Note however that lazy upating can be used (i.e., the adjustment of S's ancestors is not done immediatly), since the tree supports correctly the insert and search operations after step.

4   Experiments

We analyze the behavior of our clustered graph structure, called CMATCH ("clustered matching"), and compare it with the following competitors:

1. NAIVE is the trivial solution which stores the subscriptions in a linear structure.
2. NCMATCH ("non-clustered matching") relies on a subscriptions tree without clustering, i.e., we never introduce the lub of user's subscription during an insertion [1].

The NCMATCH implementation is mostly intended to assess experimentally the gain of the clustering in CMATCH. The impact of the number of users who registered a given subscription S is neutral, because once a subscription that matches an event is found, all the notification variants (CMATCH, NCMATCH and NAIVE) merely scan the bucket of users, sending a notification to each. For clarity we ignore the cost of this specific operation in the report of our experiments, and focus on the cost of finding the set of relevant subscriptions.

The evaluation cost is measured with respect to the following indicators: (i) number of terms comparisons, and (ii) the number of nodes visited for the tree-based solutions CMATCH and NCMATCH. We analyze successively the two major operations: the insertion of new subscriptions ("subscribe") and the search of the subscriptions that match an event ("notify").

4.1   Experimental Setting

The structure has been implemented in Java on a Pentium IV processor (3,000MHz) with 1,024MB of main memory. The implementation conforms to the specifications given in the previous section, expect for the following optimization used in CMATCH and NCMATCH. During the top-down traversal of the subscriptions tree, a same term comparison may have to be carried out repeatedly. Consider the three subscriptions $S_1 = \{\text{OOL, QuickSort, BubbleSort}\}$, $S_2 = \{\text{C++, QuickSort, BubbleSort}\}$, and $S_3 = \{\text{OOL, MergeSort, BubbleSort}\}$. The tree for these subscriptions is shown in Figure 4.

If an event e refines S_1, we know for sure that for each term in S_1, we found a subsumed term in e. We must now evaluate $e \sqsubseteq S_2$ and $e \sqsubseteq S_3$. If one or several of the terms in S_1 are also present in S_2 and S_3, it is useless to search again for a subsumed term in e.

We maintain, at each node N in the graph, a mask of bits which indicates the terms shared by N and one of its ancestors. This is illustrated in Figure 4. The parent node is the subscription $\{\text{OOL, QuickSort, BubbleSort}\}$. The two children share respectively with their parent the terms BubbleSort and QuickSort (left child), and BubbleSort (right child). A bit is set to 1 if the corresponding term is shared with the parent, or to 0. During the matching process we need

---

[1] However, in order to obtain a tree, the subscription $\{root_T\}$, containing the root term of the taxonomy, is always inserted.

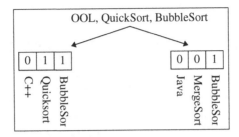

Fig. 4. Avoiding redundant comparisons

to evaluate the term comparison only for the 0-bit terms. This saves 2/3 of the comparisons for the left child in Figure 4, and 1/3 for the right child.

Our experimental setting simulates a Digital Library storing a set of scientific documents described by terms from the ACM Computing Classification System [1] taxonomy. The taxonomy contains 1,316 terms, and its maximal depth is 5.

We also implemented a subscriptions generator. The generator takes as input the average size (number of terms) of the subscriptions, the standard variation, and the cardinality of the subscriptions set. Each subscription is generated as follows: the generator picks up randomly a term and includes this term in the subscription. The process iterates until the subscription contains the required number of terms. Generated subscriptions are reduced: newly generated terms which subsume or are subsumed by one of the terms of the current subscription are ignored. We produced several sets of distinct subscriptions, with a cardinality ranging from 30,000 to 180,000.

4.2   Cost of Subscriptions

The cost of an insertion for the NAIVE approach is negligible, since it consists only in an insertion in a linear structure, performed in constant time. We focus therefore on the comparison of CMATCH and NCMATCH.

Table 1 summarizes the structural properties of the subscriptions tree. The average number of terms in a subscription is 5, with a close clustering around this value (i.e., most subscriptions have 4, 5, 6 or 7 terms). The table gives, for

Table 1. Structure of the subscriptions tree for different subscriptions datasets

Nb subscr	CMATCH			NCMATCH		
	Depth	Insertion level	Avg. fanout	Depth	Insertion level	Avg. fanout
30,000	11	7.04	2.82	4	2.63	655
60,000	13	7.71	2.86	4	2.63	362
90,000	13	7.79	2.88	4	2.63	285
120,000	13	7.82	2.89	5	2.63	239
150,000	13	8.14	2.90	5	2.64	205
180,000	13	8.05	2.92	5	2.65	183

each dataset, the depth of the subscriptions tree, the average insertion level for a new subscription and the average fanout.

The most striking feature is the very large fanout of nodes for the non-clustered solution NCMATCH. This clearly relates to the large number of terms in the taxonomy, which reduces the probability to find a refinement relationship between two *submitted* subscriptions.

As a result one obtains a tree with a few levels, where the refinement relation is sparsely represented. It can be expected that with a very large taxonomy, the tree degenerates to an almost linear structure, where all the subscriptions tend to be a child of the root node. Clearly such a structure looses all the benefits of the approach, since the insertion is costly, yet the amount of filtering remains very low.

On the other hand, the CMATCH structure which clusters the subscriptions and represents these clusters as subtrees rooted by their *lub* achieves a quite significant reduction of the fanout.

Table 2. Cost of insertions for different subscriptions datasets

Nb subscr	CMATCH		NCMATCH	
	V. nodes	Comp.	V. nodes	Comp.
30,000	36	137	10,803	34,493
60,000	45	178	24,072	76,598
90,000	52	188	36,649	116,571
120,000	56	188	49,425	157,352
150,000	64	226	62,425	198,680
180,000	68	229	77,424	212,272

Table 3. Cost of insertions for a fixed subscriptions dataset (size = 120,000 subscriptions) and a variable new subscription size

New subscr size	CMATCH		NCMATCH	
	V. nodes	Comp.	V. nodes	Comp.
4	51	154	44,797	133,108
5	53	186	45,727	139,853
6	66	256	47,327	149,938
7	77	321	50,502	163,951

Table 1 shows that the average number of children of a node is two orders of magnitude lower for CMATCH with respect to NCMATCH. This is clearly a quite desirable property since it reduces both the cost of insertions and the cost of search operations. We made several experiments that vary the size of the subscriptions, and the obtained results show that the above conclusions still hold.

Table 2 shows the cost of inserting a subscription in an existing subscriptions tree. We measure the number of nodes visited by the insertion algorithm, and the number of terms comparisons. Table 2 (illustrated by the curves of the Figure 5), gives the figures for different subscriptions datasets of the subscriptions tree and Table 3 (illustrated by the curves of the Figure 6), shows these measures for a fixed subscriptions dataset (size=120,000 subscription).

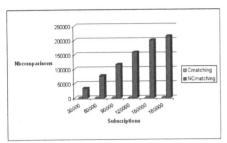
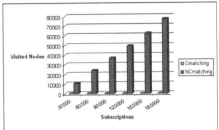

Number of comparisons	Number of nodes visited

Fig. 5. Cost of insertions for different subscriptions datasets

As expected from the trees properties, the gain of the clustered solution is quite impressive. In particular the small number of comparisons which are necessary to insert a new subscription shows that the structure can support a high ratio of updates. This constitutes an important property for a publish/subscribe system.

The results of Table 2 also definitely confirm that the clustering is essential to exploit the refinement relation and filter out most of the comparisons which are made by the non-clustered NCMATCH structure. Again, this is related to the size of the taxonomy. With a smaller one, the refinement relation would be more represented in the subscriptions tree, with a probable reduction of the gap between CMATCH and NCMATCH. However, the main conclusion of our analytical and experimental study on that matter is the high benefit, in all cases, of the lub-based clustering approach.

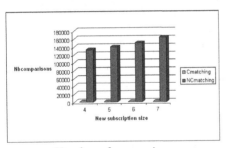
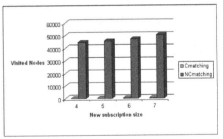

Number of comparisons	Number of nodes visited

Fig. 6. Cost of insertions for a fixed subscriptions dataset (size = 120,000 subscriptions) and a variable new subscription size

4.3   Cost of Notifications

We now turn our attention to the notification process. The results obtained for the three solutions NAIVE, NCMATCH and CMATCH are given in Table 4 and in Table 5 for our 6 subscriptions datasets. We compare both the average number

Table 4. Cost of notifications for different subscriptions datasets (for one event)

Nb subscriptions	CMATCH		NCMATCH		NAIVE	
	Visited nodes	Comp.	Visited nodes	Comp.	Visited nodes	Comp.
30,000	374	1,266	15,893	30,998	30,000	69,120
60,000	473	1,585	31,474	60,027	60,000	133,822
90,000	553	1,821	47,156	90,400	90,000	197,571
120,000	631	2,074	62,543	119,124	120,000	261,807
150,000	690	2,250	77,728	147,452	150,000	324,898
180,000	747	2,421	92,880	175,190	180,000	387,004

Table 5. Cost of notifications for a fixed tree size = 120,000 subscriptions (for one event)

Nb subscriptions	CMATCH		NCMATCH		NAIVE	
	Visited nodes	Comp.	Visited nodes	Comp.	Visited nodes	Comp.
4	350	866	54,590	88,262	120,000	219,178
5	500	1,433	57,502	99,319	120,000	245,020
6	729	2,411	68,392	133,678	120,000	271,644
7	946	3,587	70,046	156,092	120,000	311,642

of nodes visited and the average number of terms comparisons (the latter being more representative of the actual cost) for processing a single event (note that an event generates several notifications in general).

For the naive solution, the number of visited nodes is equal to the number of subscriptions. The gain of the non-clustered solution is not very significant since about 50% of the computation are saved. This is easily explained by the shape of the tree, with a small number of levels and a large fanout. The notification algorithm must visit all the nodes that refine the incoming event, and test all the children of these nodes.

The CMATCH algorithm benefits strongly from the clustering. The lubs play their role of filters and allow to get rid of most of the irrelevant computations. The number of comparisons is more than 100 times smaller than with NAIVE. Table 4 and Table 5 summarize these properties respectively for different subscriptions datasets and for a fixed subscription dataset (=120,000) and variable subscriptions sizes.

Finally Table 6 presents "normalized" cost for one notification (recall that the previous analysis considered the processing cost of one event). We simply divided the number of operations by the number of notifications triggered by an event.

Table 6 unveils an important aspect of the behavior of the algorithms, also clearly illustrated by the curves of the Figure 7. Whereas NCMATCH and NAIVE (the latter not shown on the curve because of the very large values of its figures) exhibit a linear degradation of their computation costs with respect to the size of the subscriptions set, the performances of CMATCH decrease very slowly. Actually the cost of processing required to deliver a notification turns out to be almost constant, and independent from the number of subscriptions. This shows

Table 6. Cost of notifications for different subscriptions datasets (for one notification)

	CMATCH		NCMATCH		NAIVE	
Nb subscriptions	Visited nodes	Comp.	Visited nodes	Comp.	Visited nodes	Comp.
30,000	8.90	30.14	378	738	714	1,645
60,000	9.85	33.02	655	1,250	1,250	2,787
90,000	10.63	35.02	906	1,738	1,730	3,799
120,000	11.27	37.04	1,116	2,127	2,142	4,675
150,000	11.69	38.14	1,317	2,499	2,542	5,506
180,000	12.25	39.69	1,522	2,871	2,950	6,344

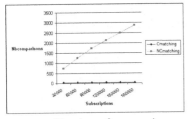
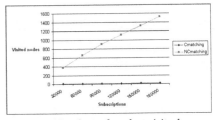

Number of comparisons   Number of nodes visited

Fig. 7. Cost of notifications for different subscriptions datasets

that the pruning effect of the tree is quite effective and removes almost all the unnecessary computations.

5   Related Work

An important amount of work has been devoted in the recent past to designing matching algorithms for large-scale systems. These works apply mostly to relational databases, where subscriptions are seen as conjunctions of predicates (i.e. of relational variables). Their main common ideas can be summarized as follows: (i) if a predicate appears in many subscriptions, avoid its repeated evaluation, and (ii) try to filter out large sets of subscriptions by evaluating first the most selective predicates. A representation approach is that of [8,15,16] which decomposes the matching process in two steps. The first one organizes the subscriptions in the form of a specific structure (e.g. : a tree), whereas the second one uses this specific structure to filter the entering events. Other proposals [3,5,11,19] compare first the attributes values of the subscriptions to the values of the event. The subscriptions matching the event are obtained by counting the number of attributes for which the test is successful.

These algorithms apply mainly to subscriptions made up of attributes whose domains are atomic (integer, string) but do not consider structured domains. [8] partially solves this problem by taking into account a relationship "kind-of". Their algorithm is efficient for the attributes which take only atomic values. However, it does not work for the attributes that can handle multiple values. The same remark holds for systems based on XML [4,6,12] or on RDF [20].

In [13,14], a new model for the publish/subscribe systems: the "state-persist" model. The main contribution of this new model is that it takes into account the notifications history. In [13] the issue of the attributes over tree domains is raised, without further explanation. [14] clearly states that the approach applies mainly to simple domains.

In summary we are not aware of a publish/subscribe technique that considers a keyword-based query language and a subsumption relations over the terms. As shown by our experimental results, the solution proposed in the present paper presents a reasonable storage cost (at most twice the number of subscriptions) and achieves a nice trade-off between the performances of subscriptions (insertion in the structure) and notification (search in the structure).

6 Conclusion

We presented in this paper a simple solution for the notification process for digital libraries, and more generally for applications that rely on keyword-based subscriptions over tree-structured domains (e.g., a taxonomy). Our main contribution is the proposal of a matching algorithm that exploits the specificities of this setting. Our experimental results confirm our original intuition that the proposed tree structure reduces significantly the cost of the notification process.

Future work aims at (i) enriching the structure of a description by adding to it a set of prerequisites necessary for the understanding of the document, (ii) enriching the structure of a subscription by adding to it a set of user backgrounds. After notifying a user about a new document, we plan to check if his backgrounds cover the prerequisites of the document. If yes then there is nothing to add to the notification, else we must inform the user that some prerequisites are needed for the new document and we must propose a set of documents from the repository (if they exist) that match those prerequisites. We are currently studying the influence of such extensions on our notification system.

References

1. The ACM computing classification system. www.acm.org/class
2. Agrawal, R., Borgida, A., Jagadish, H.V.: Efficient management of transitive relationships in large data and knowledge bases. In: Proc. of the Intl. Conf. on Management of Data (SIGMOD), pp. 253–262 (1989)
3. Aguilera, M.K., Strom, R.E., Sturman, D.C., Astley, M., Chandra, T.D.: Matching events in a content-based subscription system. In: Proc. of the ACM Intl. Symposium on Principles of Distributed Computing (PODC), pp. 53–61. ACM Press, New York (1999)
4. Altinel, M., Franklin, M.J.: Efficient filtering of xml documents for selective dissemination of information. In: Proc. of the Intl. Conf. on Very Large Data Bases (VLDB), pp. 53–64 (2000)
5. Campailla, A., Chaki, S., Clarke, E.M., Jha, S., Veith, H.: Efficient Filtering in Publish-Subscribe Systems Using Binary Decision. In: Proc. of the Intl. Conf. on Software Engineering (ICSE), pp. 443–452 (2001)

6. Chirita, P.-A., Idreos, S., Koubarakis, M., Nejdl, W.: Publish/subscribe for rdf-based p2p networks. In: Bussler, C.J., Davies, J., Fensel, D., Studer, R. (eds.) ESWS 2004. LNCS, vol. 3053, pp. 182–197. Springer, Heidelberg (2004)

7. Christophides, V., Plexousakis, D., Scholl, M., Tourtounis, S.: On labeling schemes for the semantic web. In: Proc. Intl. Conf. on World Wide Web (WWW), pp. 544–555 (2003)

8. Fabret, F., Jacobsen, H.A., Llirbat, F., Pereira, J., Ross, K.A., Shasha, D.: Filtering algorithms and implementation for very fast publish/subscribe systems. In: Proc. of the Intl. Conf. on Management of Data (SIGMOD), pp. 115–126 (2001)

9. Frej, H.B., Rigaux, P., Spyratos, N.: User notification in taxonomy based digital libraries. In: Intl. Symposium on Design of Communication (2006)

10. Furer, M., Raghavachari, B.: Approximating the minimum degree spanning tree to within one from the optimal degree. In: Proc. of the ACM-SIAM Symposium on Discrete Algorithms, ACM Press, New York (1992)

11. Gough, J., Smith, G.: Efficient recognition of events in a distributed system. In: Proc. of the Australasian Computer Science Conference (1995)

12. Hou, S., Jacobsen, H.-A.: Predicate-based filtering of xpath expressions. In: Proc. of the IEEE Intl Conf. on Data Engineering (ICDE), p. 53. IEEE Computer Society Press, Los Alamitos (2006)

13. Leung, H.K.Y.: Subject space: a state-persistent model for publish/subscribe systems. In: Proc. of the Conf. of the IBM Centre for Advanced Studies on Collaborative research (CASCON) (2002)

14. Leung, H.K.Y., Jacobsen, H.-A.: Efficient matching for state-persistent publish/subscribe systems. In: Proc. of the Conf. of the IBM Centre for Advanced Studies on Collaborative research (CASCON), pp. 182–196 (2003)

15. Pereira, J., Fabret, F., Llirbat, F., Preotiuc-Pietro, R., Ross, K.A., Shasha, D.: Publish/subscribe on the web at extreme speed. In: Proc. of the Intl. Conf. on Very Large Data Bases (VLDB), pp. 627–630 (2000)

16. Pereira, J., Fabret, F., Llirbat, F., Shasha, D.: Efficient matching for web-based publish/subscribe systems. In: Proc. of the Conf. on Cooperative Information Systems, pp. 162–173 (2000)

17. Pettie, S., Ramachandran, V.: An optimal minimum spanning tree algorithm. Journal of the ACM 49(1), 16–34 (2002)

18. Rigaux, P., Spyratos, N.: Metadata inference for document retrieval in a distributed repository. In: Maher, M.J. (ed.) ASIAN 2004. LNCS, vol. 3321, pp. 418–436. Springer, Heidelberg (2004)

19. Tryfonopoulos, C., Koubarakis, M., Drougas, Y.: Filtering algorithms for information retrieval models with named attributes and proximity operators. In: Proc. of the ACM Intl. Conf. on Research and development in information retrieval (SIGIR), pp. 313–320. ACM Press, New York (2004)

20. Wang, J., Jin, B., Li, J.: An ontology-based publish/subscribe system. In: Proc. of the ACM/IFIP/USENIX Intl. Conf. on Middleware, pp. 232–253 (2004)

Quete: Ontology-Based Query System for Distributed Sources

Haridimos Kondylakis[1,2], Anastasia Analyti[1], and Dimitris Plexousakis[1,2]

[1] Institute of Computer Science, FORTH-ICS, Greece
[2] Department of Computer Science, University of Crete, Greece
{kondylak,analyti,dp}@ics.forth.gr

Abstract. The exponential growth of the web and the extended use of database management systems in widely distributed information systems has brought to the fore the need for seamless interconnection of diverse and large numbers of information sources. Our contribution is a system that provides a flexible approach for integrating and transparently querying multiple data sources, using a reference ontology. Global semantic queries are automatically mapped to queries local to the participating sources. The query system is capable of handling complex join constructs and of choosing the appropriate attributes, relations, and join conditions to preserve user query semantics. Moreover, the query engine exploits information on horizontal, vertical, and hybrid fragmentation of database tables, distributed over the various data sources. This optimization improves system's recall and boosts its effectiveness and performance.

Keywords: Ontology-based data integration, mediation systems, query processing, table fragmentation rules.

1 Introduction

Data Integration is one of the key problems for the development of modern information systems. The exponential growth of the web and the extended use of database management systems has brought to the fore the need for seamless interconnection of diverse and large numbers of information sources. In order to provide uniform access to heterogeneous and autonomous data sources, complex query mechanisms have to be designed and implemented [16]. The design and implementation of a query mechanism is non-trivial because of the heterogeneity of the various components that are going to be queried.

In this paper, we describe an ontology-based mediator system, called *Quete*, that provides a flexible approach for transparently integrating and querying multiple relational data sources. In particular, it provides full location, language, and schema transparency for users, and integrates dynamically heterogeneous (and possibly overlapping) relational data sources in evolving environments. A common *reference ontology* is used across integration domains and the data source-to-ontology annotation process, which follows the Local-as-View approach [3], is performed only once per data source.

Y. Ioannidis, B. Novikov, and B. Rachev (Eds.): ADBIS 2007, LNCS 4690, pp. 359–375, 2007.
© Springer-Verlag Berlin Heidelberg 2007

The motivation for this work was the integration of four database systems, in order to meet the needs of the PrognoChip project [13]. The aim of the project is to identify classification and prognosis molecular markers for breast cancer, through DNA microarray technology. Specifically, our task was to integrate two Clinical Information Systems that store clinical information about patients of two different hospitals and two Genomic Information Systems that store information on DNA microarray experiment settings and results. The objective was to provide a transparent layer that could enhance knowledge extraction and data exchange between these systems. Specifically, this layer should accept ontology-based queries from tools and users, transparently break these queries into local subqueries based on metadata, send the subqueries to the constituent databases, and integrate the returned results.

Our system is an extension of a preliminary and incomplete version of Unity [10, 6], that provides the data source-to-ontology annotation mechanism and a local subquery formation algorithm. In particular, for each relational data source, a local annotator annotates (the interesting to the user) table attributes with paths over the reference ontology, called *semantic names*. We extended the local subquery formation algorithm that is provided by Unity, such that system's recall is increased with no sacrifice in precision. Additionally, we implemented the composition of the local subquery. A novel feature of our system is that horizontal, vertical, and hybrid fragmentation rules about underlying schemata can be declared and used, increasing system's recall and improving performance. In particular, we consider table fragmentation rules, during (i) the formation of the local subqueries, further extending Unity's algorithm, and (ii) during the formation of the result composition plan. This assures that local subqueries are formed and composed in such a way that final results, presented to the user, are as if there was no table fragmentation. Further, system's performance is optimized by eliminating local subqueries and avoiding joins in the result composition plan that are certain to return empty results. We want to note that our approach to data integration is by no means restricted to biomedical informatics. On the contrary, it is completely domain independent.

The rest of the paper is organized as follows: Section 2 reviews related work on mediator-based data integration. In Section 3, our architecture is described, providing details about the data source-to-ontology mappings, the processing of the user semantic query, and the incorporation of the table fragmentation rules into the system. In Section 4, preliminary experimental evaluation of Quete is provided. Finally, Section 5 concludes the paper and gives directions for further research.

2 Related Work

A *mediator* is a system that is responsible for reformulating, at runtime, a user query on a single mediated schema into a composition of subqueries over the local source schemas [7]. To achieve this, a mapping is required that captures the relationship between the local source descriptions and the mediator schema. Specifying this correspondence is a crucial step, as it influences both how difficult query reformulation is and how easily new sources can be added to or removed from the integration system. The two main approaches for establishing the mapping between each source schema and the global schema are the *Global-as-View* (GAV) and the *Local-as-View* (LAV) approaches (for an overview, see [3]). In short, in the GAV approach, each mediator relation is defined in terms of the data source relations, while in the LAV approach, data source relations are defined in terms of the mediator schema. Older projects that follow the GAV approach are TSIMMIS [4] and DISCO [17], while Information Manifold [8] follows the LAV approach.

Ontologies can be used as the global schema and it seems that database integration is currently evolving towards this direction. By accepting an ontology as a point of common reference, naming conflicts are eliminated and semantic conflicts are reduced. Below, we review a few recent ontology-based data integration projects.

In BACIIS [2] and TAMBIS [15], a single conceptualization is provided trying to capture the information in the system data sources. User queries are built and results are returned in terms of this global conceptual schema. However, any change in the sources may require the modification of the global domain conceptualization. Additionally, in TAMBIS, the integration process is restricted to combine data from sources that contain different types of information for the same semantic entity. Thus, the potential overlapping aspect of the sources or the probable incompleteness of some of them is ignored. Moreover, BACIIS only integrates Web Databases and mappings are based on text parsing of web pages.

In ONTOFUSION [1], separate conceptual schemas are used to describe the semantics of each data source. Every concept in a physical database is mapped to a virtual schema. Virtual schemas are ontologies representing the structure of the database at a conceptual level. Then, the various virtual schemas corresponding to the distinct databases are merged into new, unified virtual schemas that can be accessed by the users in order to form their queries. This approach adds more complexity to the whole task, but is otherwise promising.

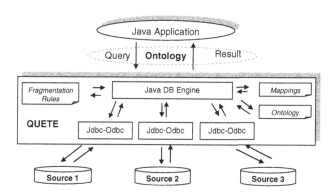

Fig. 1. The integration architecture of Quete

3 Integration Architecture

Quete is an ontology-based mediator system capable of integrating relational databases in evolving environments (participating data sources can change their schema, their semantics, their portion of shared data, or exit without any concern about the other data sources). The architecture of the system is shown in Figure 1. A user (or Java application) issues a semantic query to a *central site*, that is expressed in terms of a reference ontology. Then, the system decomposes the semantic query to one or more local (SQL) subqueries, by taking into account the ontology, the data source-to-ontology mappings, and the table fragmentation rules. The resulting subqueries are issued to the underlying databases, whose answers are sent back to the central cite. Then, the system integrates the results and the final answer is returned to the user. We have to note that no translational or wrapper software is required for the individual data sources, as the central site communicates directly with the data sources, using ODBC protocols.

3.1 Reference Ontology and Semantic Name Formulation

In order to issue semantic queries and to map underlying attributes to a common point of reference, an ontology needs to be employed as a general schema. This *reference ontology* is used to describe schema semantics since it provides standardized names for concepts with unambiguous definitions. The idea is to map the shared schema elements of the sources to the reference ontology.

In Quete, the reference ontology is organized as a graph of concepts that are related through two types of relationships; *"IS-A"* relationships and *"HAS-A"* relationships. *"IS-A"* relationships are used to model generalization/specialization of concepts, while *"HAS-A"* relationships are used to model component relationships. Each concept can have an associated set of attributes. Additionally, there is a special class of concepts, called *relationship concepts*, modeling generic relationships. To represent ontologies like these, we could use the web-ontology languages RDFS [5] and OWL [11], whereas it is important to understand that the exact organization of the concept hierarchy and the terms used to represent concepts is irrelevant as long as they are agreed upon. Although this is a rather simple modeling mechanism, we believe that it is adequate for capturing real-world schemas, since any general relationship can be represented as a concept.

Using a reference ontology, we can form the *semantic name of a table* as:

$$SN = [CN_{path}] = [CN_1; ...; CN_m],$$

and the *semantic name of an attribute* as: $SN = [CN_1; ...; CN_m] AN$,

where CN_i, $i = 1, ..., m$, are concepts of the reference ontology and AN is a (possibly inherited) attribute of concept CN_m. The semi-colon between concepts CN_i and CN_{i+1} means that concept CN_i *"HAS-A"* (generalization of) concept CN_{i+1}, $i=1, ..., m$-1. Intuitively, the semantic name of a schema element (table or attribute) captures its semantics w.r.t. the reference ontology.

We say that a semantic name $[CN_1; ...; CN_m]$ is *subsumed by* a semantic name $[CN'_1; ...; CN'_{m'}]$ if (i) $m' \leq m$, (ii) $CN_{m-m'+i}$ coincides with or is a specialization of CN'_i, $i=1, ..., m'$. Moreover, we say that a semantic name $[CN_{path}] AN$ is *subsumed by* a semantic name $[CN'_{path}] AN'$, if (i) $[CN_{path}]$ is subsumed by $[CN'_{path}]$, and (ii) $AN=AN'$. Intuitively, a semantic name SN is subsumed by a semantic name SN', if its semantics is the same or more specific than the semantics of SN'.

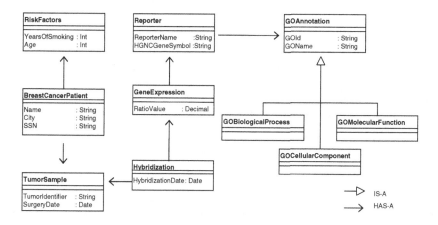

Fig. 2. An example biomedical ontology in UML

We say that two semantic names are *semantically overlapping*, if (i) their last i (for an $i > 0$) concept names (*CN*) are the same or related through the "*ISA*" relationship, and (ii) they have the same attribute name (*AN*). Additionally, we say that two schema elements are *semantically overlapping*, if their corresponding semantic names are semantically overlapping. Obviously, if a semantic name *sn* is subsumed by a semantic name *sn'* then *sn* and *sn'* are semantically overlapping.

As a running example, consider the simple biomedical ontology of Figure 2 that describes the following scenario: Each *Patient* with breast cancer is associated with some *Risk Factors*, such as smoking habits, age, etc. After undergoing surgery, her *Tumor* is removed and sent to a molecular biology laboratory for taking part in a gene expression profiling experiment, based on the (spotted) DNA microarray technology[1]. In these experiments, fragments of genes, called *Reporters*, are spotted on a microarray slide, which is *hybridized* with the cancerous tissue and a "normal" tissue. Hybridization results are analyzed to generate *Gene Expression* data, expressing through a *Ratio Value* (per spotted gene), if the gene in the cancerous tissue is over-expressed, under-expressed, or equally expressed with respect to the "normal" tissue. In other words, the gene expression profile of the tumor is compared with that of a "normal" tissue. *Gene Ontology (GO)* is a well-known ontology for the annotation of gene products in terms of the biological processes in which they participate, the particular molecular functions that they perform, and the cellular components in which they act (see http://www.geneontology.org). In particular, GO consists of 3 independent taxonomies, namely *GO Biological Process*, *GO Molecular Function*, and *GO Cellular Component*. In our example ontology, reporters are annotated with the GO terms (i.e., GO ids and their corresponding GO names) that characterize the gene products of the gene that the reporter is part of.

Note that according to our example ontology, *GeneExpression* is a relationship concept, whose instances relate a hybridization with a reporter and a ratio value, and *RatioValue* is an attribute of *GeneExpression*. Adittionally, note that (i) [*Hybridization; TumorSample*] *TumorIdentifier* is a valid semantic name, representing the

---

[1] http://www.ncbi.nlm.nih.gov/About/primer/microarrays.html

tumors participating in hybridization experiments, and (ii) [*Reporter; GOBiological-Process*] *GOName* is a valid semantic name that is subsumed by (and semantically overlaps with) [*GOAnnotation*] *GOName*.

Assume now that we want to integrate a Clinical and a Genomic database, using our example ontology. In Table 1, we give the semantic names of some of the Clinical and Genomic database schema elements (tables/attributes). Note that the atrributes *SurgicalExcision.TumorSampleId* and *Hybridization.CancerousTissue* are semantically overlapping. Of course, internal database IDs, such as *BreastCancerPatient.PatientId*, do not have semantic names.

Table 1. Clinical and Genomic database schema elements and their semantic names

Type	System Name	Semantic Name	
Table	**RiskFactors**	**[BreastCancerPatient; RiskFactors]**	
Attribute	PatientId		
Attribute	Age	[BreastCancerPatient; RiskFactors] Age	
Attribute	YearsOfSmoking	[BreastCancerPatient; RiskFactors] YearsOfSmoking	Clinical Database Schema
Table	**BreastCancerPatient**	**[BreastCancerPatient]**	
Attribute	PatientId		
Attribute	Name	[BreastCancerPatient] Name	
Attribute	City	[BreastCancerPatient] City	
Attribute	SSN	[BreastCancerPatient] SSN	
Table	**SurgicalExcision**	**[BreastCancerPatient; TumorSample]**	
Attribute	PatientId		
Attribute	TumorSampleId	[BreastCancerPatient; TumorSample] TumorIdentifier	
Attribute	SurgeryDate	[BreastCancerPatient; TumorSample] SurgeryDate	
Table	**Hybridization**	**[Hybridization]**	
Attribute	HybridizationId		
Attribute	CancerousTissue	[Hybridization; TumorSample] TumorIdentifier	
Attribute	Date	[Hybridization] HybridizationDate	
Table	**GeneExpressionData**	**[GeneExpression]**	Genomic Database Schema
Attribute	HybridizationId		
Attribute	ReporterId	[GeneExpression; Reporter] ReporterName	
Attribute	Value	[GeneExpression] RatioValue	
Table	**Reporter**	**[Reporter]**	
Attribute	ReporterId	[Reporter] ReporterName	
Attribute	GeneSymbol	[Reporter] HGNCGeneSymbol	
Table	**ReporterGOMolFun**	**[GOMolecularFunction]**	
Attribute	ReporterId	[GOMolecularFunction;Reporter] ReporterName	
Attribute	GOId	[GOMolecularFunction] GOId	
Attribute	GOName	[GOMolecularFunction] GOName	

The *data source-to-ontology annotation* phase is used to capture the data to be integrated and is performed independently in each local data source. The *Extractor* tool of Unity [10, 6] is used to extract the underlying database schema (i.e., tables, attributes, foreign keys, primary keys). The extracted schema is then stored in a specific XML file, called *X-Spec*. Then, the administrator selects the schema elements (tables/attributes) that are going to be shared, and annotates the interesting to the user schema elements with semantic names over the reference ontology. Subsequently, the final X-Spec is sent to the central site. Note that the data source-to-ontology annotation follows the Local-as-View approach [3]. Obviously, a data source is capable to change the schema, the semantics, and the portion of the data to be shared by just altering its X-Spec file.

3.2 Query Processing in Quete

After specification of the X-Spec files, the user is given the capability to issue semantic queries. The query language is an attribute-only language similar to SQL, where the *SELECT* clause contains the terms to be projected in the final results and the optional *WHERE* clause specifies selection criteria for the query. Continuing our running example, a user query, q_{user}, requesting *"the expression ratio value of genes, whose molecular function is "cell adhesion" and are expressed in (the tumor of) breast cancer patients that (i) have smoked for more than 30 years and (ii) their surgery date is the same with the date of the hybridization experiment"* is:

```
SELECT  [BreastCancerPatient]Name, [Reporter]HGNCGeneSymbol,
        [GeneExpression]RatioValue
WHERE   [RiskFactors]YearsOfSmoking>30 AND
        [Hybridization]HybridizationDate = [TumorSample]SurgeryDate
        AND [Reporter;GOMolecularFunction]GOName="cell adhesion"
ORDERBY [BreastCancerPatient]Name
```

Of course, the user must express the query terms through their semantic names. Notice that, in the semantic query, the *FROM* clause is absent. This is because, query concepts appear within semantic names. During query processing, the system automatically identifies the local tables that will be accessed and the joins that are needed to link together the various interesting to the user parts of distributed information. For the time being, Sub-selects and Union operations are not supported in the query language, while Order and Group operations are.

After submission, the user semantic query is translated to SQL subqueries that are issued in parallel to the underlying data sources. There are three major requirements in forming a local subquery:

Requirement 1: The system must identify the interesting to the user table attributes. We consider that a table attribute $T.A$ with semantic name $[CN_{path}]$ AN *is interesting to the user*, in the following cases:

Case 1: $[CN_{path}]$ AN is semantically subsumed by a semantic name *sn* appearing in the user query. Thus, the attribute $T.A$ carries the same of more specific information than that requested by the user.

Case 2: $[CN_{path}]$ AN semantically overlaps with a semantic name *sn* appearing in the user query and it exists a table T' whose semantic name $[CN'_{path}]$ is such that (i) the tables T and T' share the same primary key, and (ii) $[CN'_{path}]$ AN is subsumed by $[CN_{path}]$ AN and *sn*. Case 2 captures a common representation of the "IS-A" relationship in relational databases. Intuitively, the meaning of table T' is a specialization of that of T, and thus, T' inherits attribute A from T.

Case 3: there is a sequence of tables $(T_1, ..., T_m, T)$, such that (i) neighbour tables in the sequence can be joined based on foreign key information, and (ii) the semantic names $[CN^1_{path}], ..., [CN^m_{path}]$ of $T_1, ..., T_m$, respectively, are such that the semantic name $[CN^1_{path}; ...; CN^m_{path}; CN_{path}]^2$ AN is subsumed by a semantic name *sn* appearing in the user query. The idea behind this case is that the information requested by the user, as expressed in *sn*, can be retrieved through the appropriate joins between local tables with related semantics.

Requirement 2: As the *FROM* clause is missing from the user semantic query, the required intermediate tables, linking the tables with interesting to the user attributes (and in Cases 2 and 3, above, also the tables T' and $T_1, ..., T_m$, respectively) must be determined by the system. Additionally, the system must determine the conditions that are required to join these tables.

Requirement 3: Assume that *s*, *s'* are two local sources such that the interesting to the user attributes, identified in these sources, do not semantically overlap. Then, the local subqueries submitted to *s* and *s'* should provide the (join) attributes, called *DB link attributes*, that are needed to link the interesting to the user attributes that are provided by the two data sources.

To achieve these requirements, our system inspects the reference ontology and the XSpec files (for retrieving the data source-to-ontology mappings and information on foreign keys). The (intermediate) join tables, needed to satisfy Requirement 2 above, are decided based on an algorithm that is provided by Unity [6], which links a set of tables by forming a local join tree. A *local join tree* is a connected undirected graph whose nodes correspond to database tables and there is a link between two nodes if there is a join (based on foreign keys) between the corresponding two tables. All identified tables are joined by (i) enforcing equality on their attributes that correspond to pairs ⟨foreign key, primary key⟩, and (ii) enforcing the conditions in the *WHERE* clause of the user query that can be checked locally.

To satisfy Requirement 3, our system proceeds as follows: Let *s*, *s'* be two local sources such that the interesting to the user attributes, identified in these sources, do not semantically overlap. Let *TS* and *TS'* be the sets of tables of *s* and *s'*, respectively, that have been identified in order to satisfy Requirements 1 and 2, above. Our system inspects *TS* and *TS'*, and if it finds a pair of attributes $T.A$ and $T'.A'$ ($T \in TS$ and $T' \in TS'$) that are semantically overlapping then it includes them in the local subqueries to *s* and *s'*, respectively. These attributes will be used for joining the local subqueries to *s* and *s'*, at result composition time.

Based on the above analysis, our local subquery formation algorithm builds one SQL subquery, for each data source. This SQL subquery provides all the attributes of the local source that are interesting to the user, plus the DB link attributes (if any). We want to note that Unity considers as interesting to the user only the attributes whose semantic names appear in the user query. This is a subcase of Case 1, above. Thus, in Unity, an attribute whose semantic name semantically overlaps, but is not the same, with a semantic name in the user query is ignored. Moreover, Unity does not take care of Requirement 3.

When the results from the local subqueries are returned to the central site, they are composed based on a *result composition plan*, that is formed using Join, Union, and Projection operations, while, if needed, operations Group and Order are applied at the

---

[2] If two neighbour concepts names in $CN^1_{path}; ...; CN^m_{path}; CN_{path}$ are the same then only one is kept.

end. After execution of this composition plan on the local subquery results, final results are presented to the user.

Continuing our running example, consider a Clinical database (DB1) with tables *RiskFactors, BreastCancerPatient,* and *SurgicalExcision,* and a Genomic database (DB2) with tables *Hybridization, GeneExpressionData, Reporter,* and *ReporterGO-MolFun.* After describing the attributes of these tables using our example ontology, as shown in Table 1, the corresponding X-Spec files are sent to the central site. Assume now that the user query q_{user} is issued.

Checking the corresponding X-Spec files, it is concluded that the attributes *BreastCancerPatient.Name,* *RiskFactors.YearsOfSmoking,* *SurgicalExcision.SurgeryDate,* and *SurgicalExcision.TumorSampleId* should appear in the *SELECT* clause of the Clinical database subquery. The first attribute is needed to be presented to the user, the second and third to satisfy the condition in the *WHERE* clause of the user query, and the fourth (which is a DB link attribute) is needed to join this local subquery with that submitted to the Genomic database. The local join tree that connects the tables *BreastCancerPatient, RiskFactors,* and *SurgicalExcision* is: *RiskFactors – BreastCancerPatient – SurgicalExcision.*

Similarly, the attributes *Reporter.GeneSymbol, GeneExpressionData.Value, Hybridization.Date, ReporterGOMolFun.GOName,* and *Hybridization.CancerousTissue* should appear in the *SELECT* clause of the Genomic database subquery and the corresponding local join tree is: *Hybridization – GeneExpressionData – Reporter – ReporterGOMolFun.*

Then, the system forms the following two subqueries, q_1 and q_2, that will be issued to the local databases DB1 and DB2, respectively.

```
DB1: SELECT  BreastCancerPatient.Name, RiskFactors.YearsOfSmoking,
             SurgicalExcsion.SurgeryDate, SurgicalExcsion.TumorSampleId
     FROM    RiskFactors, BreastCancerPatient, SurgicalExcision
     WHERE   RiskFactors.YearsOfSmoking>30 AND
             RiskFactors.PatientId=BreastCancerPatient.PatientId AND
             BreastCancerPatient.PatientId=SurgicalExcision.PatientId
DB2: SELECT  Reporter.GeneSymbol, GeneExpressionData.Value,
             Hybridization.Date, ReporterGOMolFun.GOName,
             Hybridization.CancerousTissue
     FROM    Hybridization, GeneExpressionData, Reporter,ReporterGOMolFun
     WHERE   Hybridization.HybridizationId =
             GeneExpressionData.HybridizationId AND
             GeneExpressionData.ReporterId = Reporter.ReporterId  AND
             Reporter.ReporterId= ReporterGOMolFun.ReporterID AND
             ReporterGOMolFun.GOName= "cell adhesion"
```

The above subqueries q_1 and q_2 are issued in parallel to the Clinical and Genomic database, respectively, using threads. Note that, the condition [*RiskFactors*]*YearsOfSmoking*>30 (resp. [*Reporter;GOMolecularFunction*]*GOName*="cell adhesion"), appearing in the *WHERE* clause of the user query q_{user}, is checked in q_1 (resp. q_2). However, the condition [*Hybridization*]*HybridizationDate* = [*TumorSample*]*SurgeryDate* cannot be checked in a local subquery alone.

Our algorithm for forming the result composition plan is given below (Alg. 3.1). We want to note that Alg 3.1 does not have to search the XSpec files or the reference ontology, as all the information that it requires from these sources has already been retrieved during the formation of the local subqueries. In Section 3.3, we describe how Alg 3.1 is extended, in the case that knowledge on table fragmentation exists.

Algorithm 3.1. *ResultCompositionPlan(q_{user}, QS)*
Input: (i) the user semantic query, q_{user}, and (ii) the local subqueries QS = {q_1, ..., q_n}
Output: Composition plan for the local subquery results

1: *For i := 1, ..., n do* {
    Let the *projection attributes* of q_i be the attributes returned by q_i;
    } /* *End For* */

2: Let {S_1, ..., S_k} be all the *minimal* subsets of QS such that:
    (i) for each semantic name *sn* appearing in q_{user}, there is a projection attribute
        of a subquery in S_i, whose semantic name semantically overlaps with *sn*,
    (ii) for each subquery in S_i, there is a projection attribute which does not
        semantically overlap with a projection attribute of another subquery in S_i, and
    (iii) there is *global join tree* connecting all subqueries of S_i, i = 1, ..., k;

3: *For i := 1, ..., k do* {
    3.1 Join the local subqueries in S_i, by forcing equality on their semantically
        overlapping projection attributes and applying the user specified conditions,
        not already applied to the local subqueries in S_i;
    /* *Join semantically overlapping subqueries*
    and apply user specified conditions */
    3.2 Project the new subquery p_i to the attributes whose semantic names semantically
        overlap with these in the *SELECT, GROUPBY,* and *ORDER* clause of q_{user};
    } /* *End For* */
4: Union resulting subqueries p_i' (for i = 1, ..., k) by aligning their semantically overlapping
    attributes; /* *Union p_i', i = 1, ..., k and produce a single subquery* */
5: Apply to the resulting subquery, the Group and Order operations (if any) indicated in q_{user};
6: *Return* the formed result composition plan, after usual optimization techniques are
    applied;

In Alg. 3.1, input local subqueries are placed in *QS*. The *projection attributes* of each local subquery *q* are defined as the attributes returned by *q* (Step 1). Local subqueries having projections attributes that are semantically overlapping are called *semantically overlapping*. A *global join tree* is a connected undirected graph whose nodes correspond to local subqueries and there is a link between two nodes, if their corresponding local subqueries are semantically overlapping. We consider that semantically overlapping subqueries represent information that overlaps on their semantically overlapping projection attributes, and thus they should be joined.

Step 2 computes all the *minimal* subsets S_i, i= 1, ..., k, of *QS* such that: (i) for each semantic name *sn* appearing in the user query, there is a projection attribute of a subquery in S_i, whose semantic name semantically overlaps with *sn*, (ii) for each subquery in S_i, there is a projection attribute which does not semantically overlap with a projection attribute of another subquery in S_i, and (iii) there is a *global join tree* connecting all subqueries in S_i. The reasoning behind Step 2 is that the answers returned to the user should be correct and provide all requested attributes.

Then, local subqueries in S_i, i= 1, ..., k, are joined by (i) forcing equality on their semantically overlapping projection attributes, and (ii) applying the user specified conditions, not already applied to the local subqueries in S_i (Step 3.1). Each resulting subquery p_i, i= 1, ..., k, is projected to the attributes whose semantics names semantically overlap with these in the *SELECT, GROUPBY,* or *ORDER* clause of the user query (Step 3.2). Subsequently, all resulting subqueries p_i', i= 1, ..., k, are unioned (Step 4). Finally, Group and Order operators are applied, as indicated in the user query (Step 5).

Note that due to conditions (i) and (ii) of Step 2, it holds that for each $i = 1, ..., k$, $|S_i| \leq r$, where r is the number of different semantic names appearing in the user query. Additionally, note that although there is no $j \neq i$, such that $S_i \subseteq S_j$, for $i, j = 1, ..., k$, local subquery sets S_i can be overlapping. Thus, the result composition plan can be optimized by first joining common subqueries and then, proceeding to further joins that use the result. Alg. 3.1 returns the optimized result composition plan (Step 6).

Continuing our running example, the set $S_1 = \{q_1, q_2\}$, where q_1 and q_2 are the local subqueries submitted to DB1 and DB2, respectively, satisfies the conditions of Step 2 of Alg 3.1. Thus, subqueries q_1 and q_2 are joined based on the conditions (i) *SurgicalExcision.TumorSampleId = Hybridization.CancerousTissue*, and (ii) *SurgicalExcision.SurgeryDate = Hybridization.Date* (see Step 3.1). Note that condition (i) is due to the fact that attributes *SurgicalExcision.TumorSampleId*, *Hybridization.CancerousTissue* are semantically overlapping, while condition (ii) is due to the user specified condition *[Hybridization]HybridizationDate = [TumorSample]SurgeryDate*. The resulting query p_1 is then projected to the attributes *BreastCancerPatient.Name*, *Reporter.GeneSymbol*, and *GeneExpressionData.Value* that are interesting to the user (see Step 3.2) and results are ordered based on *BreastCancerPatient.Name* (See Step 5).

The composition of the local subquery results is done with the help of a central DBMS and consists of the following five steps:

1. For every subquery issued to a local data source, design the temporary table that will be constructed in the central database to store the returned results.
2. Build these tables in the central (lightweight) database.
3. Build the global SQL query that will be issued to the central database, according to the result composition plan formed by Alg. 3.1 (in the result composition plan, local subqueries are replaced by their corresponding temporary tables).
4. Store the local subquery results into the temporary tables, created in Step 2.
5. Execute the global SQL query (formed in Step 3) over the temporary tables, get the results, and present them to the final user.

The first four steps are executed in *parallel*, using threads. As the global SQL query is issued to the central database, Join, Union, Order, and Group operations are executed by the central DBMS.

3.3   Considering Table Fragmentation Rules

In Quete, we take into account information on horizontal, vertical, and hybrid table fragmentation [12], where table fragments are distributed over the various data sources. This consideration improves system's recall and optimizes performance.

In our case, horizontal fragmentation is based on a *defining condition*, which is a boolean expression (in conjunctive normal form) of simple conditions: "*Attr comparison-operator Value*", where *Attr* is an attribute, *comparison-operator* is one of $<$, $>$, $=$, $\leq$, $\geq$, $\neq$, and *Value* is a number, or string. For example, consider the table *BreastCancerPatient(PatientId, Name, City, SSN)*. A horizontal fragmentation of this table could result to two fragments. The first fragment contains tuples with $PatientId < 500$ and is stored in DB1, while the second contains tuples with $PatientId \geq 500$ and is stored in DB2. Moreover, the system supports vertical fragmentation. For example, vertical fragmentation of the table *BreastCancerPatient* could result to three fragments. The first fragment with attributes *PatientId* and *Name* is stored in DB1, the second with attributes *PatientId* and *City* is stored in DB2, and the third with attributes *PatientId* and *SSN* is stored in DB3. We also support *hybrid fragmentation*, which is a tree-structured partitioning of a table, formed by successive horizontal and vertical fragmentations, or vice-versa.

The tables that result after the application of horizontal, vertical, and hybrid fragmentation are called horizontal, vertical, and hybrid fragments, respectively. Hybrid fragments are *associated with a condition* which is the conjunction of the defining conditions of the horizontal fragmentations through which they have been derived (if any) or *true*. The rules that describe table fragmentation are called *table fragmentation rules* and stored in an XML file in the central site.

Extending our local subquery formation algorithm, we consider table fragmentation rules during the formation of the local subqueries. Specifically, assume that (N,E) is the local join tree, identified for forming a local subquery in the case that no knowledge on table fragmentation exists. Then, if an $f \in N$, refers to a table fragment, our algorithm forms one subquery q_f querying only f, and another subquery q_r querying the rest of the tables in N, keeping in the *SELECT* clause of q_r the foreign key that will connect q_f with q_r, at result composition time. Table fragmentation rules are also checked during the formation of the result composition plan. The reason is that fragments of the same table t (containing information interesting to the user) should be composed through the appropriate Join and Union operations before any further processing, so that final results (presented to the user) are the same as if t had never been fragmented.

An additional benefit from using *horizontal* and *hybrid fragmentation rules* is that the system may be able to predict that the execution of certain subqueries on table fragments will return empty results, and thus avoid their execution, optimizing performance. In Quete, horizontal and hybrid fragmentation rules are checked, each time a generated local subquery refers to a horizontal or hybrid table fragment. If the fragment's associated condition conflicts with that in the *WHERE* clause of the local subquery then the subquery is discarded.

Continuing our running example, consider another Clinical database (DB4) with the table *Breast_cancer_patient(PatientId, Name, City, SSN, Age, YearsOfSmoking)*, but with no information about surgeries. Thus, our integrated environment now includes the sources DB1, DB2, DB3, and DB4.

Assume now that a table *PatientWithBreastCancer*, containing information about breast cancer patients in the whole prefecture of Heraklion, has been (at some time) fragmented horizontally into two fragments. The first with *defining condition= <City = "Iraklion">* is stored in DB1 and the second with *defining condition= <City ≠ "Irakion">* is stored in DB4 and is the table *Breast_cancer_patient*. Assume now that the first fragment is further fragmented vertically (for reasons of perfomance) into two fragments that are the tables *BreastCancerPatient* and *RiskFactors*. Note that all tables *BreastCancerPatient*, *RiskFactors*, and *Breast_cancer_patient* are hybrid fragments of the same table, the first two with associated condition *<City = "Iraklion">* and the third with associated condition *<City ≠ "Iraklion">*.

Assume now that the user poses the semantic query q_{user} of our running example. If no knowledge about this hybrid fragmentation exists, information about tumor surgeries (and thus, about the gene expression profiles) of breast cancer patients that do not live in the city of *Iraklion* will be lost. This is because, (i) all projection attributes of the local subquery, issued to DB4, semantically overlap with a projection attribute of the local subqueries q_1 and q_3, and (ii) there is no way to link the clinical information that is stored in DB4 with the genomic information stored in DB2. If, however, knowledge about this hybrid fragmentation exists, the system can combine the table fragments to reconstruct the original table *PatientWithBreastCancer* and then, take its join with the table *SurgicalExcision* of DB1 (based on foreign key information). This way there will be no information loss, and interesting gene expression information about breast cancer patients, living in the whole prefecture of Heraklion, will be retrieved.

To support knowledge on horizontal, vertical, and hybrid table fragmentation, Alg. 3.1 is extended by adding after Step 1, the statement *QS* = *CombineFragments(QS)*. The algorithm *CombineFragments(S)* (Alg. 3.2) takes as input a set of local subqueries *S* and combines the table fragments in *S*, so as to reconstruct the original tables. Then, it joins the original tables *t* with the other subqueries issued to the local sources of the fragments of *t*. Thus, there will be no information loss due to the fact that table fragments are distributed over the various data sources.

Algorithm 3.2. *CombineFragments(S)*
Input: a set of local subqueries $S=\{q_1,...,q_n\}$
Output: a new set of subqueries that recombine fragmented tables and proceed to useful joins

1: *For i* :=1,...,*n do* {
    Let $DBs(q_i)$ be a singleton set with the local source of q_i;
    } */* End For */*
2: Let *FS* be the local subqueries in *S* that refer to table fragments;
3: Let $\{S_1, ..., S_k\}$ be maximal subsets of *FS* such that all subqueries in S_i, for *i* =1, ..., *k*, refer either (i) to vertical fragments of the same table or (ii) to hybrid fragments of the same table such that their associated conditions are not conflicting;
4: *For i* :=1, ..., *k do* {
    Join the subqueries in S_i, by forcing equality on their primary key;
    Let the *projection attributes* of the resulting subquery *q* be the union of the projection attributes of the subqueries in S_i, keeping from the common primary keys only one;
    $FS := (FS - S_i) \cup \{q\}$; $DBs(q) := \cup\{DBs(q') \mid q' \in S_i\}$;
    } */* End For */*
5: Let $\{S_1, ..., S_k\}$ be maximal subsets of *FS* such that all subqueries in S_i, for *i* =1, ..., *k*, refer either to horizontal fragments of the same table or to hybrid fragments of the same table;
6: *For i* :=1,...,*k do* {
    Take the union of the subqueries in S_i;
    Let the *projection attributes* of the resulting subquery *q* be the projection attributes of one of the subqueries in S_i;
    $FS := (FS - S_i) \cup \{q\}$; $DBs(q) := \cup\{DBs(q') \mid q' \in S_i\}$;
    } */* End For */*
7: *FS'* :=*FS*;   *S'* :=*S* - *FS*;   */* Initialize FS' and S' */*
8: *For* each subquery *q* $\in$ *FS do* {
    / Join reconstructed tables with other local subqueries */*
    For each subquery *q'* $\in$ *S* - *FS do* { */* q' is a local subquery */*
        If $DBs(q') \subseteq DBs(q)$ *then* {
        Join *q'* and *q* based on foreign key information between (i) the tables appearing in *q'* and (ii) the table fragments appearing in *q* and stored in $DBs(q')$;
        Let the *projection attributes* of the resulting subquery *q"* be the union of the projection attributes of *q* and *q'*, removing the foreign key used in the join;
        $FS' := FS' - \{q\}$; $S' := (S' - \{q'\}) \cup \{q''\}$; }
    } */* End For */*
    } */* End For */*
9: *Return(S'* $\cup$ *FS'*);

In Step 4, the vertical and hybrid fragmentation rules are considered and all (i) vertical fragments of the same table, and (ii) hybrid fragments of the same table, such that *their associated conditions are not conflicting*, are joined. Note that subquery sets S_i, for $i = 1, \ldots, k$, can be overlapping if they refer to hybrid fragments of the same table. Additionally, note that all subqueries in S_i, for $i = 1, \ldots, k$, are local and, as they refer to vertical and hybrid fragments of the same table, have common primary keys.

In Step 6, only the horizontal fragmentation rules are considered and all horizontal fragments of the same table are unioned. Additionally, in Step 3.2, all subqueries created in Step 3.1 that refer to hybrid fragments of the same table are unioned (fully reconstructing the original table). Note that in Step 6, all subquery sets S_i, for $i=1, \ldots, k$, are pairwise disjoint and all subqueries in S_i, for $i=1, \ldots, k$, have common projection attributes.

Thus, vertically fragmented tables are reconstructed through Step 4, horizontally fragmented tables are reconstructed through Step 6, and hybrid-fragmented tables are reconstructed through both steps 4 and 6.

Then, Alg. 3.2 joins the subqueries that reconstruct the original tables t with the other subqueries issued to the local sources of fragments of t. The join is based on foreign key information, identified during the formation of the local subqueries. In our example, the subquery that reconstructs the original table *PatientWithBreastCancer* is formed by (i) a join on the fragments stored in DB1 (by enforcing equality of their *PatientId* attributes), and (ii) a subsequent union of the result with the fragment stored in DB4. The resulting subquery is then joined with the second subquery submitted to DB1: SELECT PatientId, TumorSampleId, SurgeryDate FROM SurgicalExcision by forcing equality on the *PatientId* attributes of the two subqueries.

Finally, we would like to mention that considering table fragmentation rules, unnecessary joins (in Step 3.1 of Alg 3.1) between local subqueries, that refer to fragments of the same table and have conflicting associated conditions, are avoided. Thus, performance is improved.

4   Preliminary Performance Evaluation

To evaluate Quete, we performed several experiments. Here, we present the evaluation conducted for the needs of the PrognoChip project. Our resources were limited, thus we used only four machines with an Intel Pentium 4 processor on 3.4 GHz, and 2 GB of RAM. In particular, for our experiments, we used four *SQLServer* local databases (two Clinical and two Genomic). The two Clinical databases capture patient clinical information of two Hospital environments. The first Genomic database is dedicated to DNA microarray experiment settings, and the second one to the storage of gene expression profiles of tumors, participating in microarray experiments. In these two Genomic databases, there exist two horizontal fragments of the same table. The schema of each database was annotated using a reference ontology, designed by us. In our study, we issue a user semantic query that uses around 10 tables from each database. Our central database also uses *SQLServer* and resides at the same machine as one of the Clinical databases. To evaluate our system, we have built a benchmark program that loads local tables with a prefixed number of rows.

In Figure 3, the performance results of four experiments are shown. In the first two experiments (denoted by, `3DBs fragment`, `3DBs no fragment`), we used three databases (one Clinical and two Genomic), with and without considering table fragmentation rules, respectively. In the last two experiments (denoted by, `4DBs fragment`, `4DBs no fragment`), we used all four databases (two Clinical and two Genomic), again with and without considering table fragmentation rules, respectively. We see that the whole system has an acceptable performance, even with large data sets and complex queries. Moreover, knowledge on table fragmentation improves the performance of the system.

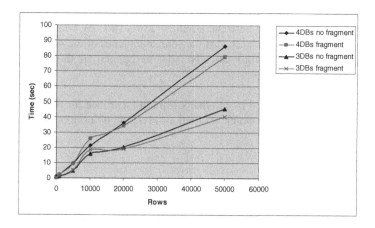

Fig. 3. Performance of Quete with and without knowledge on table fragmentation

In particular, let us define the *elapsed time* of a local subquery q, ET_q, as the time from issuing q to the corresponding local source, until results are returned to the central site. Considering table fragmentation rules, a performance gain is achieved in the case that $ET_{hq} > ET_{slq}$, where hq is a local subquery involving a conflicting horizontal fragment, and slq is the slowest local subquery that does not involve a conflicting horizontal fragment. This is because local subqueries are issued in parallel, thus the total elapsed time coincides with the elapsed time of the slowest subquery. Further, a performance gain is achieved at result composition time, since unnecessary joins between local subqueries, that refer to conflicting table fragments, are avoided.

5 Conclusions and Outlook

Quete is an ontology-based mediator system that integrates several underlying relational databases by providing the user with the capability to transparently query them. Since every source relation is defined over the reference ontology, it follows the Local-as-View approach [3], and thus, can flexibly accommodate the addition/deletion and evolution of the local sources that participate in the integrated system. Moreover, underlying sources can evolve at will, without any changes to the reference ontology. Comparing with Information Manifold [8], which also follows the Local-as-View approach, our supported data source-to-ontology mappings are quite simple. However, we sacrifice expressiveness in order to achieve performance and easiness in use (both for local annotators and users).

A novel feature of Quete is that rules concerning horizontal, vertical, and hybrid table fragmentation can be declared and used, increasing system's recall and improving performance. To the best of our knowledge, no other ontology-based mediator system takes into account table fragmentation information.

Finally, we would like to note that our work does not claim to be complete and further performance evaluation experiments need to be performed. Our future plans also include the expansion of our implementation to non-relational data sources, and in particular, web databases with limited query capabilities (through wrappers). Moreover, we plan to enhance and ease the administrator's task, by extending the schema mapping tool described in [9], such that the semantic names of the data source schema elements are generated semi-automatically and stored directly in the corresponding X-Spec files.

Acknowledgments. We would like to thank Ramon Lawrence for providing us with a preliminary version of Unity, which we used as a starting point in our research.

References

1. Alonso-Calvo, R., Maojo, V., Billhardt, H., Martin-Sanchez, F., Garcia-Remesal, M., Perez-Rey, D.: An agent- and ontology-based system for integrating public gene, protein, and disease databases. Journal of Biomedical Informatics 40(1), 17–29 (2007)
2. Ben Miled, Z., Li, N., Bukhres, O.: BACIIS: Biological and Chemical Information Integration Systems. Journal of Database Management 16(3), 73–85 (2005)
3. Calvanese, D., Lembo, D., Lenzerini, M.: Survey on methods for query rewriting and query answering using views. Technical Report D1.R5, Dipartimento di Informatica e Sistemistica, Università di Roma La Sapienza (April 2001)
4. Garcia-Molina, H., Hammer, J., Ireland, K., Papakonstantinou, Y., Ullman, J., Widom, J.: Integrating and Accessing Heterogeneous Information Sources in TSIMMIS. In: Procs. of the AAAI Symposium on Information Gathering, pp. 61–64 (1995)
5. Klyne, G., Carroll, J.J.: Resource Description Framework (RDF): Concepts and Abstract Syntax. W3C Recommendation (10 February 2004), Available at http://www.w3.org/TR/rdf-concepts/
6. Lawrence, R., Barker, K.: Multidatabase Querying by Context. In: Procs. of the 20th DataSem 2000, pp. 127–136. Czech Republic (October 2000)
7. Lenzerini, M.: Data Integration: A Theoretical Perspective. In: Procs. of the 21st ACM SIGACT-SIGMOD-SIGART Symposium on Principles of Database Systems (PODS 2002), pp. 233–246 (2002)
8. Levy, A.Y., Rajaraman, A., Ordille, J.J.: Querying heterogeneous information sources using source descriptions. In: Procs. of the 22nd VLDB Conference, pp. 251–262 (1996)
9. Manakanatas, D., Plexousakis, D.: A Tool for Semi-Automated Semantic Schema Mapping: Design and Implementation. In: Procs. of the Intern. Workshop on Data Integration and the Semantic Web (DisWeb 2006), in conjunction with CAiSE'06, pp. 290–306 (2006)

10. Mason, T., Lawrence, R.: Dynamic Database Integration in a JDBC Driver. In: Procs. of the 7th International Conference on Enterprise Information Systems - Databases and Information Systems Integration Track, Miami, FL (2005)
11. McGuinness, D.L., van Harmelen, F.: OWL Web Ontology Language Overview. W3C Recommendation (10 February 2004), Available at http://www.w3.org/TR/owl-features/
12. Ozsu, T., Valduriez, P.: Principles of Distributed Database Systems, 2nd edn. Prentice-Hall, Englewood Cliffs, NJ (1999)
13. Potamias, G., Analyti, A., Kafetzopoulos, D., Kafousi, M., Margaritis, T., Plexousakis, D., Poirazi, P., Reczko, M., Tollis, I.G, Sanidas, E., Stathopoulos, E., Tsiknakis, M., Vassilaros, S.: Breast Cancer and Biomedical Informatics: The PrognoChip Project. In: Procs. of the 17th IMACS world Congress Scientific Computation, Applied Mathematics and Simulation, Paris, France (2005)
14. Ramakrishnan, R.: Database Management Systems. McGraw-Hill Companies, Inc. (1997)
15. Stevens, R., Baker, P., Bechhofer, S., Ng, G., Jacoby, A., Paton, N.W., Goble, C.A., Brass, A.: TAMBIS: Transparent Access to Multiple Bioinformatics Information Sources. Bioinformatics 16(2), 184–186 (2000)
16. Sujansky, W.: Heterogeneous Database Integration in Biomedicine. Journal of Biomedical Informatics 34(4), 285–298 (2001)
17. Tomasic, A., Raschid, L., Valduriez, P.: Scaling Access to Heterogeneous Data Sources with DISCO. IEEE transactions on Knowledge and Data Engineering 10(5), 808–832 (1998)

Author Index

Lecture Notes in Computer Science

Sublibrary 3: Information Systems and Application, incl. Internet/Web and HCI

For information about Vols. 1– 4277
please contact your bookseller or Springer

Vol. 4537: K.C.-C. Chang, W. Wang, L. Chen, C.A. El-lis, C.-H. Hsu, A.C. Tsoi, H. Wang (Eds.), Advances in Web and Network Technologies, and Information Management. XXIII, 707 pages. 2007.

Vol. 4531: J. Indulska, K. Raymond (Eds.), Distributed Applications and Interoperable Systems. XI, 337 pages. 2007.

Vol. 4526: M. Malek, M. Reitenspieß, A. van Moorsel (Eds.), Service Availability. X, 155 pages. 2007.

Vol. 4524: M. Marchiori, J.Z. Pan, C.d.S. Marie (Eds.), Web Reasoning and Rule Systems. XI, 382 pages. 2007.

Vol. 4519: E. Franconi, M. Kifer, W. May (Eds.), The Semantic Web: Research and Applications. XVIII, 830 pages. 2007.

Vol. 4518: N. Fuhr, M. Lalmas, A. Trotman (Eds.), Comparative Evaluation of XML Information Retrieval Systems. XII, 554 pages. 2007.

Vol. 4508: M.-Y. Kao, X.-Y. Li (Eds.), Algorithmic Aspects in Information and Management. VIII, 428 pages. 2007.

Vol. 4506: D. Zeng, I. Gotham, K. Komatsu, C. Lynch, M. Thurmond, D. Madigan, B. Lober, J. Kvach, H. Chen (Eds.), Intelligence and Security Informatics: Biosurveillance. XI, 234 pages. 2007.

Vol. 4505: G. Dong, X. Lin, W. Wang, Y. Yang, J.X. Yu (Eds.), Advances in Data and Web Management. XXII, 896 pages. 2007.

Vol. 4504: J. Huang, R. Kowalczyk, Z. Maamar, D. Martin, I. Müller, S. Stoutenburg, K.P. Sycara (Eds.), Service-Oriented Computing: Agents, Semantics, and Engineering. X, 175 pages. 2007.

Vol. 4500: N.A. Streitz, A. Kameas, I. Mavrommati (Eds.), The Disappearing Computer. XVIII, 304 pages. 2007.

Vol. 4495: J. Krogstie, A. Opdahl, G. Sindre (Eds.), Advanced Information Systems Engineering. XVI, 606 pages. 2007.

Vol. 4480: A. LaMarca, M. Langheinrich, K.N. Truong (Eds.), Pervasive Computing. XIII, 369 pages. 2007.

Vol. 4471: P. Cesar, K. Chorianopoulos, J.F. Jensen (Eds.), Interactive TV: A Shared Experience. XIII, 236 pages. 2007.

Vol. 4469: K.-c. Hui, Z. Pan, R.C.-k. Chung, C.C.L. Wang, X. Jin, S. Göbel, E.C.-L. Li (Eds.), Technologies for E-Learning and Digital Entertainment. XVIII, 974 pages. 2007.

Vol. 4443: R. Kotagiri, P. Radha Krishna, M. Mohania, E. Nantajeewarawat (Eds.), Advances in Databases: Concepts, Systems and Applications. XXI, 1126 pages. 2007.

Vol. 4439: W. Abramowicz (Ed.), Business Information Systems. XV, 654 pages. 2007.

Vol. 4430: C.C. Yang, D. Zeng, M. Chau, K. Chang, Q. Yang, X. Cheng, J. Wang, F.-Y. Wang, H. Chen (Eds.), Intelligence and Security Informatics. XII, 330 pages. 2007.

Vol. 4425: G. Amati, C. Carpineto, G. Romano (Eds.), Advances in Information Retrieval. XIX, 759 pages. 2007.

Vol. 4412: F. Stajano, H.J. Kim, J.-S. Chae, S.-D. Kim (Eds.), Ubiquitous Convergence Technology. XI, 302 pages. 2007.

Vol. 4402: W. Shen, J.-Z. Luo, Z. Lin, J.-P.A. Barthès, Q. Hao (Eds.), Computer Supported Cooperative Work in Design III. XV, 763 pages. 2007.

Vol. 4398: S. Marchand-Maillet, E. Bruno, A. Nürnberger, M. Detyniecki (Eds.), Adaptive Multimedia Retrieval: User, Context, and Feedback. XI, 269 pages. 2007.

Vol. 4397: C. Stephanidis, M. Pieper (Eds.), Universal Access in Ambient Intelligence Environments. XV, 467 pages. 2007.

Vol. 4380: S. Spaccapietra, P. Atzeni, F. Fages, M.-S. Hacid, M. Kifer, J. Mylopoulos, B. Pernici, P. Shvaiko, J. Trujillo, I. Zaihrayeu (Eds.), Journal on Data Semantics VIII. XV, 219 pages. 2007.

Vol. 4365: C.J. Bussler, M. Castellanos, U. Dayal, S. Navathe (Eds.), Business Intelligence for the Real-Time Enterprises. IX, 157 pages. 2007.

Vol. 4353: T. Schwentick, D. Suciu (Eds.), Database Theory – ICDT 2007. XI, 419 pages. 2006.

Vol. 4352: T.-J. Cham, J. Cai, C. Dorai, D. Rajan, T.-S. Chua, L.-T. Chia (Eds.), Advances in Multimedia Modeling, Part II. XVIII, 743 pages. 2006.

Vol. 4351: T.-J. Cham, J. Cai, C. Dorai, D. Rajan, T.-S. Chua, L.-T. Chia (Eds.), Advances in Multimedia Modeling, Part I. XIX, 797 pages. 2006.

Vol. 4328: D. Penkler, M. Reitenspiess, F. Tam (Eds.), Service Availability. X, 289 pages. 2006.

Vol. 4321: P. Brusilovsky, A. Kobsa, W. Nejdl (Eds.), The Adaptive Web. XII, 763 pages. 2007.

Vol. 4317: S.K. Madria, K.T. Claypool, R. Kannan, P. Uppuluri, M.M. Gore (Eds.), Distributed Computing and Internet Technology. XIX, 466 pages. 2006.

Vol. 4312: S. Sugimoto, J. Hunter, A. Rauber, A. Morishima (Eds.), Digital Libraries: Achievements, Challenges and Opportunities. XVIII, 571 pages. 2006.

Vol. 4306: Y. Avrithis, Y. Kompatsiaris, S. Staab, N.E. O'Connor (Eds.), Semantic Multimedia. XII, 241 pages. 2006.

Vol. 4302: J. Domingo-Ferrer, L. Franconi (Eds.), Privacy in Statistical Databases. XI, 383 pages. 2006.

Vol. 4299: S. Renals, S. Bengio, J.G. Fiscus (Eds.), Machine Learning for Multimodal Interaction. XII, 470 pages. 2006.

Vol. 4295: J.D. Carswell, T. Tezuka (Eds.), Web and Wireless Geographical Information Systems. XI, 269 pages. 2006.

Vol. 4286: P.G. Spirakis, M. Mavronicolas, S.C. Kontogiannis (Eds.), Internet and Network Economics. XI, 401 pages. 2006.

Vol. 4282: Z. Pan, A. Cheok, M. Haller, R.W.H. Lau, H. Saito, R. Liang (Eds.), Advances in Artificial Reality and Tele-Existence. XXIII, 1347 pages. 2006.

Vol. 4278: R. Meersman, Z. Tari, P. Herrero (Eds.), On the Move to Meaningful Internet Systems 2006: OTM 2006 Workshops, Part II. XLV, 1004 pages. 2006.